D1329738

Environmental Behaviour
of
Agrochemicals

*Progress in Pesticide Biochemistry
and Toxicology*

Volume 9

Environmental Behaviour
of
Agrochemicals

Edited by

T. R Roberts

Corning Hazleton, Harrogate, UK

and

P. C. Kearney

Water Resources Research Center, University of Maryland, USA

A Wiley-Interscience Publication

JOHN WILEY & SONS

CHICHESTER · NEW YORK · BRISBANE · TORONTO · SINGAPORE

Copyright © 1995 by John Wiley & Sons Ltd.
Baffins Lane, Chichester
West Sussex PO19 1UD, England
Telephone: *National* Chichester (01243) 779777
International +44 1243 779777

All rights reserved.

No part of this book may be reproduced by any means,
or transmitted, or translated into a machine language
without the written permission of the publisher.

Other Wiley Editorial Offices

John Wiley & Sons, Inc., 605 Third Avenue,
New York, NY 10158-0012, USA

Jacaranda Wiley Ltd, 33 Park Road, Milton,
Queensland 4064, Australia

John Wiley & Sons (Canada) Ltd, 22 Worcester Road,
Rexdale, Ontario M9W 1L1, Canada

John Wiley & Sons (SEA) Pte Ltd, 37 Jalan Pemimpin #05-04,
Block B, Union Industrial Building, Singapore 2057

Library of Congress Catalog Card Number: 83-647760
ISSN 0887-6142

British Library Cataloguing in Publication Data:

A catalogue record for this book is available from the British Library

ISBN 0 471 95301 6

Typeset in 10/12pt Times by Keytec Typesetting Ltd, Bridport, Dorset.
Printed and bound in Great Britain by Biddles Ltd, Guildford, Surrey.
This book is printed on acid-free paper responsibly manufactured from sustainable forestation,
for which at least two trees are planted for each one used for paper production.

QP82.2
.P4P76
1981
vol. 9
copy 1

Contributors to Volume 9

L. Bergström
Department of Soil Sciences, Division of Water Quality Management, Swedish University of Agricultural Sciences, Box 7072, 750 07 Uppsala, Sweden

C. D. Brown
Soil Survey and Land Research Centre, Cranfield University, Silsoe, Bedfordshire, MK45 4DT, UK

B. Burgoa
USDA, Agricultural Research Service, PO Box 748, Tifton Georgia, 31793, USA

A. D. Carter
Soil Survey and Land Research Centre, Shardlow, Derby, DE72 2GN, UK

C. V. Eadsforth
Shell Research Ltd, Sittingbourne, Kent, ME9 8AG, UK

A. T. Eldefrawi
School of Medicine, University of Maryland, 655 West Baltimore Street, Baltimore, Maryland, 21201–1559, USA

M. E. Eldefrawi
School of Medicine, University of Maryland, 655 West Baltimore Street, Baltimore, Maryland, 21201–1559, USA

P. R. Fisk
Shell Research Ltd, Sittingbourne, Kent, ME9 8AG, UK

J. M. Hollis
Soil Survey and Land Research Centre, Cranfield University, Silsoe, Bedfordshire, MK45 4DT, UK

N. Jarvis
Department of Soil Sciences, Division of Water Quality Management, Swedish University of Agricultural Sciences, Box 7072, 750 07 Uppsala, Sweden

K. R. Rogers
US Environmental Protection Agency, Environmental Monitoring Systems Laboratory, PO Box 93478, Las Vegas, NV 89193, USA

M. H. Russell
Research Association, DuPont Agricultural Products, Experimental Station, Wilmington, DE 19880–0402, USA

CONTRIBUTORS

A. W. TAYLOR — *Consultant in Conservation and Environmental Sciences, 7302 Baylor Avenue, College Park, Maryland, 20740, USA*

R. D. WAUCHOPE — *USDA, Agricultural Research Service, PO Box 748, Tifton, Georgia, 31793, USA*

A. P. WOODBRIDGE — *Shell Research Ltd, Sittingbourne, Kent, ME9 8AG, UK*

Contents

Preface

The theme for Volume 7 of 'Progress in Pesticide Biochemistry and Toxicology', published in 1990, was the environmental fate of pesticides and the contents ranged from pesticides in surface and ground water, soil and in fish. Such is the interest in the behaviour of agrochemicals in the environment that this volume covers a similar range of topics, chosen to complement the earlier volume. I am delighted that Philip Kearney agreed to join me as guest editor for the book.

We make no apology for devoting another volume to environmental issues, since in the areas of agrochemical development and re-registration, the movement of pesticides away from the site of application continues to be of critical importance in environmental safety assessment.

The first chapter covers the all important physicochemical properties of chemicals and summarizes current methods of estimation of these properties, both theoretical and experimental. This is followed by four chapters on pesticide mobility in soil. In Chapter 2, the report of a specialist industry group is presented. The agrochemical industry trade association GIFAP/ECPA commissioned its experts to prepare this review with recommended approaches to assessing pesticide mobility in soils. The group published a short paper available directly from GIFAP in Brussels and the full paper is published in this volume. The following chapter addresses pesticide mobility from the more specific perspective of soils and their influence on pesticide mobility and this complements the more general chapter on hydrogeology in Volume 7 of this series. Finally, on leaching of pesticides through soil, we include a timely comparison of computer simulation models and their applications. The level of interest in modelling has increased dramatically in recent years with the development and 'validation' of models by US and European scientists. Interest in this approach from regulators has also increased. In Chapter 5, the potential run-off of pesticides from soil surfaces into surface water is discussed, including experimental and theoretical approaches.

The volatilization of pesticide residues (Chapter 6) is also an important contribution since there continues to be controversy over the significance of long-range transport of agrochemicals.

The final two chapters cover analytical methodologies for agrochemicals in environmental matrices with Chapter 8 on biosensors and Chapter 9 on current methodologies for analysis of natural waters. There is no doubt that *in situ* measurement presents a highly desirable alternative to sampling and subsequent analysis, and the review of research on biosensors covers this aspect.

This volume will be the last in the 'Progress in Pesticide Biochemistry and Toxicology' series in its current format, since it will be integrated with the new Wiley series on 'Agrochemicals and Plant Protection'. This new series will build on the success of the 'Progress' volumes and have a broader scope, to cover important aspects such as discovery, structure–activity relationships, formulation technologies, biotechnology, regulatory issues and integrated pest management, in addition to the important areas of safety assessment already within the present scope. Together with Professor Junshi Miyamoto of Japan I look forward to developing this new project, covering all the major areas of interest with agrochemicals.

Harrogate   T. R. ROBERTS
November 1994

CHAPTER 1

Estimation of physicochemical properties: theoretical and experimental approaches

P. R. Fisk

Environmental Behaviour of Agrochemicals
Edited by T. R. Roberts and P. C. Kearney　　© 1995 John Wiley & Sons Ltd

INTRODUCTION

Background

The prediction of the values of physicochemical properties is of widespread use in chemistry and chemical engineeering, and a wide range of sources has been drawn upon in this review. The main emphasis will be on compounds used in crop protection, but the information should be of use in other disciplines, particularly the growing area of environmental risk assessment of chemicals in general. Many of the principles developed in the sphere of crop protection are now being applied more widely.

The provision of the values of physicochemical properties of crop protection compounds is required at all stages of progression of a substance, from discovery to commercial maturity. Whilst there is varying emphasis from one organization to another, there are some common features in the way new substances are developed. The key challenge is how to reduce the number of options from the infinite number of structures that can be devised and prepared. A combination of experience, cost information, biochemical and physicochemical data, computational chemistry, the suitability of formulation technology, and the environmental and toxicological profile are used in this process. The utilization of physicochemical data varies between every organization and individual involved. However, a pattern is shown in Table 1, which will be amplified upon below. Here, and throughout the text, symbols will be used that are defined in the glossary.

Discovery

Whilst there is usually little choice at the regulatory end of the progression, there is much more debate at the earlier stages. At the discovery level, the chemist will still be exploring lines of synthesis. Predicted values of physicochemical properties are widely used at this stage, alongside input from molecular modelling programmes and biochemistry (aimed at characterizing the possible active site or sites). Partitioning properties are useful in cases where an optimum lipophilicity is believed to exist. Advice can be given as to how that lipophilicity can be achieved. Inevitably, the range of data examined is less at this stage, when there are many candidates.

Lead optimization

During this phase, biological screening data will be confirmed and the dose response of targets to members of a candidate class of compound will be measured. The presentation of the compound as a formulation now becomes important, and physicochemical data are useful at this point. Solubility is of increasing importance because the range of formulation strategies used is restricted by the reduction in the number of solvents considered acceptable (on health and safety grounds). In recent years a new emphasis has emerged in many companies: the evaluation of candidates on the basis of environmental properties. Environmental testing is too expensive to leave until the choice of compound has been narrowed. Therefore, assessment of potential benefits and problems can be made, even on the basis of calculation only.

Table 1 The uses of physicochemical data during progression from discovery to developement

Level	Purpose of physicochemical data	Progressive data requirement
Discovery	• Input to QSAR programmes • Basic chemical data • Evaluation of competitor products	$\log P_{ow}$, pK_a Hydrolysis, water solubility, melting
Lead optimization	• Assistance to formulation programmes • Early environmental assessment • Lab–field correlation	Solubility in solvents, thermal stability Hydrolysis and photolysis rates, soil adsorption Volatility
Development	• Full range of data required for commercialization	

Partition properties, solubility, estimation of hydrolytic stability and bio-degradability can be used with environmental fate models to build up a picture of the likely behaviour of a compound in the environment. This has served to bring calculated physicochemical data to the fore.

Commercialization

At the commercial stage the demand is driven by regulatory requirements and the safety standards of the particular manufacturer. At this level, there is now a high degree of harmony between the requirements of the European Community and the US Environmental Protection Agency, leading organizations in the regulatory areas. It is sufficient to state that comprehensive data sets are required.

Uses of predicted property values

A predicted value of a physicochemical property is of use even when measurement itself may be a viable option, because:

- it can help decide which properties of which substances under consideration are of most urgency and potential value to measure.
- it can provide an approximate value which enables the experimental design to be optimized efficiently without the need for preliminary experiments.
- calculations can be used to build upon several measured values and thereby give very reliable predictions for a whole series.
- it can show that measurement could be impossible or not useful and therefore no attempt to measure need be made.

Several of the uses of the data described will be examined later in the report. The properties described in the report are those of relevance to areas described in Table 1. They may be grouped as follows.

Properties of the compound

- Melting point
- Boiling point
- Vapour pressure
- Water solubility
- Thermal stability

Properties of solutions of the compound

- Octanol/water partitioning
- Dissociation constant

- Diffusion coefficient
- Henry's constant
- Hydrolysis rate and photolysis rate

The emphasis in this review will necessarily be on single, known substances, not mixtures. Also, because the majority of crop protection compounds are complex, multifunctional substances, there will not be extensive discussion of methods that are appropriate only to simpler structures such as hydrocarbons.

The review will describe first some general principles, then describe methods for each property in turn. Application of some of the methods to a test set will be made. Some specific areas where physicochemical property data are used will be exemplified.

In the absence of a relevant published review of measurement methods, Appendix 1 on this matter has been included for completeness.

Appendix 2 deals with the area of experimental methods of estimating properties from related, simpler (or more cheaply obtained) ones. A well known example is the estimation of octanol/water partition coefficient by reverse phase high performance liquid chromatography (HPLC), which will be reviewed below.

An important factor in this review is consideration of the accuracy desired for a particular purpose compared to what is achievable. At the discovery phase, a large number of closely related structures will be under consideration. For example, the absolute value of a partition coefficient may often, in these circumstances, be less important than accurate assessment of the differences of the values for the members of the series. Differences in $\log P_{ow}$ need to be known to within 0.1–0.2 log units for use in QSAR. However, for an environmental assessment, it is sufficient to be able to say with high confidence that the value lies within a range, say 3.5–4.5; although apparently a wide range, it is sufficient for that purpose. There may, however be regulatory end-points that need to be followed; for example, the system for labelling substances in the EC has the $\log P_{ow}$ less or greater than 3 as a parameter.

The use of data estimation methods for such regulatory needs has been discussed in general (Hart, 1991), and will be returned to later in the chapter (see 'Uses of predicted data').

GENERAL PRINCIPLES

Prior review literature

Lyman *et al.* (1982) have provided a practical and comprehensive handbook. It has been widely used and is regularly quoted by other authors. The present paper reviews progress in the field since that authoritative work was published. A further update (Lyman, 1985) is helpful also. Whilst the book is invaluable, it applies largely to substances somewhat simpler than the majority of current pharmaceuticals and crop protection compounds, which restricts its usefulness. However, it was undoubtedly a stimulus for further activity, particularly for those involved with the 'optimization' phase described in Table 1.

Briggs (1990) has reviewed the behaviour of pesticides in soil on the basis of their physical and chemical properties and included a brief survey of predictive methods. This is an update of an earlier review (Briggs, 1981). Another relatively recent review (Chung, 1985) has an emphasis on safety and does not introduce any new approaches beyond the earlier reviews. More recently, Yalkowsky and Mishra (1992) have proposed a scheme for prediction of a wide range of properties in a coordinated way, but without comparison with other approaches.

There are two further reviews that are part of commercially available services. One is the computer program CHEMEST (1991, and Boethling *et al.*, 1988) which is heavily based on the work of Lyman mentioned above. The Fraunhofer Institute has compiled a review of methods for some properties of relevance (Klein, 1992).

Reviews for particular properties are described below.

Principles and concepts which apply to several property-prediction methods

This section provides a classification of the types of methods that are used, and will be referred back to in the main part of the report. Lyman (1985) provides further details on some of these. The next section lists techniques used to apply the methods described in this section; not every approach is applied to the different properties in the same way. The purpose of these two sections is to provide a framework of understanding so that the particular method can be seen in its context.

Readers interested only in the final results may prefer to move to the section on 'Methods for prediction of properties' where particular properties are described.

A number of generalizations may be more regarding how physicochemical property prediction is approached. Several classes of method may be identified:

(1) Methods which use correlation of property A with property B (C, etc.).
(2) Methods which use accepted fundamental equations as a basis.
(3) Methods which use molecular fragment constants.

(4) Statistical methods which use no obvious theory-derived basis, but which use realistic inputs and statistics to turn these into usable relationships.

These categories will now be expanded upon.

Correlation methods

Whilst chemists have always noted obvious relationships between properties, perhaps the best known use of this approach is the Collander equation (Collander, 1951) relating partition coefficients (K) in different solvent systems

$$\log K_1 = a \log K_2 + b$$

where K_1 and K_2 refer to different solvents, e.g. octanol and cyclohexane. Another early use of this kind of approach was made in the relationship of boiling point to molar refraction (Meissner, 1949).

Occasionally, these approaches merge with those based on theory. An example is the Wilke–Chang approach to the prediction of molecular diffusion coefficients in solvents (Wilke and Chang, 1955), which is described in the section on 'Diffusion coefficient':

$$D = 7.4 \times 10^{-8} \, (xM)^{0.5} T / \eta V^{0.6}$$

In this example the desired property is related to several known properties; it is also an extension of the Stokes–Einstein equation, which has a sound theoretical basis.

As applies to many methods, these approaches also yield better predictive equations when the 'training set' on which the equation is established is a set of analogous substances; the equation then only applies to further analogues. In some cases, it is possible then to attempt to rationalize the meaning of the coefficients for different sets of analogues.

Methods based on fundamental equations or physical models

Fundamental equations: There are very few useful examples in this category, particularly for pesticides which are, by their nature, complex substances in the main, frequently outside the scope of ideal equations. An example is the Watson modification of the Clausius–Clapeyron equation to predict vapour pressure and its temperature dependence (see 'Methods for prediction of properties'). Even before this can be made useful for purposes of predictions, it requires a host of empirical constants and still requires a boiling point as an input.

Physical models: Even the most fundamental equations are based on underlying models and assumptions. Therefore, model-based approaches may be considered as attempts to develop new understanding of physico-chemical processes. Their testing, interpretation and evaluation forms part of their acceptance into the body of scientific knowledge. An example is the use of the Hammett equation,

$$\log k - \log k_0 = \sigma\rho$$

based originally on the relative rates of hydrolysis of esters (see 'Description of electronic effects using linear free energy relationships').

The σ constants that result can be interpreted in terms of molecular structure, and ρ values describe the nature of the transition state in the reaction. Whilst there is an element of empiricism in the development of linear free energy relationships (LFER), once the initial step had been made they can only be described as a major success. Finer examination has led to extensions to σ constants of many types and the search for an underlying understanding at a deeper level (see later, 'Description of electronic effects using linear free energy relationships'). The LFER has passed from being a hypothesis to being an accepted scientific principle.

It could be argued that some fragment methods are based on physical models and that some other methods are in fact a 'blend'. For example, fragment methods for partition coefficient start with the 'model' of a molecule having fragments that differ in their relative contributions to the overall lipophilicity of the molecule. The resulting π values make chemical sense, thereby supporting the model. However, the range of correction terms needed, and the empirical methods needed to generate them, are beyond easy rationalization in every case.

Fragment methods

Fragment methods view a molecule as composed of specified parts, which contribute individually to the compound property. The development and use of fragment methods proceeds as follows:

(1) Decide what constitutes a fragment.
(2) Decide upon a set of test data ('training set').
(3) Use multiple regression or another technique to establish how

$$\text{Property} = \Sigma \text{ (fragment values)}$$

and thus the values for the fragments examined.
(4) Examine the outliers and consider what interactions between fragments may have been ignored.
(5) Establish an approach to dealing with the interactions and define a new method.

$$\text{Property} = \Sigma \text{ (fragment values)} + \Sigma \text{ (correction terms)}$$

to establish new fragment values and corrections terms.

(6) Test the method on data outside the training set, then repeat steps 4 and
    5.

Step 1 is perhaps the most crucial and one which does not always receive
full attention as to why one approach was preferred to another. Further
details of various published methods are given under the relevant properties.
Similarly, the scientific meaning of the fragment values and correction terms
should be considered and preferably understood. It is sometimes the case for
methods to require correction terms which are significant compared to the
fragment values, even for simple structures. The user is thereby warned that
the method may not be robust. It is essential to know the limitations of any
method.

Fragment-based methods are particularly useful when reliable data are
available for close analogues of the test structure; extension from the known
result is made, and the errors are thereby considerably reduced.

Statistical methods

A wide range of statistical methods is being applied to study of QSAR for
pesticides. As yet, this battery of methods is not being widely used on
physical property prediction. However, it is important to note that if
physicochemical properties are inputs into models of, for example, environmental fate, there are alternatives. The quantum mechanical molecular
descriptors could be used instead. Whilst it may in future be possible to
predict environmental fate from molecular descriptors, the need to relate
any such model to an easily visualized 'common currency' of physicochemical properties may be felt by some.

It is beyond the scope of this review to describe the methods and their
validation in detail. A useful review is available (Livingstone, 1989). The
methods may be divided into two classes, often referred to as 'supervised'
and 'unsupervised' learning. In the latter, the techniques used are more free
to explore relationships between variables, and are therefore less likely to
produce change effects. Examples include principal components analysis,
factor analysis and cluster analysis. Multiple regression, discriminant analysis and partial least squares are examples of supervised learning.

Tools used in property-prediction methods

The following sections describe generally applicable tools that are used by
the methods in various ways. The purpose of this section is to simplify the
description of the methods themselves, given in 'Methods for prediction of
properties'.

Structure description

The methods use a variety of descriptions of chemical structure, from none at all (e.g. for most property correlation methods) through to molecular orbital calculations. Table 2 describes this progression.

SMILES notation: SMILES is a widely used notation system for molecular structure. It lives up to its name (Simplified Molecular Input Line Entry System) because of its ease of use. It has been developed and described by Weininger (1988).

Generally, atoms are represented by their atomic symbols. Hydrogen need never be added (the programs interpret them). Table 3 illustrates some features.

Table 2 Descriptions of structure used in property prediction

Structure description	Example
List of general structural type	'Alkane', 'phenolic'
Count of atoms present	Ethane = $2 \times C, 6 \times H$
Count of bonds present	Toluene $3 \times C$ (alkane) to H $5 \times C$ (aromatic) to H $1 \times C$ (aromatic) to C (alkane) $6 \times C$ (aromatic) to C (aromatic)
Description of structure without stereochemical input	SMILES notation (see below) in which the structure is uniquely described
Description of shape	Chemical graph theory
Molecular modelling methods	Shape and electronic structure are described (see 'Molecular modelling methods')

Table 3 Examples of the use of SMILES notation

Feature	Example	SMILES
Single bonds	Ethane	CC
Double bond	2-Butene	CC=CC
Ring	Methylcyclohexane	CC1CCCCC1
Aromatic ring	Pyridine	c1ccccn1
Branching	Propan-2-ol	CC(O)C
Halogens	Dichloromethane	ClCCl
Bicyclic systems	Naphthalene	c1ccc2ccccc2c1
Non-fused ring	Diphenylether	c1ccccc1Oc2ccccc2
Nitro group	Nitroethane	CCN(=O)(=O)

For rings, the numbers indicate where the ring starts and stops. There is not always a unique SMILES string for a particular structure, but the program interprets them uniquely.

Molecular connectivity

Connectivity is an example of a topological index, that is, a relationship between three-dimensional chemical structure and the property of interest. This may be considered as a subset of chemical graph theory (Sabljic and Trinajstic, 1981). Molecular connectivity has been particularly advanced by Kier and Hall; see Hall (1990), Kier (1990), Hall and Kier (1991) for recent reviews. At the simplest level, the number of neighbours of each non-hydrogen atom is counted (δ values), and an index derived from those numbers. The first connectivity index is called χ and where

$$(1\chi)^{-2} = \Sigma(\delta_i \delta_j)$$

The summation is over the bonds, and where $\delta_{i,j}$ are the number of atoms in the skeleton bonded to atoms i and j.

Other more complex indices are similarly defined, as $^m\chi_t$ where structure fragments t containing m bonds are counted. At the next level, there is a valence index which takes account of atom type by a count of the number of core and valence electrons. The technique is usually applied by multiple regression. A property is described by: property $-\Sigma$ (structure index)$_i$ + constant (structure index)$_i = \Sigma_j$ (structure elements)$_j$.

Connectivity index calculations are very complex and the computer program MOLCONN-X (1991) has been found to be useful.

Linear solvation energy relationships

This approach has been applied to many properties, including physicochemical data, chromatographic data and biological data (Taft et al., 1985). It is essentially a regression where the derived property is a linear function of parameters V, π^*, β, α, defined as V, molecular volume; π^*, a dipolarity/polarizability term; β, proton acceptor ability; α, proton donor ability.

They are obtained from several sources. V may be based on measured molar volumes or calculated from molecular modelling; π^* is found from work on solvatochromic parameters (i.e. the effect of solvents on spectra of model substances); β and α are derived from experiments measuring hydrogen bond strength (infrared or ultraviolet spectroscopy). The strengths of the linear solvation energy relationships (LSER) lie in their use of specific, relevant measurements. A weakness is the π^* factor being influenced by several essentially different molecular properties. A further problem has been that the values for V, π^*, β and α for a new substance have to be obtained. The development of the method has been described by Kamlet, Abboud and Taft in many papers; see for example Taft *et al.* (1985).

Whilst the attempt to use interpretable parameters is laudable, the method results in equations of very similar form to other methods and is occasionally restrictive in the availability of π^*, β and α values. It is a rapidly developing field, although mostly in the hands of one research group.

Molecular modelling methods

A review of this topic is beyond the scope of this report, but some background is given because some modelling is being used in descriptions of molecular shape, volume and area. For example, Rohrbaugh and Jurs (1987) compare their shape index to physicochemical descriptors of molecules. Molecular volume and area calculated by modelling methods are being used, particularly in prediction of solubility and partition coefficient. This is because, in any fundamental understanding of these properties, it is necessary to consider the cavity that has to form in water to accommodate the solute.

Kollman has produced several useful reviews of molecular modelling, aimed at drug design, but which are also generally relevant (Kollman, 1985, 1987; Kollman and Merz, 1990).

'Molecular modelling' means theoretical chemistry applied to 'real chemistry', necessarily first involving acceptance of the principles of quantum mechanics. Then approximations are required. Table 4 lists some of the approaches being used.

The major emphasis of modelling is to be able to gain insight into molecular behaviour, especially by comparing molecules. Examination of molecular similarity also uses statistical methods, the development of

Table 4   Methods in molecular modelling

Ab initio methods	Quantum mechanical method using molecular orbital theory—computed wave functions with minimal approximation
Molecular dynamics	Simulations of molecular motion
Molecular mechanics	Neglect electrons and look at movement of nuclei ('ball and springs' approach)
Conformational analysis and searching	Calculation of relative energy of the possible conformations of a molecule, aiming to find the minimum energy
Probe methods	Calculation of the response of a molecule to a hypothetical probe placed at points of a 3-D grid

comparative molecular field analysis (CoMFA) being a powerful recent example (Cramer *et al.*, 1988). CoMFA is a numerical representation of the electrostatic and van der Waal's energies of a molecule (both in magnitude and orientation); it is a probe method.

The BC(DEF) method

This method, developed by Cramer (1980a, 1980b) is an example of the use of principal components analysis. Cramer collated P_{ow}, MR, BP and V data for a set of substances, and calculated the five principal components describing other properties such as water solubility. The method was then extended to provide fragment contributions to B, C, D, E and F. Generally, B ('bulk') and C ('cohesiveness') are the most statistically significant terms. A more recent paper (Cramer, 1983) gives a more detailed account of the method. He also outlines restrictions on the value of the method: first, the BC(DEF) values are of most use for simple compounds, and second, they are predictive for properties that are non-stereospecific. These restrictions probably explain why the method has not been extensively developed.

Description of electronic effects using linear free energy relationships

The insight of Hammett greatly advanced mechanistic understanding in organic chemistry. His formulation started with the Hammett equation:

$$\log k - \log k_0 = \sigma \rho$$

where k_0 is the rate constant for the reaction with the substituent = hydrogen; k is the rate constant for the reaction with the substituent in place; σ is a constant describing the substituent; ρ is a substituent describing the reaction itself.

It was rapidly extended into other areas, including pK_a correlations. The history of this subject has been usefully reviewed by Shorter (1978) who charts the extension into various types of σ constant and the need for para-σ constants, the decomposition of σ into inductive and resonance effects, and the development of descriptors of steric effects by Taft. A deeper review is available from Ehrenson *et al.* (1973), and more recent ones from Taft and Topsom (1987) and Charton (1981, 1987, 1989), who also indicate the way a more unified approach is replacing a plethora of special cases.

Use of σ constants has been common in QSAR and plays a part in some property-prediction methods.

Activity coefficients and UNIFAC

Description of solubility and partitioning processes is frequently made in terms of molar concentrations. For strict thermodynamic reasons, activity rather than concentration should be used for real solutions. However, only in the case where authors have developed methods based on group activity coefficients (i.e. a fragment method) does this become important in the context of this review.

It is most simply illustrated for the case of solubility in water, which is evaluated in 'Methods for prediction of properties'. For a liquid solute, by definition

$$\gamma_w x_w = a_w$$

and when a liquid solute is in equilibrium with water,

$$\gamma_w x_w = \gamma_0 x_0$$

If water is poorly soluble in the solute then $x_0 = 1$ and $\gamma_0 = 1$, and $x_w = 1/\gamma_w$.

Solid solutes are described below.

UNIFAC is a model in which

$$\ln \gamma_i = \ln \gamma_i^{(c)} + \ln \gamma_i^{(R)}$$

where $\gamma_i^{(c)}$ is the combinational part (entropy, size, shape) and $\gamma_i^{(R)}$ is the 'residue', presumed to be concerned with intermolecular interactions. $\gamma_i^{(c)}$ is then expressed in terms of group volumes and areas. $\gamma_i^{(R)}$ is determined by difference, for the groups present. The approach has been developed by Fredenslund et al., (1977).

There are two main compilations of UNIFAC data (Magnussen et al., 1981; and Gmehling et al., 1982), which have been extended (Macedo et al., 1982; Tregs et al., 1987; Hansen and Jurs, 1991).

An anomaly in UNIFAC has been that data based on air–water equilibria have given better predictions of water solubility. A new set of groups has recently been introduced by Yalkowsky and Mishra (1992) termed AQUAFAC. This is discussed in detail below.

CHEMEST

CHEMEST is an on-line or PC-based system for estimating physical/chemical properties important in the assessment of the environmental fate of chemicals. An investigation of its validity was made by Boethling et al. (1988). This program is referred to several times, as appropriate, in 'Methods for prediction of properties'. The validation was undertaken with 170 chemicals from seven classes (alcohols, aldehydes and ketones, amines, carbamates, esters, ethers and phenols).

CHEMEST is a computerized version of a selection of the methods of Lyman et al. (1982).

METHODS FOR PREDICTION OF PROPERTIES

The first two sections of this chapter have described the general principles behind the methods. This section provides a summary of some of the notable methods described in the literature, with an emphasis on the results of the last 10 years.

Melting point

CHEMEST

In CHEMEST, two methods are available:

(1) Lorenz and Herz (Gold and Ogle, 1969). The melting point T_M is estimated from the boiling point T_B by $T_M = 0.5839 T_B$. Although useful, such a method relies on the estimation of boiling point (see 'Boiling point').
(2) Grain and Lyman (1983). The manual for CHEMEST describes this method in which

$$T_M = 0.474 T_b' + b\rho_b - 28$$

where $T_b' = 10^3/(10^{-(4/m^{1/2})})$ and ρ_b is the liquid density at the normal boiling point; b is a constant related to chemical class; m is molar mass $(g\,mol^{-1})$. The method is usually restricted to substances for which the ρ_b value may be estimated. The Boethling review (Boethling *et al.*, 1988) used method 1 only and found a mean error of 36 °C for 141 substances.

Method of Abramowitz

This method (Abramowitz and Yalkowsky, 1990) uses boiling point as an input, along with three other properties describing the symmetry and shape and substitution pattern. It was developed as an extensive series of rigid aromatic compounds.

Methods specific to individual compound classes

Charton and Charton (1984) have given a useful treatment of the problem of estimating melting point, producing an equation that can correlate for specific subsets.

    Tsakanikas and Yalkowsky (1988) applied a similar approach to flexible hydrocarbons. Dearden (1991) use descriptors α and β (described in 'General principles of prediction of values'), shape descriptors and molar refractivity to describe the melting of anilines.

Boiling point

Methods available in CHEMEST

The seven methods available in CHEMEST have been extensively tested as described by Rechsteiner (1982). They are:

(1) Meissner—inputs required: parachor and molar refraction;
(2) Lydersen—fragment and structural factors;
(3) Miller—a variant of method 2;
(4) Ogata—size;
(5) Somayajulu—total atomic number;
(6) Kinney—fragment method;
(7) Stiel—number and arrangement of C atoms.

They are chosen according to the scheme shown in Figure 1.

In the review by Boethling *et al.* (1988), the Meissner method was used for 179 organic chemicals, with a mean error of 23 °C, and underestimation for twice as many compounds as there were overestimations. Carbamates were particular badly estimated; it is reasonable to infer from this that such approaches will fail for complex substances. This problem is made more severe when it is considered that few useful data exist for complex

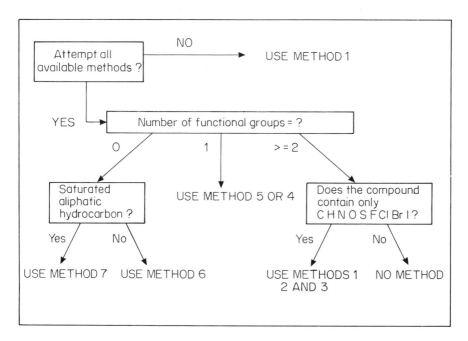

Figure 1   Scheme showing means of choosing methods available in CHEMEST

substances, owing to decomposition occurring before boiling. This is important with respect to prediction of vapour pressure.

Yalkowsky (Simamora *et al.*, 1993) has published a method which could be of use to predict melting and boiling points of certain crop protection compounds in that it focuses on rigid aromatic molecules. This is a new fragment-based method.

Method of Chen and Maddox

Chen and Maddox have produced an interesting new approach to the prediction of boiling point (Lai, 1987). Their approach is to assume that each methylene group contribution differs from the preceding one by a constant ratio (r_c). Thus for n methylene groups, in alkanes

$$T_B = a + b_c(1 - r_c^n)/(1 - r_c)$$

where a and b_c are constants. For functional classes such as alcohols and halogenated substances, additional b constants and terms of the form $(1 - r_f^n)/(1 - r_f)$ are added appropriately.

The method has been applied to 1169 organic compounds with an average error of 1.3%. The paper gives some useful worked examples.

The approach is typical of the pragmatic approach to these problems taken by chemical engineers. However, at the fundamental level, the r_c^n term is simply a way of accounting for the tendency of molecules, for statistical reasons and because of intermolecular van der Waals forces, to 'curl up' into near-spherical shapes. With respect to the transfer from liquid to vapour, each additional methylene adds less than the previous one to the intermolecular forces restricting evaporation.

This paper needs to be followed up because it is not a complete solution to the problem of boiling point prediction. The procedure is complex for multifunctional molecules and most potential users would benefit from an automated version. However, the insight provided by allowing for chain-length-dependent fragment values is a powerful one, and should be considered for use in other methods. The results in the paper apply to 13 common chemical classes, which further shows the usefulness of the method. It shows that there has to be some treatment of molecular size, probably in both liquid and gas phases, presumably by molecular modelling, if progress is to be made for the prediction of boiling point. The results of Jurs (see 'Methods specific to particular compound classes') suggest that the liquid state needs to be the focus of attention.

Methods specific to particular compound classes

Jurs has published a series of papers covering the prediction of boiling point for specific classes of compounds. There are alkenes (Hansen and Jurs,

1987), alcohols (Smeeks and Jurs, 1990) and furans, tetrahydrofurans and thiophenes (Stanton *et al.*, 1991; Egolf and Jurs, 1993). In the most recent two papers, a wide variety of molecular descriptors was investigated, based on molecular modelling (partial atomic charges), topological indices (connectivity), geometry (surface area) and physical properties (polarizability, molar refractivity, molecular weight). In the study of furans, 244 compounds were considered initially, with 80% of the variance explained and a standard deviation of the regression (s) of 28 °C. This last figure has to be borne in mind in that the range covered was 50–350 °C; at the lower end, 28 °C is clearly significant. A reduced 'training set' gave better regression statistics ($s = 13$ °C). The model was extended to include 11 descriptors. Thiophenes initially required a different model, although inclusion of neural networks methods has improved the statistical analysis.

The previous two papers from Jurs used topological descriptors only. For the alkenes, the best model found used eight descriptors and had an estimation error of 2 °C. For the alcohols, correlation from models containing between two and seven descriptors were produced. The two-descriptor model was based on the connectivity index $^1\chi$ and the charge on the oxygen atom; the estimation error was 4 °C. For the seven-descriptor model various charges and several topological indices were used; the estimation error was 2 °C.

It is clear from this that whilst interesting, potentially useful results have been developed, but there is no clear advantage over the older, simpler methods. Cramer's BC(DEF) approach does not work as well as the Meissner method, and is unlikely to be developed further. An approach that works for all classes is needed. Possibly the newer modelling methods such as CoMFA may help, alongside dynamics simulations.

Vapour pressure

This section does not include papers for which the aim was to obtain vapour pressure correlations over very wide temperature ranges (and pressures greater than atmospheric pressure).

Methods in CHEMEST and other methods based on boiling point

CHEMEST includes three methods described by Lyman *et al.* (1982). The most useful is the modified Watson correlation, for liquids and solids. This is now a recommended method in the recent update of the EC Dangerous Substances Directive (Official Journal of the European Communities, 1992). However, it is applicable only to values $>10^{-5}$ Pa. The review by Boethling *et al.* (1988) shows an average error of almost 2 log units, which is also typical of experience in the author's work.

The other more recent papers utilize this approach (Twu, 1983; Guthrie, 1986).

The method of Yalkowsky (see 'Approach based on thermodynamic equations') is also an example of this method. Mackay *et al.* (1982) have introduced a method which uses a modified Clausius–Clapeyron equation with the result:

$$\ln P_V = -(4.4 + \ln T_B)(1.803(T_B/T - 1) - 0.803 \ln (T_B/T))$$
$$- 6.8(T_M/T - 1)$$

Methods of Briggs

Briggs (1992) has outlined a 'rule of thumb' approach whereby an organic molecule is considered in terms of the equivalent carbon atom count. That is, heteroatoms are excluded and each counts for a number of carbons. Thus, for example, a complex substance may be considered as being equivalent to $C_{14}H_{30}$, which has a known boiling point and vapour pressure. This approach is relatively undeveloped in the open literature and, whilst rather empirical, it should not be scorned. It is comparable to the concept of retention index in gas chromatography, where retention is described in terms of retention time relative to hydrocarbons.

Briggs (1981, 1990) also describes a fragment method for boiling point (and from there, vapour pressure).

Solvatochromatic parameters

Banerjee *et al.* (1990) gives:

$$\log P_V = 7.82 - 7.29(V_I/100) - 6.41\pi^* + 3.25(\pi^*)^2 - 0.01(T_M - 25)$$

with $n = 53$ and $r = 0.98$.

UNIFAC parameters

Banerjee *et al.* (1990) gives:

$$\log P_V = 6.94 - (2.25V_u + 4.23 \log \gamma_c) - 0.577 \log \gamma_R - 0.01(T_M - 25)$$

with $n = 118$ and $r = 0.95$.

V_u is related to γ_c (defined in 'General principles of prediction of values').

Approach based on thermodynamic equations

Mishra and Yalkowsky (1991) use standard expressions for the enthalpies of phase changes, and then use approximations for the entropies of vaporization and melting. These approximations are claimed to be superior to Trouton's and Walden's rules respectively. The final result, which gives an estimation error of ±0.2 log units, is:

$$\ln P_V = - ([T_M - T]/T) (8.5 - 5.0 \log \sigma + 2.3 \log \Phi)$$
$$- ([T_B - T]/T) (10 + 0.08 \log \Phi)$$
$$+ ([T_B - T]/T - \ln [T_B/T]) (-6 - 0.9 \log \Phi)$$

where σ is a rotational symmetry number and Φ is a conformational flexibility number; Φ and σ have been described (Tsakanikas and Yalkowsky, 1988).

Method using Henry constant and water solubility

It is apparent that many property predictions are interdependent, as clearly described by Lyman (1985). For the prediction of vapour pressure of complex substances, such as typical crop protection compounds and pharmaceuticals, the need for a boiling point input is very restrictive.

Therefore, the author has investigated the use of a method for predicting vapour pressure from water solubility and Henry's constant; the Henry constant is obtained from the fragment method described by Meylan and Howard (1991), obtained in 'Henry's constant'. The water solubility is either predicted or already available.

The utility of the approach is described in 'Tests of some of the methods'.

Methods specific to particular classes of compound

The methods for predicting vapour pressure given above have been applied to a wide range of substances. All of them would have improved success if tailored to particular subclasses. There are other methods published that also illustrate this principle:

- Rordorf (1986, 1989)—dibenzodioxins and dibenzofurans;
- Hoshino *et al.* (1985)—substituted benzenes;
- Burkhard (1985)—halogenated aromatic hydrocarbons (group contribution method);
- D'Souza and Teja (1987)—carboxylic acids (uses the idea of an effective carbon number; Ambrose and Sprake, 1970).

Also, Burkhard *et al.* (1985) have published a review of 11 predictive methods for polychlorinated biphenyls (PCBs). Of the general methods, that of Mackay *et al.* (1982) was found to be the best. Methods using a PCB training set were, as expected, better.

Octanol/water partition coefficient

Prediction of octanol/water partition coefficient is the best known area to those involved in QSAR and environmental studies. It is only necessary here to point out some recent developments.

Fragment-based methods

The work of Hansch and Leo, Rekker and others, over the last 25 years has been of fundamental importance in the whole field of structure–property prediction; see Hansch and Leo (1979). However, some new areas have been examined:

(1) Taylor and co-workers have developed fragment values for solvent systems other than octanol, and described circumstances when these may be more relevant (Leahy *et al.*, 1989, 1991).
(2) Howard and Meylan have developed a similar system to CLOGP, called LOGKOW (1992). The difference lies in the fragment definition; they include the hydrogen atom rather than treating it separately. The program is commercially available, as is Rekker's PROLOGP.

Potential rivals to the position of CLOGP have to vie with the following statistics for reliable log P_{ow} values (Leo, 1992).

$$\log P_{ow} = 0.903 \ (\pm 0.005) \ \text{CLOGP} + 0.209 \ (\pm 0.12)$$

with $n = 7800$, $r = 0.982$, $s = 0.398$.

There is an unexpected departure from unit gradient and zero intercept. The literature for LOGKOW claims a better performance than CLOGP; this is discussed further in 'Tests of some of the methods', but it is clear that the relationship of LOGKOW to log P_{ow} is nearer to the expected line of unit gradient and zero intercept.

Methods based on surface area and volume

Several papers have been published using the linear solvation energy approach, for example Kamlet *et al.* (1988) and Leahy (1986). Whilst the correlations achieved are impressive, the number of substances is still relatively small (about 100).

There is a large number of other papers, each describing correlations between solvent-accessible surface area and the octanol/water partition coefficient. Amongst these, the work of Dunn is of particular note (Dunn *et al.*, 1987), because a wider range of solvent systems was studied, using principal components analysis. These workers found it necessary to add the specifically bonded water of hydration to the molecules before calculating the surface area.

Methods using molecular modelling techniques

Some surface area methods use these calculations (see 'Methods based on surface area and volume'). This general approach has been surprisingly quiet. Bodor and Huang (1992) have examined prediction of the partition coefficient of 302 compounds using inputs of properties derived from molecular orbital (AM1) calculations. AM1 was used to optimize the geometry and then provide dipole movement, O and N atom charges, surface area, 'ovality' and other descriptors.

Crippen (Ghose *et al.*, 1988) used calculations of charge density to validate choices and definition of fragments. To avoid the need for any 'correction factors', if possible, these workers defined only atom types (i.e. the atom and its environment), and then predicted some partition coefficients. This interesting approach awaits further development. It requires many atom types to be defined, and may not offer any advantages unless it results in a significant reduction in the number of fragments needing to be defined.

Solubility in water

Yalkowsky and Banerjee (1992) have recently produced a comprehensive and readable review. Therefore, this section on water solubility will be shorter than the topic itself merits. The main approaches to water solubility estimation described are: predictions from partition coefficient, linear solvation energy relationships (LSER) and prediction of activity coefficient.

The authors strongly support the generality of the $\log P_{ow}$-based prediction, and also the thermodynamic soundness of the activity coefficient route; they also attack the LSER school, because they feel that the π^* term is poorly defined. They do, however, acknowledge that the LSER method has insights into the processes involved in solubility.

There is little doubt that the prediction of water solubility (S_W) from $\log P_{ow}$ is the most accessible method for most purposes. The equation most widely used is:

$$\log S_W = -1.123 \log P_{ow} + 0.686 - 0.0099(T_M - 25)$$

where S_W is in mol/l and T_M is in °C.

This equation is used in CHEMEST and has been reviewed by Boethling *et al.* (1988), who showed an error factor of about 4, for 155 substances. Its application to some new substances is described in 'Tests of some of the methods'. A modified form is used for acids:

$$\log S_W = -0.650 \log P_{ow} + 0.289 - 0.0099(T_M - 25)$$

The LSER equation is:

$$\log S_W = 0.54 - 0.0332V + 0.46\pi^* + 5.17\beta$$

(Taft et al., 1985).

It is also improved by the same 'correction' for the solid phase to be turned into a hypothetical supercooled liquid, and the use of calculated volumes (Leahy, 1986).

The use of activity coefficient requires some estimation of the coefficient, e.g. by group contributions.

The AQUAFAC approach (Myrdal et al., 1993) is a new group contribution method. It has been applied to 12 atoms or groups in up to three environments (sp^3, sp^2, or two sp^2). It is successful for the compound classes represented by the groups used.

Henry's constant

There are several useful methods, some of which have been reviewed by Lyman (1985). Suntio et al. (1988) give some useful background and data.

It should also be noted that it is possible to use the approximation

$$h = P_V/S_W$$

where vapour pressure and water solubility can be estimated separately.

Hine and Mookerjee

Hine and Mookerjee (1975) have developed a fragment method, which has been extended by Meylan and Howard (1991), who introduced a 'bond' method, wherein the bonds in a substance are considered as fragments. This method is particularly useful for low functionality compounds or for extending a series of which one or more members is already characterized. The initial compilation included 59 chemical bonds; the program HENRY from Syracuse Research has extended the list to 94 (Meylan and Howard, 1992).

Connectivity index method

Connectivity indices were used by Nirmalakhandan and Speece (1988), who also required input of a fragment method for polarizability. The method may be criticized because it has not been applied to difficult structures.

Acid dissociation constant

As described in 'General principles of prediction of values', this area is foundational in the development of linear free energy relationships. The comprehensive review by Perrin *et al*. (1981) is still the most useful work. A very wide range of structures is described, including cases of heterocyclic ring and aliphatic compounds. In the latter case, a ΔpK_a method is used to account for the inductive effects of neighbouring groups. Prediction of pK_a is usually possible to within one unit, and normally to less than 0.5; for series, the relative errors may be as low as 0.2 units. Quantum mechanical approaches are unable to compete with such errors, because it is so difficult to model the solvation processes involved. Gas phase proton affinities have been studied with some success (Mautner *et al.*, 1986), but these cannot be directly compared to pK_a values.

Rate of hydrolysis

Knowledge of hydrolysis is of primary importance for two distinct reasons. First, measurements of partition coefficient and solubility in water become impossible for a substance unstable over the test duration. Second, hydrolysis is an important process in environmental fate and metabolism of chemicals. The basic principles of hydrolysis mechanisms are well known, and so the following is a brief outline. Neely (1985) has provided a useful review.

Hydrolysis is an example of nucleophilic reactions, which may proceed *via* several mechanisms, for example:

$$S_N1 \quad R\text{---}X \rightarrow [R^+] + X^-$$
$$\downarrow$$
$$ROH$$

$$S_N2 \quad R\text{---}X + OH^- \rightarrow ROH + X^-$$

$$B_{AC}2 \quad R.CO.X \underset{k_{-1}}{\overset{k_1}{\rightleftharpoons}} \left[\begin{array}{c} OH \\ | \\ R\text{---}C\text{---}X \\ | \\ OH \end{array} \right] \overset{k_2}{\rightarrow} \begin{array}{c} R.CO.OH \\ \\ +X^- \end{array}$$

Square brackets indicate an intermediate, usually short-lived. For crop protection compounds, a mechanism such as $B_{AC}2$ or similar ones (addition–eliminations) occur quite frequently. S_N1 and S_N2 reactions usually follow pseudo-first-order kinetics in both disappearance of the substrate and appearance of products. However, if k_{-1} is significant then other rate laws apply. For multistep reactions where rate constants such as k_1 and k_2 are of similar magnitude, then the appearance of products will not follow first-order kinetics and the intermediate may also be detectable. Hydrolysis is a complex process, and not easily susceptible to theoretical treatments such as quantum mechanical ones. This is because several species are involved: starting substance, intermediates and products and associated transition states. Solvent water is intimately involved in all stages. The energy differences between the various states are small. Overall, linear free energy relationships at present provide a more useful framework. A recent review by Taft and Topsom (1987) on the origin of electronic effects is particularly interesting for its insights. For a practical use, the relationship between rate constant and σ and/or steric parameters for the substituents still represents the approach to use.

Estimation of rate constants is important, because a complete set of measured values can be costly. There is a scattering of recent literature; the original work by Hammett (1940) is of little practical use for new classes, at least before some measurements have been made. Katagi (1992) gives a well written account of some of the issues to be considered, and covers organophosphorus and carbamate pesticides.

Another report (IUPAC, 1991), includes a brief discussion of carboxylic acid esters, carbamates, phosphates, acetates and ketals, phthalates and alkyl halides.

However, for a new structural class some measurements are essential, using a good protocol such as OECD method 111. The standard compilations of literature (Harris, 1982) are of relatively little use because they cover simple functionalities only, and frequently high cosolvent levels were used. Therefore, for a compound at the development stage, measurements have to be made. At the discovery level, if a series of, for example, 20 analogues are to be considered, three or four at least will need to be tested in order to establish a valid Hammett relationship (linear free energy relationship).

Translation of laboratory data into field or *in vivo* effects has to take into account temperature, pH and composition of the media. The effect of temperature can only be considered on the basis of measurements and the use of the Arrhenius equation. pH is covered by consideration that the pseudo-first-order rate constant is described by

$$k_1 = k_N + k_A[H^+] + k_B[OH^-]$$

where N refers to attack by water, A by proton and B by hydroxyl (specific acid–base catalysis). An example of this has been given (Takahashi *et al.*, 1985), describing hydrolysis of cypermethrin.

Other than specific acid–base catalysis, the composition of the medium can also give changes in rate of up to an order of magnitude, for any and each of the following:

- general acid or base catalysis,
- specific catalysis by anions,

- metal ion effects,
- ionic strength.

General catalysis is, for example, the participation in the reaction by ions that can receive a proton from an intermediate. Specific catalysis is the direct involvement of an ion in a step of the reaction. Metal ion effects occur primarily due to complex formation. Ionic strength will most strongly affect reactions that involve charged intermediates.

Diffusion coefficient

Wilke and Chang (1955) presented an equation for the prediction of molecular self-diffusion coefficients (D). In this

$$D = 7.4 \times 10^{-8} \, (xM)^{0.5} T / \eta V^{0.6}$$

where D is diffusion coefficient ($cm^2 \, s^{-1}$); M is molar mass of solvent ($g \, mol^{-1}$); V is molar volume ($cm^3 \, mol^{-1}$); x is solvent association parameter; and η is solvent viscosity (cP). Minor developments on this model have been made (see Tucker and Nelken, 1982).

The author has investigated the original data set used by Wilke and Chang, using different inputs. These were: CMR_A, calculated molar refractivity of solute; M_B, molar mass of solvent ($g \, mol^{-1}$); M_A, molar mass of solute ($g \, mol^{-1}$); and η_B, viscosity of the solvent (cP). The result was:

$$\log(D/T) = -5.95 + 0.18 \log M_B - 0.93 \log \eta_B - 0.42 \log(CMR_A)$$

M_A was not needed.

This expression is very similar to that of Wilke and Chang. The principal differences are as follows:

1. Measured molar volumes are replaced by calculated CMR values from the Pomona programs (Medchem, 1987).
2. This model does not use an arbitrary x factor.
3. This model, however, has certain chemical classes as outliers; these are polar molecules. These outliers may be explained as being caused by the tendency of water to hydrogen bond to the diffusing molecule. Therefore, the effective CMR_A is increased.

The full description of this new method will be published (Fisk and Jonathan, 1995).

TESTS OF SOME OF THE METHODS

Several properties relevant to prediction of environmental fate are considered in this section.

The methods used are as follows, which have been detailed under 'Methods for prediction of properties':

- Log P_{ow}: CLOGP (version 3.53), and LOGKOW;
- Water solubility: CHEMEST, using measured log P_{ow} and melting point as inputs;
- Henry's constant: HENRY program (version 2), 'bond' method;
- Vapour pressure: calculated from the water solubility and Henry's constant.

These were chosen for reasons of accessibility.

Application to recent discoveries

The substances chosen are some of those recently discovered (BCPC, 1991, 1992), and their structures are illustrated in Figure 2. The results are collated in Table 5, and expressed graphically in Figure 3 (log P), Figure 4 (water solubility), and Figure 5 (Henry's constant).

Application to established products

The substances have been chosen to represent a diversity of functional groups. Data were obtained from *The Pesticide Manual* (1983) or from the Hansch and Leo database (Medchem, 1987). The results obtained are shown in Table 6. Inspection of the table shows that the predictions are generally successful. They are much superior to the values predicted for the recently invented compounds, reflecting the fact that the programs have taken in the past data to some extent.

Octanol/water partition coefficient

In several cases, CLOGP reported a missing fragment. This is not uncommon in fields such as pesticide discovery, where novelty is essential to the process. The novelty frequently brings combinations of new and existing fragments that the method cannot deal with. However, it is possible to make progress despite this, depending on the purpose of the calculations. In discovery work, the researcher may be interested in relative values and therefore the absolute correctness of a new fragment need not be a concern. The researcher may make an estimate of the value and proceed. In the context of environmental fate, it is usually sufficient to obtain an order of magnitude value. The approach used here is to modify the structure to the closest reasonable analogue that CLOGP can handle.

In the series shown in Table 5, for the structures that CLOGP can handle, it is of interest to note that the correction factors were always important:

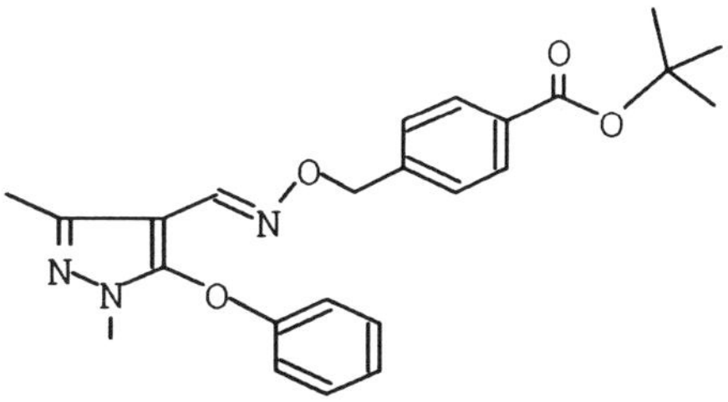

(1) Fenpyroximate

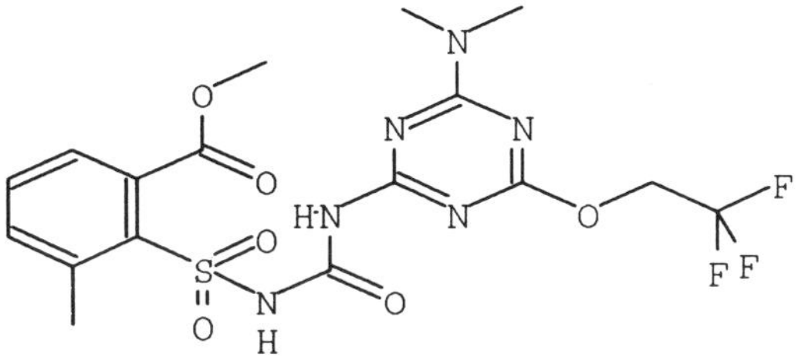

(2) DPX-66037

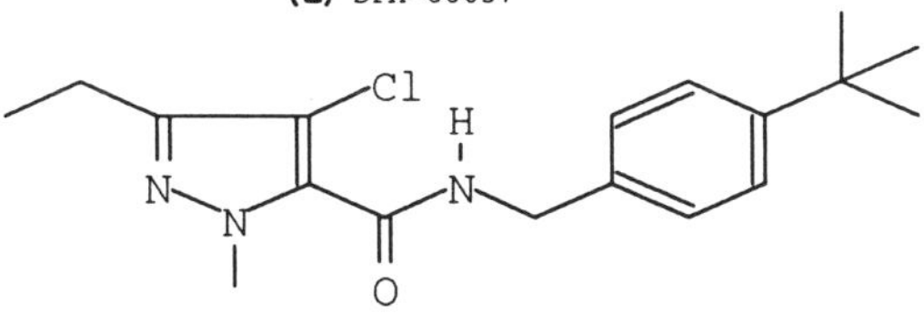

(3) MK-239

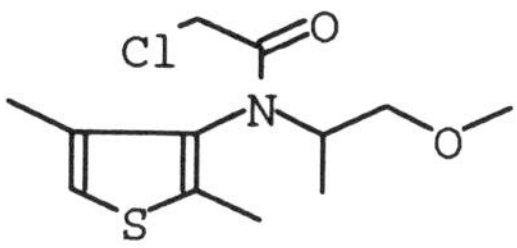

(4) SAN 5824

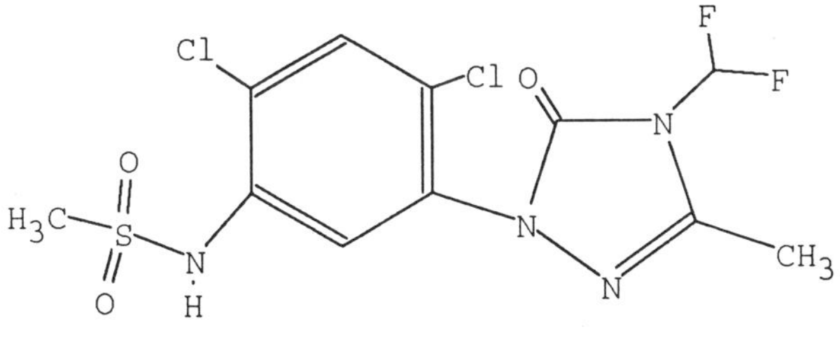

(5) F6285

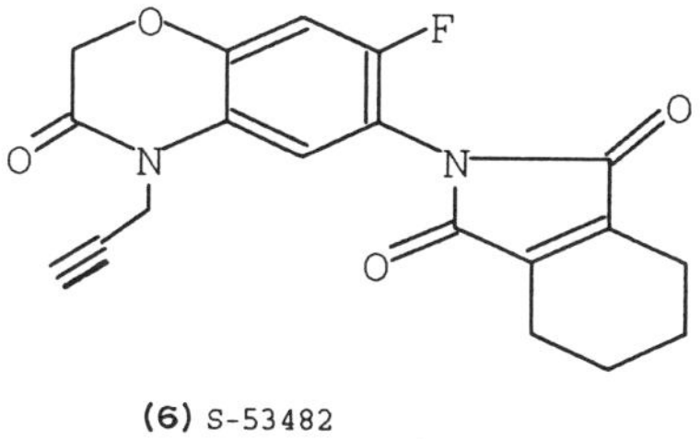

(6) S-53482

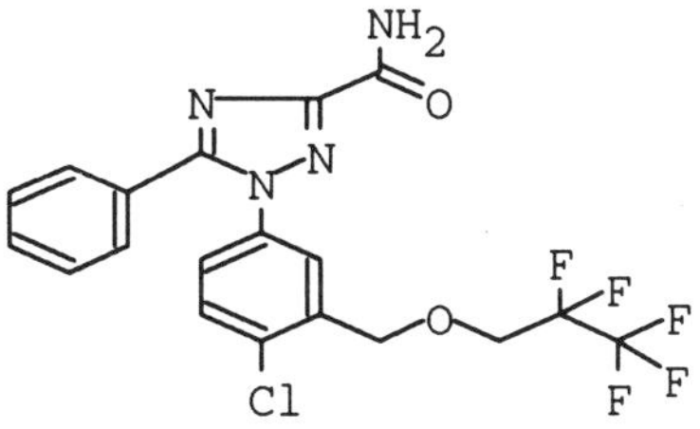

(7) Flupoxam

Figure 2

(8) NC-319

(9) Fipronil

(10) MAT 7484

(11) CGA 215944

(12) Fenazaquin

(13) RH 5992

(14) Flufenprox

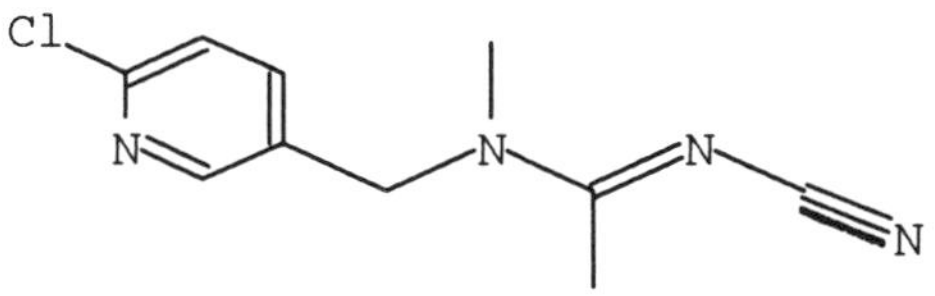

(15) NI-25

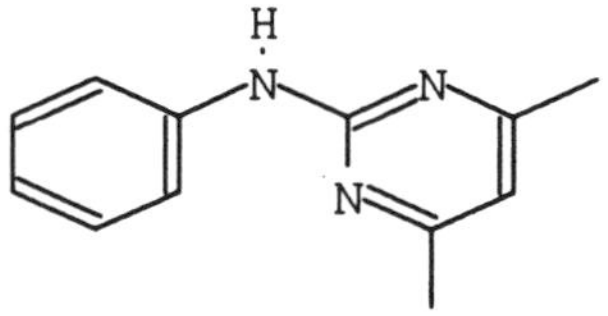

(16) Pyrimethanil

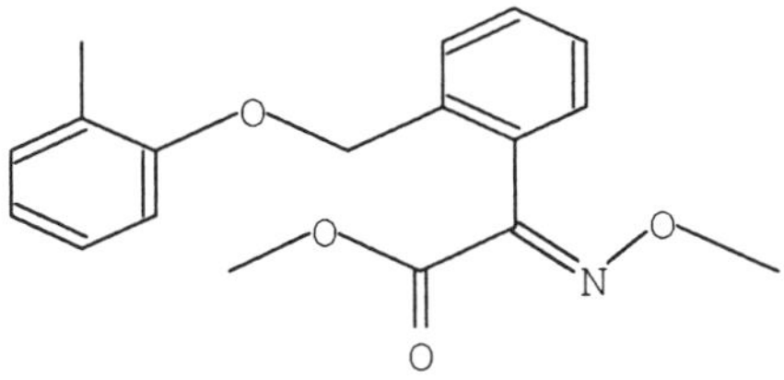

(17) BAS 490F

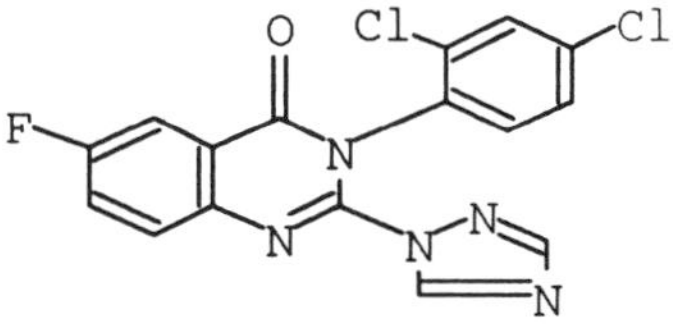

(18) Fluquinconazole

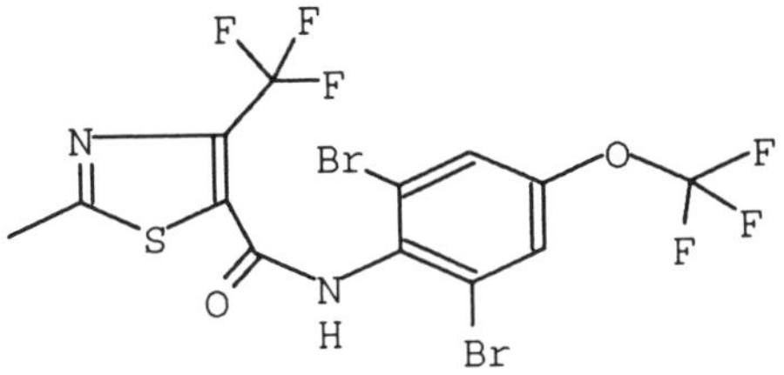

(19) MON 24000

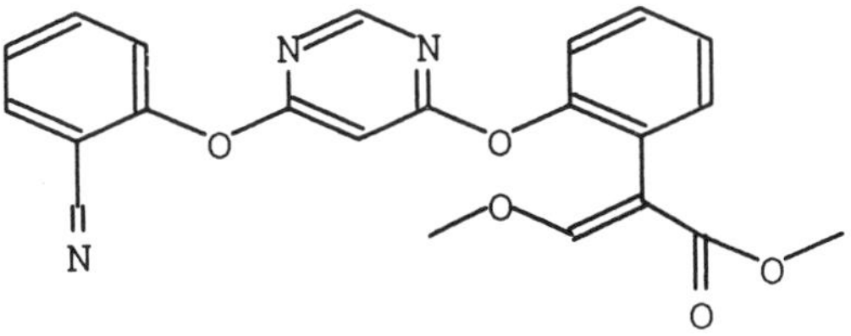

(20) ICIA 5504

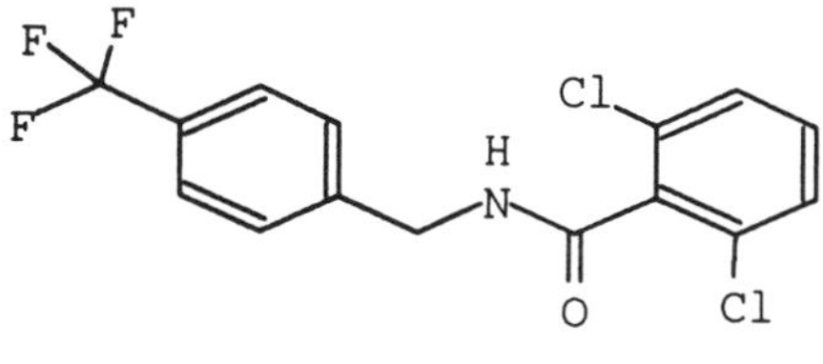

(21) XRD-563

Figure 2   (cont) Chemical structures of recent discoveries

Table 5 Results for recent discoveries

Compound	Henry's constant (Pa m³ mol⁻¹)		Water solubility (mg l⁻¹) at 25 °C		CLOGP	LOGKOW	log P
	Bond method	Measured	Calculated	Measured			
1 Fenpyroximate	4.98E − 06	2.09E − 0.1	0.00556	0.015	6.96	4.42	
2 DPX-66037	1.27E − 09		10200	110	4.37	1.94	
3 MK-239	1.63E − 06	1.74E − 02	4.7	2.8	4.83	5.76	4.61
4 SAN 582	6.16E − 05	8.62E − 03	372	1174	3.17	2.57	
5 F6285	5.09E − 07	6.60E − 08	4090	780		1.93	
6 S-53482	5.24E − 09	6.35E − 02	280	1.79	1.81	1.34	
7 Flupoxam	6.32E − 10	1.34E+03	10	1	3.71	4.45	
8 NC-319	9.75E − 11	4.50E − 09	945	36	1.69	− 0.35	
9 Fipronil	1.52E − 04	8.14E − 05	1.3	2	4.41	7.19	4
10 MAT 7484	1.41E − 02	2.20E − 01	4.5	5.5	3.58	4.19	4.93
11 CGA 215'944	3.59E − 08	7.80E − 08	3250	270		0.89	0.2
12 Fenazaquin	4.31E − 03	4.90E − 01	0.97	0.1	5.57	5.76	5.51
13 RH 5992	8.74E − 06		0.16			4.79	
14 Flufenprox	1.01E − 02	2.40E − 02	0.0757	0.0025	6.64	8.08	
15 NI-25	7.01E − 03	5.30E − 08	262	4200		2.55	
16 Pyrimethanil	2.49E − 01	3.62E − 03	312	121	3.68	3.19	2.5
17 BAS 490F	2.90E − 03	3.65E − 04	45	2		3.68	3.4
18 Fluquinconazole	1.51E − 08	2.41E − 06	10	1	2.65	2.83	3.2
19 MON 24000	6.24E − 08		2	1.6	4.48	6.34	4.1
20 ICIA 5504	8.11E − 09	4.03E − 04	255	10		1.57	2.64
21 XRD-563	1.90E − 04		40	3.5	4.12	5.16	

Table 6 Results for established products

Compound	Henry's constant (Pa m³ mol⁻¹)		Water solubility (mg l⁻¹)		Vapour pressure (Pa)		CLOGP	LOGKOW	log P
	Bond method	Measured	Calculated	Measured	Calculated	Measured			
Permethrin	2.94E − 02	5.11E+02	0.24	0.2	1.80E − 05	2.61E − 01	6.86	7.43	6.15
Parathion	3.04E − 02	6.07E − 02	71	24	7.41E − 03	5.00E − 03	3.47	3.73	3.83
Cyanazine	1.93E − 07	2.82E − 07	147	171	1.18E − 07	2.00E − 07	1.72	2.51	2.22
Linuron	1.22E − 03	6.15E − 03	65	81	3.17E − 04	2.00E − 03	3.18	2.91	3.2
Triadimefon	1.22E − 04	1.13E − 04	301	260	1.25E − 04	1.00E − 04	3.03	2.94	2.77
Dicamba	3.55E − 04	1.53E − 04	2040	6500	3.27E − 03	4.50E − 03	2.46	2.14	2.21
Oxamyl	3.55E − 06	2.43E − 05	355000	280000	5.74E − 03	3.10E − 02		− 1.2	− 0.48
Malathion	8.51E − 05	1.21E − 02	3590	145	9.25E − 04	5.30E − 03	2.14	2.29	2.36

2 log units is quite usual. Corrections for proximity of groups were common. These account for the fact that two or more adjacent polar groups are 'sharing' some of the solvating molecules of water and are therefore able to affect the solvent structure less than might be expected. Through bond effects (σ–ρ corrections) are usually, but not always, less significant. The significance of a 'missing fragment' has to be seen in this context.

The program LOGKOW gave a result in every case, although inspection of the data shows that the quality of the prediction was less good where comparison was possible. These generalizations apply because the definition of fragments is less demanding in LOGKOW than in CLOGP.

The results shown in Figure 3 are not inspiring! However, a fair test is to consider a larger database of measured values available at Sittingbourne Research Centre. For over 400 compounds, over the range of log P of -3 to 6, the average deviation was approximately one log unit. For the same database, the performance of LOGKOW was broadly similar, but superior in the important respect that it could deal with many more compounds than CLOGP could, just as was found in the current test set.

The differences between the results from two programs are illustrated in

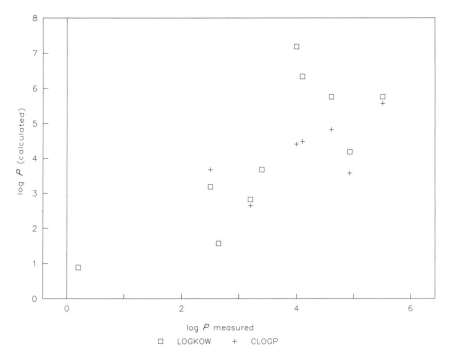

Figure 3   Measured versus calculated log P values

Table 7

LOGKOW:

Type	Num	LOGKOW fragment description	Coeff	Value
Frag	3	—CH$_3$ (aliphatic carbon)	0.5473	1.6419
Frag	2	—CH$_2$ (aliphatic carbon)	0.4911	0.9822
Frag	14	Aromatic carbon	0.2940	4.1160
Frag	2	Aromatic nitrogen	−0.7324	−1.4686
Frag	2	—O— (oxygen, one aromatic attach)	−0.4664	−0.4664
Frag	1	—*tert* carbon (3 or more carbon attach)	0.2676	0.2676
Factor	1	*ortho*-alkoxy (thio) to 1 aromatic nitrogen	0.4549	0.4549
Const		Equation constant		0.2290
			Log $K_{ow}=$	5.7604

CLOGP:

Class	Type	Log(P) contribution description	Comment	Value
Fragment	1	Aromatic nitrogen (Type 2)	MEASURED	−1.120
Fragment	2	Aromatic nitrogen (Type 2)	MEASURED	−1.120
Fragment	3	Ether	MEASURED	−0.610
Isolating	Carbon	6 Aliphatic isolating carbon(s)		1.170
Isolating	Carbon	14 Aromatic isolating carbon(s)		1.820
Fusion	Carbon	1 Extended aromatic *iso*-C(s)		0.100
Fusion	Carbon	1 Extended hetero-aromatic *iso*-C(s)		0.310
Exfragment	Branch	2 chain and 0 cluster branch(es)	(Chain)	−0.260
Exfragment	Hydrog	22 Hydrogen(s) on isolating carbon(s)		4.994
Exfragment	Bonds	6 chain and 0 alicyclic (net)		−0.720
Electronic	Sigrho	6 potential interactions; 4.25 used	Within ring	1.010
Result	v3.4	All fragments measured	CLOGP	5.574

Table 7, which shows an output from each program for calculation of the log P of fenazaquin.

Water solubility

Water solubility predictions used the measured log P_{ow} values (where available) in preference to calculated ones, otherwise CLOGP was used in preference to LOGKOW. Compounds with no melting point stated were assumed to be liquids. If a range of melting point was stated, the lowest temperature was used. The regression statistics for the test set were as follows:

$$\log S_{W(calc)} = 0.865 \log S_{W(meas)} + 0.824$$

with $n = 21$, $r^2 = 0.720$, $s = 0.86$.

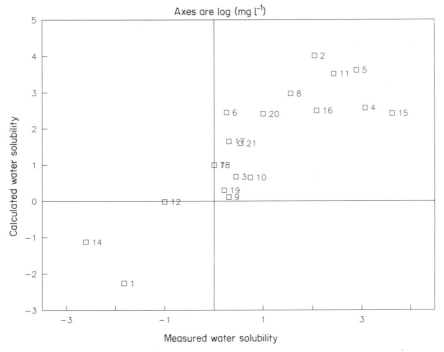

Figure 4   Measured versus calculated water solubility. Axes are $\log(\text{mg l}^{-1})$

These data show that there is a tendency to overestimate the solubility by the prediction based on partition coefficient and melting point, as shown by the non-zero intercept. A possible cause of this is that the method fails for what are, in general, substances of greater complexity than the ones upon which the method was based. Consider first a simple compound of $\log P = 1.9$, with a melting point of 32 °C, such as p-cresol. From Table 5 and Figure 2, we may compare it with substances which have higher melting points (>100 °C), several ring systems and perhaps six or more functionalities. From the equation shown in 'Solubility in water', the melting point correction term is only one log unit for a melting point of 125 °C; this may be inadequate for complex compounds. It should be noted that phenomena concerned with the solvation in water should normally be accounted for in the $\log P$ term; however, in the partition process the possibility for unexpected solvation of either or both water and octanol molecules by a compound cannot be discounted. In $\log P$ prediction, such effects (such as unusually strong hydrogen bonds, or steric hindrance) may mutually cancel because partitioning involves two phases. In contrast, solubility involves one phase only. There is clearly scope for further examination of the correlation of water solubility with $\log P_{\text{ow}}$.

Henry's constant

For this series, only one substance was sufficiently simple for the 'group method' to be successful. Therefore results from the 'bond method' were used throughout. For this method, the correction factors are still somewhat undeveloped, but are typically two to three orders of magnitude. Therefore, it is not surprising that results are disappointing (see Figure 5). Again the data may be compared to larger sets.

For the database available at Sittingbourne, for about 100 compounds with known water solubility and vapour pressure, the results showed a line of unit gradient for predicted vs measured values (measured meaning the estimate based on the ratio of water solubility to vapour pressure). The average error was ±2 log units.

Another database was also examined, published by the USDA (Wauchope *et al.*, 1992). The results showed a deviation from a line of unit gradient for predicted vs measured values (measured meaning the estimate based on the ratio of water solubility to vapour pressure). Also, the average error was ±3 log units. The cause of the rather poor performance of the

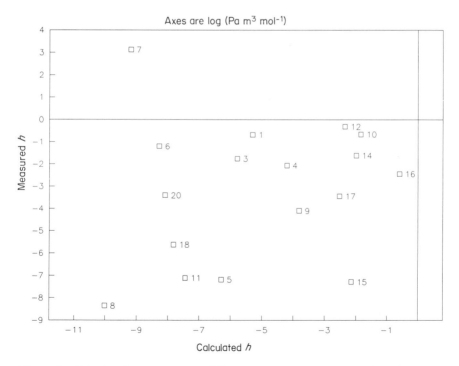

Figure 5   Calculated versus measured Henry constant. Axes are $\log (\mathrm{Pa\,m^3\,mol^{-1}})$

method was probably twofold:

(1) The data were from many sources and probably of varying quality.
(2) The 'bond method' needs too many correction factors to make it work well.

Vapour pressure

In the absence of reliable boiling points (if achievable) for crop protection compounds, the method of estimating vapour pressure from Henry's constant (calculated) and water solubility (measured or calculated) has been used. The results are not shown here as it is obvious from the previous sections how poor they are. It is not surprising that this approach can give large propagated errors. It illustrates well the need for more work on this important property.

USES OF PREDICTED DATA

It is beyond the scope of this review to cover all these topics in depth, so brief surveys of the way physicochemical properties are used predictively will be given.

Soil adsorption

The soil adsorption coefficient is an equilibrium constant for partition of a compound between soil and water. It is useful in modelling the environmental fate of chemicals, and is expressed as K_{oc}. This is the equilibrium constant for the compound between soil organic matter and water. It has been usefully reviewed by Lyman (1982). The K_{oc} value has been the subject of considerable effort in predicting it, with P_{ow} as the most common basis. However, there are also methods based on connectivity (for example, Bahnick and Doucette, 1988).

Gerstl (1990) has provided the most comprehensive review to date. In this paper approximately 600 compounds are discussed, from a variety of chemical classes. Statistical analysis of the data was made according to

$$\log K_{oc} = a \log P_{ow} + b$$

and was more reliable than equivalent relationships based on solubility in water.

Table 8 shows some of the relevant data.

Professor Gerstl has kindly provided his data set, and a repeat statistical

Table 8   Soil adsorption coefficient predicted from Pow

Group	a	b	r^2
Amides	0.253	1.776	0.808
Carbamates	0.433	0.919	0.863
Organophosphorus pesticides	0.689	0.530	0.672
Triazines	0.586	0.826	0.894
Triazoles	0.583	0.969	0.799
Ureas	0.545	0.943	0.713
All compounds	0.679	0.663	0.831

analysis was performed. The result for all classes combined was

$$\log K_{oc} = 0.656 \log P_{ow} + 0.736$$

$$0.018 \qquad\qquad 0.059 \qquad \text{(standard errors)}$$

resulting in the estimation error in $\log K_{oc}$ of ± 1 unit.

Bioaccumulation factor

Connell (1988) has provided a wide-ranging and authoritative review. Bioaccumulation factor (K_B) is best predicted by equations of the form

$$\log K_B = a \log P_{ow} + b$$

Equations based on water solubility tend to be less good. The values of a and b vary with substance type and, more importantly, with the organism under consideration.

In recent years the limits of this approach have been considered. At high $\log P_{ow}$ values (>5), K_B tends to reach a limiting value. Also, for high molecular weights (>600), some consider that membrane permeability is too slow; these arguments are described by Geyer et al. (1987), Barron (1990) and Binstein et al. (1993).

A further exception has been considered by Anliker and Moser (1987), who have pointed out that some substances have very low solubility in both water and lipids and are therefore exceptions. A simple partition model for the organism and substance is inappropriate for these.

Approaches other than those based on partition coefficient have also been reported. These are based on various methods, as follows:

- quantum mechanical descriptors (Saito et al., 1992);
- solvatochromic parameters (Hawker, 1990);
- topological indices (Sabljic, 1990).

These methods show some good correlations; although they may have conceptual advantages over correlations based on the partition coefficient, they offer no practical advantages yet.

Rate of volatilization from water

The Henry's constant of a substance is a measure of its equilibrium between an ideal solution phase and the vapour phase. It is a dimensionless equilibrium constant, termed h herein.

The maximum amount of substance (of known Henry's constant) lost to a known volume of vapour space from a known volume of solution can be calculated by assuming ideal gas behaviour and the volume of a mole of substance at the temperature and pressure of the system. Thus, under thermodynamically ideal conditions, not approaching saturation of the water, in a closed system:

$$\text{percentage in air} = 100(1 + h(V_{\text{air}}/V_{\text{water}}))$$

Henry's constants are frequently cited with dimensions $\text{Pa}\,\text{m}^3\,\text{mol}^{-1}$, and will need to be scaled to dimensionless values before the above equation can be used.

With respect to open systems, the rate of volatilization is much more complex to assess. This subject has been reviewed by Thomas (1982) and Mackay (1985). It should rarely be an issue in the laboratory, but in outdoor testing such as mesocosms, streams or the environment proper (e.g. movement of agrochemicals from exposed water courses), it has to be considered. In summary, a diffusion model for movement of molecules from the water phase to the air phase is set up. Assuming the worst case, the phases are considered as being well mixed. The result obtained is:

$$C_{\text{w}} = (C_{\text{A}}/h) + (C_{\text{w0}} - (C_{\text{A}}/h))\exp(-kat/V_{\text{w}})$$

where C_{w} is the concentration in water; C_{A} is the concentration in air; h is the dimensionless Henry's constant; C_{w0} is the concentration in water at zero time; k is a mass transfer coefficient; a is the area of the gas–liquid interface; t is time; V_{w} is the volume of the solution; and

$$1/k = (1/k_{\text{w}}) + (1/k_{\text{A}}h)$$

where k_{w} is the mass transfer coefficient in water, typically $10^{-5}\,\text{m}\,\text{s}^{-1}$; and k_{A} is the mass transfer coefficient in air, typically $10^{-3}\,\text{m}\,\text{s}^{-1}$. Mass transfer coefficients are derived from a knowledge of diffusion coefficients and the hydrodynamics found in the test system.

Quantitative structure–activity relationships

This section includes several areas of application of physicochemical data that are part of the understanding of the mode of action of crop protection compounds.

Mobility of compounds in plants and insects

This is a large and difficult subject, but some aspects are relevant here. The penetration of compounds through the insect and plant cuticles has received much attention. Reviews are available; see Schoenherr and Reiderer (1989) for plants, and Ford *et al.* (1981) for insects. An example of the correlation with $\log P$ is given by Chamel (Chaumat *et al.*, 1992). Within plants, compounds may move within xylem or phloem systems. Grayson and Kleier (1990) have developed a model of plant phloem which describes a critical pH gradient and critical $\log P_{ow}$ range.

Design of crop protection compounds

The introduction described the interplay of factors affecting the design of crop protection compounds. The subject has been reviewed (Graham-Bryce, 1984), and reviewed with respect to photosystem II inhibition by Draber (1987). Some examples of studies in which particular physicochemical properties played a significant part are given below:

- Benzoylphenoxyurea larvicides: Nakagawa *et al.*, 1984; $\log P_{ow}$, σ, size;
- Thiolane oxime carbamate insecticides: Kurtz and Durden, 1987; $\log P_{ow}$, size;
- Pyrethroid insecticides: Ford *et al.*, 1985; $\log P_{ow}$.

Toxicology and QSAR

There has been relatively little use of QSAR or physicochemical properties in the prediction of *in vivo* toxicity, or even in *in vitro* testing of mutagenicity. In studies of mammalian toxicology, substances are usually presented to the test animal for ingestion, thereby allowing access of the substance to the interior of the animal. The transport of the substance may still be important, however, in that interior barriers have to be crossed. There has been an emphasis on rule-based systems, probably because severe toxic effects tend to be associated with certain functional groups. Hansch has made some contributions on the analysis of *in vitro* data, using $\log P_{ow}$ as a descriptor; see Hansch (1991) and Kumar Debnath *et al.* (1994). A rare example of analysis of acute toxicity data is given by do Amaral *et al.* (1991).

Ecotoxicology and QSAR

The toxicity of chemicals to aquatic organisms has been widely studied, and the general existence of a correlation of toxicity with $\log P_{ow}$ has been examined in depth. This correlation occurs because of the narcotic mechanism of toxicity. However, many agrochemicals have specific modes of action of toxicity to aquatic organisms, and the log P_{ow} based correlations break down in such cases. A useful review of the general mechanism and exceptions to it has been provided by Lipnick (1985), and a recent conference proceedings includes some discussion of crop protection compounds (Hermens and Opperhuizen, 1991). Donkin and Widdows (1990) have provided a wide-ranging and useful review from a biological perspective.

Models of environmental fate

Models of environmental fate are growing in importance in the assessment of risks associated with the use of crop protection compounds and industrial chemicals. Roberts (1990) has given a useful overview of the environmental fate of pesticides.

Regional models based on dividing a notational 'world' into compartments are useful, because they allow the user to focus on the most important issues. Thus, in the fugacity approach (Mackay, 1991; Mackay et al., 1992), the compartments are soil, water, biota in water, sediments in water, and air. At the simplest level, inputs of soil adsorption coefficient, water solubility, Henry's constant and bioaccumulation are needed. The system is allowed to come to equilibrium, and concentrations in the compartments are calculated. At higher levels, degradation and rate of movement between compartments are used as inputs. The latter requires knowledge of diffusion coefficients.

The Mackay approach can be useful for the study of crop protection compounds, by indicating which compartment is most important, and possible further examination by more detailed models. For example, consider a hypothetical series of stable substances of variable log P_{ow}, but with other properties fixed as follows:

$$S_W = 50\,\text{mg/l}, \; P_V = 10^{-5}\,\text{Pa}, \; \text{MW} = 300$$

The value of S_W is not critical; this is because long-term environmental processes are assumed to involve partitioning under conditions where no phase (i.e. compartment) achieves saturation. Figure 6 shows the amount of substance in the soil and water compartments (the amount in air being small) calculated according to a Mackay Level 1 model, as it depends upon log P_{ow}. The balance between water and soil swings strongly around log $P_{ow} = 3$. A similar analysis shows $P_V > 10^{-3}$ Pa can result in significant volatilization (log P_{ow} set at 3).

For crop protection compounds, volatility may be low and the substance deliberately applied to plants and soil. Models of the compound's movement in the soil are appropriate. For example, PRZM (pesticide root zone model, Carsel et al., 1984) calculates the seasonal movement, depth and leaching

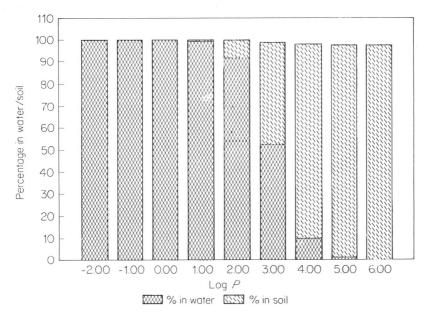

Figure 6 Percentage in water/soil versus log P for a typical pesticide

potential. It requires many inputs, including data such as diffusion coefficients, water solubility and soil adsorption.

CONCLUSIONS

This review has shown that significant efforts have been made over many
years to provide methods for the estimation of values of physicochemical
properties. With respect to some of the particular problems posed by crop
protection compounds, there remains much progress to be made.

For members of a series for which some measured values are known, the
methods are capable of providing values that are suitable for most purposes.
However, if values are required for complex structures, there are very few
methods that are robust enough to give even the correct order of magnitude.
Many of the methods use inputs of related physicochemical properties, and
therefore there is the strong possibility that errors will be propagated at each
stage.

What of the future? An ideal method would be robust, easy to use and
readily available. It would apply to any compound type so that no
measurements on a 'training set' would be needed. Furthermore, it would be
adjustable to account for the end-use of the data—the needs of QSAR are

different to prediction of environmental fate. The author's view is that the way forward lies in a combination of the best of the current technologies with an advancement in the use of quantum mechanical and statistical techniques. Some useful methods, such as the LSER-based equations, could become more useful if they were developed to apply to a wider set of compounds. It would be highly desirable for a project to run that emphasized the measurement of key values of physicochemical properties to allow the extension of the current techniques.

It is the author's view that the following are useful and will remain so:

• CLOGP and LOGKOW, for structures not complicated by stereochemical factors;
• prediction of vapour pressure from boiling point;
• prediction of water solubility from log P_{ow} and melting point;
• estimation of dissociation constant, Henry's constant and diffusion coefficients by current methods.

These are unlikely to be improved upon easily.

The properties that are the most difficult to deal with are:

• melting point,
• boiling point,
• rate of hydrolysis,
• log P_{ow} for complex or novel structures.

It is to be hoped that efforts will continue in these areas.

APPENDIX 1: RAPID ESTIMATION METHODS FOR PHYSICOCHEMICAL PROPERTIES

Octanol/water partition coefficient

Estimation of P_{ow} by chromatographic retention behaviour has been made for over 25 years. Despite this, papers claiming minor novelties continue to appear! There are two main approaches, HPLC and TLC.

High performance liquid chromatography

A procedure suitable for log P_{ow} prediction for regulatory purposes has been described by Eadsforth (1986) and developed by OECD (1991). The system used is as follows:

• reverse phase HPLC, octadecylsilica, methanol: water normally at 75:25 (isocratic elution).
• UV and refractive index detection is normal.
• calibration with suitable reference substances.

- statistical analysis of the $\log P_{ow} - \log k^1$ relationship (k^1 is capacity factor), including 95% confidence range of the predicted $\log P_{ow}$; ± 0.5 is typical.

If the data are not required for regulatory purposes, significant variants on the above are frequently needed. For example, in QSAR the range of P_{ow} of the test substances examined may be very small, and a different solvent regimen may help to achieve $\pm$ of 0.2 log units.

Changing the column may produce good results, but it is suspected that k^1 from some column types, for example, phenyl–silica, is giving information about many factors, not just hydrophobicity.

Thin-layer chromatography

This is very similar in principle to the HPLC method (and predates it). It is essential to use reverse phase plates, such as octadecylsilica. The main advantage over HPLC is that highly insoluble substances can easily be 'spotted' onto the plate. They may well stay on the baseline, which is easily demonstrated. Also, it may not always be easy to elute such substances from HPLC columns, with potentially expensive consequences.

Vapour pressure

It is well known that gas chromatography retention times correlate with vapour pressure; see papers by Bidleman (e.g. Hinckley *et al.*, 1990). It is essential to calibrate the system with suitable reference substances, although these may not be members of a homologous series, depending on the accuracy of the estimation that is required. Both isothermal and temperature-programmed conditions may be used. Columns packed with OV-101 are suitable; estimations of vapour pressures in the range 10^4–10^{-6} Pa are possible.

Water solubility

Exactly the same approach as for P_{ow} may be used. However, it is essential to use closely analogous reference substances; see Tomlinson and Hafkensheid (1986).

APPENDIX 2: RELIABLE MEASUREMENT OF PHYSICOCHEMICAL PROPERTIES

The purpose of this appendix is to summarize what is involved in the measurement of the properties that have been described. For regulatory work, the appropriate guidelines should always be consulted.

Melting point

Although the capillary method is very familiar, good practice dictates that several procedures should be adopted:

- appropriate heating rate (3 °C per minute) from 10 °C below the melting point.
- observation of all stages of melting.

Differential scanning calorimetry and differential thermal analysis both offer significant advantages in the determination of melting and boiling points. Thermal instability or phase transitions prior to melting can be detected. Results are frequently more accurate than conventional methods, particularly for impure samples.

Boiling point

Several methods are available:

- ebulliometer.
- dynamic method: measurement of the vapour recondensation temperature in the reflux whilst boiling.
- distillation method.
- Siwoloboff method (see below).
- DSC and DTA as above.

The dynamic method is useful for reduced pressure measurements.

The Siwoloboff method is straightforward and reliable. It involves dipping a capillary (fused at the upper end) containing an air bubble, into a heated tube containing the sample. At the boiling point the capillary produces a stream of bubbles.

Vapour pressure

It is frequently necessary to perform measurements at temperatures above the required temperature, and to extrapolate the data, typically using $\log(P_V)$ vs T^{-1}, to obtain the desired result, with appropriate statistical analysis.

For pressures of 10 Pa or higher, the 'static method' is easy to use. It is particularly useful for substances containing volatile impurities. It involves balancing the pressure generated by the substance with pressure from a vacuum pump plus nitrogen gas supply.

For pressures of 10 Pa and below, the gas saturation method is particularly useful for substances of purity greater than 90% not containing volatiles, and for which an analytical method is available. It involves passing a carrier gas over the substance and trapping the substance that volatilizes.

Solubility in water

It is essential to know whether the test substance is stable in water over the test interval used. Two methods are available, the choice depending on the solubility range, and in both it is essential to demonstrate that equilibrium has been reached. Good thermostatting is advisable.

For solubility greater than $10 \, \text{mg} \, l^{-1}$, the substance is stirred in a flask, at an excess of at least five times the solubility. Excess test substance should be separated from the solution via centrifugation; avoid filtration unless it is demonstrated that the filters do not adsorb dissolved substance. Analytical methods should be appropriate; that is, unless the substance is of very high purity, chromatographic analysis is preferred.

For solubility less than $10 \, \text{mg} \, l^{-1}$, the column elution method should be used. A film of substance is coated onto an inert support bed in a column. Water is allowed to equilibrate and then slowly passed over the support. Samples of eluate are collected and analysed.

It is essential to record the pH at the start and end of the tests.

Octanol/water partition coefficient

The measurement is not as easy as its frequency of use might suggest. The basic 'shake-flask' approach is the method of choice. The use of column methods (Mirrlees *et al.*, 1976; Miyake and Terada, 1978) analogous to that described for water solubility, or the 'filter-probe' method (Tomlinson *et al.*, 1985), can be useful but are not proven over the whole range. The steps below should ensure success:

(1) Do not test unstable substances or surfactants.
(2) Use at least three concentrations of the test substance in the phase used to initally 'load' the material. Choose appropriate phase volumes.
(3) Do not approach the limit of solubility in either phase.
(4) Always use presaturated octanol and water.
(5) Measure the pH of the water at the start and end of the test.
(6) After shaking or rapidly stirring the test substance for at least 24 hours, centrifuge the mixture in a thermostatted centrifuge. Two stages are sometimes required.
(7) Separate the phases *via* a funnel. Do not contaminate the device (e.g. a syringe) used for sampling a phase with droplets of the other phase.
(8) Use a validated chromatographic analytical method, after a concentration step if necessary. Concentration may be achieved by normal methods such as extraction cartridges or solvent extraction. 'Validated' means the inclusion of blanks, controls and tests of recovery efficiency.

If the above procedure is followed then results in the range $-4 < \log P_{ow} < 6$ can be achieved, dependent on analytical sensitivity. Values in the literature outside this range are probably suspect because they have not been obtained by the shake-flask method.

Acid dissociation constant

Measurement of K_a of an acid RH or a base B starts with a definition: for

$$RH + H_2O \rightleftharpoons R + H_3O^+$$

$$K_a = [R^-]\,[H_3O^+]/[RH]$$

or, for bases,

$$H_2O + BH^+ \rightleftharpoons B + H_3O^+$$

$$K_a = [B]\,[H_3O^+]/[BH^+]$$

i.e. all K_a values are expressed as loss of a proton.

   Traditionally, K_a values have been measured by acid or base titration. For many substances this approach is not possible due to lack of solubility. Albert and Serjeant (1971) clearly describe the alternatives. In summary, these are as follows.

UV/visible spectrometry

This is appropriate when the two forms (RH/R^- or BH^+/B) have significantly different UV spectra. Other spectrometric methods such as nuclear magnetic resonance (in D_2O) are normally unsuitable due to insufficient water solubility.

Variation of S_W or P_{ow} with pH

This method makes use of the enhanced hydrophilicity of the ionized form. It is a very powerful method; see Schaper (1979) for a review, which also deals with multiple ionization.

Hydrolysis

Testing of the rate of hydrolysis of a substance may be performed as follows:

(1) Develop an analytical method for the quantification of the substance. It should be sensitive enough to follow at least two half-lives, and account for products also, if possible. High performance liquid chromatography is often appropriate.
(2) Prepare deoxygenated, sterile buffered solutions at the required pH values and temperatures.
(3) Dose the test substance into the buffer, using less than 1% of cosolvent, at a concentration of less than half the saturated concentration or 0.1 M, whichever is the lower.
(4) Analyse samples at suitable time intervals.
(5) Interpret the data according to standard kinetic models.

GLOSSARY OF SYMBOLS

π^*	polarizability term in LSER
α	hydrogen bond acceptor term in LSER
β	hydrogen bond donor term in LSER
P_{ow}	octanol/water partition coefficient
CMR	calculated molar refractivity from MEDCHEM
T_B	boiling point
V	molar volume
k	rate constant
σ	Hammett sigma constant
ρ	Hammett rho constant
T_M	melting point
P_V	vapour pressure
pK_a	$-\log_{10}$ (acid dissociation constant)
K	general equilibrium constant
π	fragment value used in log P estimation
$m_{\chi t}$	general connectivity index
B,C,D,E,F	principal components of a method (Cramer, 1980a, 1980b)
γ	activity coefficient
x	mole fraction
a	activity
S_W	solubility in water
T	temperature
h	Henry's constant
r	regression coefficient

NOTE ADDED IN PROOF DECEMBER 1994

Syracuse Research extended their range of software available, to include melting point, boiling point, vapour pressure, atmospheric oxidation, hydrolysis, water solubility and soil adsorption.

REFERENCES

Abramowitz, R., and Yalkowsky, S. H. (1990). 'Melting point, boiling point, and symmetry', *Pharm. Res.*, **7**, 942–947.

Albert, A., and Serjeant, E. P., (1971) 'The Determination of Ionization Constants', Chapman and Hall, London.

do Amaral, A. T., Miyazaki, Y., Stachissini, A. S., Caprara, L., and Oliveira, L. (1991). 'A QSAR study of the acute toxicity of a series of hydrochlorides of *N,N*-[(dimethylamino)ethyl] 4-substituted benzoates', in *'QSAR: Rational Approaches to the Design of Bioactive Compounds'* (eds C. Silipo and A. Vittoria), Elsevier, Amsterdam.

Ambrose, D., and Sprake, C. H. S. (1970). 'Thermodynamic properties of organic oxygen compounds', *J. Chem. Thermodynamics*, 1970, **2**, 631.

Anliker, R., and Moser, P. (1987). 'The limits of bioaccumulation of organic pigments in fish: their relation to the partition coefficient and the solubility in water and octanol', *Ecotox. Environ. Safety*, **13**, 43–52.

Bahnick, D. A., and Doucette, W. J. (1988). 'Use of molecular connectivity indices to estimate soil sorption coefficients for organic chemicals', *Chemosphere*, **17**, 1703–1715.

Banerjee, S., Howard, P. H., and Lande, S. S. (1990). 'General structure–vapor pressure relationships for organics', *Chemosphere*, **21**, 1173–1180.

Barron, M. G. (1990). 'Bioconcentration—will water-borne organic chemicals accumulate in aquatic animals? *Environ. Sci. Technol.*, **25**, 536–539.

BCPC (1991). *Proceedings 1991 British Crop Protection Conference*. British Crop Protection Council, London.

BCPC (1992). *Proceedings 1992 British Crop Protection Conference*. British Crop Protection Council, London.

Binstein, S., Devillers, J., and Karcher, W. (1993). 'Nonlinear dependence of fish bioconcentration on *n*-octanol/water partition coefficient', *SAR and QSAR in Environmental Research*, **1**, 29–39.

Bodor, N., and Huang, M-J. (1992). 'An extended version of a novel method for the estimation of partition coefficients', *J. Pharm. Sci.*, **81**, 272.

Boethling, R. S., Campbell, S. E., Lynch, D. G., and La Veck, G. D. (1988). 'Validation of CHEMEST, an online system for the estimation of chemical properties', *Ecotox. Environ. Safety*, **15**, 21–30.

Briggs, G. G. (1981). 'Relationships between chemical structure and the behaviour and fate of pesticides', *Proceedings 1981 British Crop Protection Conference—Pests and Diseases*, pp. 701–710.

Briggs, G. G. (1990). 'Predicting the behaviour of pesticides in soil from their physical and chemical properties', *Philos. Trans. R. Soc. London, B.* **329**, 375–382.

Briggs, G. G. (1992). 'Environmental behaviour as a guide to the synthesis of pesticides', presented at '*Environmental Fate of Chemicals: Prediction and Measurement*', a Royal Society of Chemistry and SETAC symposium, September 1992.

Burkhard, L. P. (1985), 'Estimation of vapor pressures for halogenated aromatic hydrocarbons by a group-contribution method', *Ind. Eng. Chem. Fundam.*, **24**, 119–120.

Burkhard, L. P., Andren, A. W., and Armstrong, D. E. (1985). 'Estimation of vapor pressures for polychlorinated biphenyls: a comparison of eleven predictive methods', *Environ. Sci. Technol.*, **19**, 500–507.

Carsel, R. F., Smith, C. N., Mulkey, L. A., Dean, J. D., and Jowise, P. P. (1984). *Users Manual for the Pesticide Root Zone Model (PRZM): Release 1*. USEPA EPA-600/3-84-109. US Govt. Print Office, Washington, DC.

Charton, M. (1981). 'Electrical effect substituent constants for correlation analysis', *Prog. Phys. Org. Chem.*, **13**, 119–252.

Charton, M. (1987). 'A general treatment of electrical effects', *Prog. Phys. Org. Chem.*, **16**, 287–316.

Charton, M. (1989). 'Directing and activating effects of doubly bonded groups', in *The Chemistry of Doubly-bonded Functional Groups* Vol. 2, Pt. 1, 239–298 (ed. S. Patai), pp. John Wiley, Chichester.

Charton, M., and Charton, B. I. (1984). 'The estimation of melting points', in *QSAR in Design of Bioactive Compounds* (ed. M. Kuchar), pp. 41–51. Prous, Barcelona.

Chaumat, E., Chamel, A., Taillandier, G., and Tissut, M. (1992). 'Quantitative relationships between structure and penetration of phenylurea herbicides through isolated plant cuticles', *Chemosphere*, **24**, 189–200.

CHEMEST (1991). Supplied to the author by Technical Database Services, New York.

Chung, P. T. (1985). 'Estimation of physical and chemical properties', *Proc. Tech. Semin. Chem. Spills 2nd*, pp. 75–90, Environ. Prot. Service, Ottawa.

Collander, R. (1951). 'The partition of organic compounds between higher alcohols and water', *Acta Chem. Scand.*, **5**, 774–780.

Connell, D. W. (1988). 'Bioaccumulation behaviour of persistent organic chemicals with aquatic organisms', *Rev. Environ. Contam. Toxicol.*, **101**, 117–154.

Cramer III, R. D. (1980a). 'BC(DEF) parameters. 1. The intrinsic dimensionality of intermolecular interactions in the liquid state', *J. Am. Chem. Soc.*, **102**, 1837–1849.

Cramer III, R. D. (1980b). 'BC(DEF) parameters. 2. An empirical structure-based scheme for the prediction of some physical properties', *J. Am. Chem. Soc.*, **102**, 1849–1859.

Cramer III, R. D. (1983). 'BC(DEF) coordinates 3. Their acquisition from physical property data, *Quant. Struct.-Act. Relat Pharmacol., Chem. Biol.*, 2(1), 7–12.

Cramer III, R. D., Patterson, D. E., and Bunce, J. D. (1988). 'Comparative molecular field analysis (CoMFA). 1. Effect of shape on binding of steroids to carrier proteins', *J. Am. Chem. Soc.*, **110**, 5959–5967.

Dearden, J. C. (1991). 'The QSAR prediction of melting point, a property of environmental relevance', in *QSAR in Environmental Toxicology—IV* (eds J. L. H. Hermens and A. Opperhuizen), p. 68, Elsevier, Amsterdam.

Donkin, P., and Widdows, J. (1990). 'Quantitative structure–activity relationships in aquatic invertebrate toxicology', *Rev. Aquat. Sci.*, **2**, 375–398.

Draber, W. (1987). 'Can quantitative structure–activity analyses and molecular graphics assist is designing new inhibitors of photosystem II?', *Z. Naturforsch.*, **42c**, 713–717.

D'Souza, R., and Teja, A. S. (1987). 'The prediction of the vapor pressures of carboxylic acids', *Chem. Eng. Commun.*, **61**, 13–22.

Dunn III, W. J., Koehler, M. G., and Grigoras, S. (1987). 'The role of solvent-accessible surface area in determining partition coefficients', *J. Med. Chem.*, **30**, 1121–1126.

Eadsforth, C. V. (1986). 'Application of reverse-phase HPLC for the determination of partition coefficients', *Pestic. Sci.*, **17**, 311–325.

Egolf, L. M., and Jurs, P. C. (1993). 'Prediction of boiling points of organic heterocyclic compounds using regression and neural network techniques', *J. Chem. Inf. Comput. Sci.*, **33**, 616–625.

Ehrenson, S., Brownlee, R. T. C., and Taft, R. W. (1973). 'A generalized treatment of substituent effects in the benzene series', *Prog. Phys. Org. Chem.*, **10**, 1–80.

Fisk, P. R., and Jonathan, P. (1995). submitted to *Chem. Eng. Sci.*

Ford, M. G., Greenwood, R., and Thomas, P. J. (1981). 'The kinetics of insecticide action. Part I: the properties of a mathematical model describing insect pharmacokinetics', *Pestic. Sci.*, **12**, 175–198.

Ford, M. G., Greenwood, R., Leake, L. D., Szydlo, R. M., and Turner, C. H. (1985). 'The relationship between the physicochemical properties of pyrethroid insecticides and their speed of neurotoxic action', *Pestic. Sci.*, **16**, 673–683.

Fredenslund, A., Gmehling, J., and Rasmussen, P. (1977). 'Vapor–Liquid Equilibria Using UNIFAC'. Elsevier, Amsterdam.

Gerstl, Z. (1990). 'Estimation of organic chemical sorption by soils', *J. Contamin. Hydrol.*, **6**, 357–375.

Geyer, H. J., Scheunert, I., and Korte, F. (1987). 'Correlation between the bioconcentration potential of organic environmental chemicals in humans and their *n*-octanol/water partition coefficients', *Chemosphere*, **16**, 239–252.

Ghose, A. K., Pritchett, A., and Crippen, G. M. (1988). 'Atomic physicochemical parameters for three dimensional structure directed QSAR III: modelling hydrophobic interactions', *J. Comp. Chem.*, **9**, 80–90.

Gmehling, J., Rasmussen, P., and Fredunslund, Aa. (1982). 'Vapor-liquid equilibria by UNIFAC group contribution. Revision and extension 2'. *Ind. Eng. Chem. Process Des. Dev.*, **21**, 118–127.

Gold, P., and Ogle, G. (1969). 'Estimating thermophysical properties of liquids. Part 4—Boiling, freezing and triple point temperatures', *Chem. Eng.*, 119.

Graham-Bryce, I. J. (1984). 'Optimization of physicochemical and biophysical properties of pesticides', 185–207, in '*Pesticide Synthesis through Rational Processes*' (eds P. S. Magee, G. K. Kohn and J. J. Menn), ACS Symposium Series 255, American Chemical Society, Washington, DC.

Grain, C. F. (1983). 'Vapour pressure', in '*Handbook of Chemical Property Estimation Methods*' (eds W. J. Lyman, W. F. Reehl, and D. H. Rosenblatt), Chapter 14, McGraw-Hill, New York.

Grain, C. F., and Lyman, W. J. (1983). 'Enhancement to CHEMEST program: estimation of melting points', EPA Contract 68-01-6271, Arthur D. Little, Inc., Cambridge, MA.

Grayson, B. T., and Kleier, D. A. (1990). 'Phloem mobility of xenobiotics. IV. Modeling of pesticide movement in plants', *Pestic. Sci.*, **30**, 67–79.

Guthrie, P. J. (1986). 'Estimation of heats of vaporization for non-associating liquids from the boiling points at various pressures', *Can. J. Chem.*, **64**, 635–640.

Hall, L. H. (1990). 'Computational aspects of molecular connectivity and its role in structure–property modeling', in *Computational Chemical Graph Theory* (ed. D. H. Rouvray), Chapter 8, pp. 202–233, Nova Press, New York.

Hall, L. H., and Kier, L. B. (1991). 'The molecular connectivity chi indexes and kappa shape indexes in structure–property modeling', in *Rev. Comput. Chem.* (1991), **2**, 367–422.

Hammett, L. P. (1940). 'Physical Organic Chemistry', McGraw-Hill, New York.

Hansch, C., and Leo, A. J. (1979). *Substituent Constants for Correlation Analysis in Chemistry and Biology*, John Wiley, New York.

Hansch, C. (1991). 'New perspectives in QSAR' 3–10, in *QSAR: Rational Approaches to the Design of Bioactive Compounds*' (eds C. Silipo, and A. Vittoria), Elsevier, Amsterdam.

Hansen, P. J., and Jurs, P. C. (1987). 'Prediction of olefin boiling points from molecular structure', *Anal. Chem.*, **59**, 2322–2327.

Harris, J. C. (1982). 'Rate of hydrolysis', in *Handbook of Chemical Property Estimation Methods* (eds W. J. Lyman, W. F. Reehl and D. H. Rosenblatt), Chapter 7, McGraw-Hill, New York.

Hart, J. W. (1991). 'The use of data estimation methods by regulatory authorities', *Sci. Total Environ.*, **109/110**, 629–633.

Hawker, D. (1990). 'Description of fish bioconcentration factors in terms of solvatochromic parameters', *Chemosphere*, **20**, 467–477.

Hermens, J. L. H., and Opperhuizen, A. (eds) (1991). *QSAR in environmental toxicology—IV*, Elsevier, Amsterdam. Previously published in *Sci. Total Environ.*, **109/110**.

Hinckley, D. A., Bidleman, T. F., Foreman, W. T., and Tuschall, J. R. (1990). 'Determination of vapour pressures for nonpolar and semipolar organic compounds from gas chromatographic retention data', *J. Chem. Eng. Data*, **35**, 232–237.

Hine, J., and Mookerjee, P. K. (1975). 'The intrinsic hydrophobic character of organic compounds. Correlations in terms of structural contributions', *J. Org. Chem.*, **40**, 292–298.

Hoshino, D., Zhu, X-R., Nagahama, K., and Hirata, M. (1985). 'Prediction of vapor pressures of substituted benzenes by a group-contribution method', *Ind. Eng. Chem. Fundam.*, **24**, 112–114.

IUPAC (1991). 'The use of quantitative structure–activity relationships for predicting rates of environmental hydrolysis processes', *Pure Appl. Chem.*, **63**, 1667–1676.

Kamlet, M. J., Doherty, R. M., Abraham, M. H., Marcus, Y., and Taft, R. W. (1988). 'Linear solvation energy relationships. 46. An improved equation for correlation and prediction of octanol–water partition coefficients of organic nonelectrolytes', *J. Phys. Chem.*, **92**, 5244–5255.

Katagi, T. (1992). 'Application of molecular orbital calculations to the estimation of environmental and metabolic fates of pesticides', in *Rational Approaches to Structure, Activity and Ecotoxicology of Agrochemicals* (eds W. Draber, and T. Fujita), Chapter 21, CRC Press, Boca Raton, Florida.

Kier, L. B. (1990). 'Indexes of molecular shape from chemical graphs', in *Computational Chemical Graph Theory*, (ed. D. H. Rouvray), Chapter 6, pp. 152–174, Nova Press, New York.

Klein, W. (1992). *SAR Programm*, Fraunhofer Institute, D-5948 Schmallenberg, Germany.

Kollman, P. (1985). 'Theory of complex molecular interactions: computer graphics, distance geometry, molecular mechanics, and quantum mechanics', *Acc. Chem. Res.*, **18**, 105–111.

Kollman, P. (1987). 'Molecular modeling', *Ann. Rev. Phys. Chem.*, **38**, 303–316.

Kollman, P., and Merz, K. M. (1990). 'Computer modeling of the interactions of complex molecules', *Acc. Chem. Res.*, **23**, 246–252.

Kumar Debnath, A., Shusterman, A. J., Lopez de Compadre, R. L., and Hansch, C. (1994). 'The importance of the hydrophobic interaction in the mutagenicity of organic compounds', *Mutat. Res.*, **305**, 63–72.

Kurtz, A. P., and Durden, J. A. (1987). 'Quantitative structure–activity relationships in insecticidal oxathiolane and dithiolane oxime carbamates and related compounds', *J. Agric. Food Chem.*, **35**, 115–121.

Lai, W. Y., Chen, D. H., and Maddox, R. N. (1987). 'Application of a non-linear group-contribution model to the prediction of physical constants. 1. Predicting normal boiling points with chemical structure', *Ind. Eng. Chem. Res.*, **26**, 1072–1079.

Leahy, D. E. (1986). 'Intrinsic molecular volume as a measure of the cavity term in linear solvation energy relationships: octanol–water partition coefficients and aqueous solubilities', *J. Pharm. Sci.*, **75**, 629–636.

Leahy, D. E., Taylor, P. J., and Wait, A. R. (1989). 'Model solvent systems for QSAR Part 1. Propylene glycol dipelargonate (PGDP). A new standard solvent for use in partition coefficient determination', *Quant. Struct.–Act Relat.*, **8**, 17–31.

Leahy, D. E., Morris, J. J., Taylor, P. J., and Wait, A. R. (1991). 'Membranes and their models: towards a rational choice of partitioning system', 75–82 in *QSAR: Rational Approaches to the Design of Bioactive Compounds* (eds C. Silipo and A. Vittoria) Elsevier, Amsterdam.

Leo, A. J. (1992). '30 years of calculating $\log P_{oct}$', *QSAR 92 Meeting, Duluth*, p. 125.

Lipnick, R. L. (1985). 'Validation and extension of fish toxicity QSARs and interspecies comparisons for certain classes of organic chemicals', *Pharmacochem. Libr., 8 (QSAR Toxicol. Xenobiochem.)*, 39–52.

Livingstone, D. J. (1989). 'Multivariate quantitative structure–activity relationship methods which may be applied to pesticide research', *Pestic. Sci.*, **27**, 287–304.

LOGKOW (1992). Syracuse Research Corporation, New York.

Lyman, W. J. (1982b). 'Adsorption coefficient for soils and sediments', in *Handbook of Chemical Property Estimation Methods* (eds W. J. Lyman, W. F. Reehl and W. F. Rosenblatt), Chapter 4, McGraw-Hill, New York.

Lyman, W. J. (1985). 'Estimation of physical properties', in *Environmental Exposure from Chemicals* (eds W. B. Neely and G. E. Blau), Chapter 2, pp. 13–48, CRC Press, Boca Raton, Florida.

Lyman, W. J., Reehl, W. F., and Rosenblatt, D. H. (1982). *Handbook of Chemical Property Estimation Methods*, McGraw-Hill, New York. This book has recently been republished by the American Chemical Society.

Macedo, E. A., Weidlich, U., Gmehling, J., and Rasmussen, P. (1982). 'Vapor–liquid equilibria by UNIFAC group contribution. Revision and extension 3', *Ind. Eng. Chem. Process Des. Dev.*, **21**, 676–678.

Mackay, D. (1985). 'Air/water exchange coefficients', in *Environmental Exposure from Chemicals* (eds W. B. Neely and G. E. Blau), Chapter 5, pp. 91–108, CRC Press, Boca Raton, Florida.

Mackay, D. (1991). *Multimedia Environmental Models: the Fugacity Approach*, Lewis Publishers, Chelsea, Michigan, USA.

Mackay, D., Bobra, A., Chan, D. W., and Shiu, W. Y. (1982). 'Vapour pressure correlations for low-volatility environmental chemicals', *Environ. Sci. Technol.*, **16**, 645–649.

Mackay, D., Paterson, S., and Shiu, W. Y. (1992). 'Generic models for evaluating the regional fate of chemicals', *Chemosphere*, **24**, 695–717.

Magnussen, T., Rasmussen, P., and Fredunslund, Aa. (1981). 'UNIFAC parameter table for prediction of liquid–liquid equilibria', *Ind. Eng. Chem. Process Des. Dev.*, **20**, 331–339.

Mautner, M. M.-N., Liebman, J. F., and delBene, J. E. (1986). *J. Org. Chem.*, **51**, 1105–1110.

Medchem (1987). *Medchem Software manual*, Medicinal Chemistry Project, Pomona College, Claremont, California.

Meissner, H. P. (1949). 'Critical constants from parachor and molar refraction', *Chem. Eng. Prog.*, **45**, 149–153.

Meylan, W. M., and Howard, P. H. (1991). 'Bond contribution method for estimating Henry's law constants', *Environ. Toxicol. Chem.*, **10**, 1283–1293.

Meylan, W., and Howard, P. (eds) (1992). *Users Guide for HENRY Program*, Syracuse. Research Corporation, Syracuse, New York.

Mirrlees, M. S., Moulton, S. J., Murphy, C. T., and Taylor, P. J. (1976). 'Direct measurement of octanol–water partition coefficients in liquid–liquid chromatography', *J. Med. Chem.*, **19**, 615–618.

Mishra, D. S., and Yalkowsky, S. H. (1990). 'Estimation of entropy of vaporization: effect of chain length', *Chemosphere*, **21**, 111–117.

Mishra, D. S., and Yalkowsky, S. H. (1991). 'Estimation of vapor pressure of some organic compounds', *Ind. Eng. Chem. Res.*, **30**, 1609–1612.

Miyake, K., and Terada, H. (1978). 'Preparation of a column with octanol-like properties for high-performance liquid chromatography', *J. Chromatogr.*, **157**, 386–390.

MOLCONN-X (1991). Hall Associates Consulting, Quincy, MA 02170.

Myrdal, P., Ward, G. H., Simamora, P., and Yalkowsky, S. H. (1993). 'AQUA-FAC: aqueous functional group activity coefficients', *SAR and QSAR in Environmental Research*', **1**, 53–61.

Nakagawa, Y., Iwamura, H., and Fujita, T. (1984). 'Quantitative structure–activity studies of benzoylphenylures larvicides', *Pestic. Biochem. Physiol.*, **21**, 309–325.

Neely, W. B. (1985). 'Hydrolysis', in *Environmental Exposure from Chemicals* (eds W. B. Neely and G. E. Blau), Chapter 7, pp. 157–176, CRC Press, Boca Raton, Florida.

Nirmalakhandan, N. N., and Speece, R. E. (1988). 'QSAR model for predicting Henry's constant', *Environ. Sci. Technol.*, **22**, 1349–1357.

OECD (1991). 'OECD Guidelines for Testing of Chemicals, OECD, Paris.

Official Journal of the European Communities, L383A, 29th December 1992.

Perrin, D. D., Dempsey, B., and Serjeant, E. P. (1981). '*pK_a Prediction for Organic Acids and Bases*', Chapman and Hall, London.

The Pesticide Manual (1983). 7th Edition Eds. C. R. Worthing and S. B. Walker, British Crop Protection Council, London.

Rechsteiner, C. E. (1982). 'Boiling point', in *Handbook of Chemical Property Estimation Methods* (eds W. J. Lyman, W. F. Reehl and D. H. Rosenblatt), Chapter 12, McGraw-Hill, New York.

Roberts, T. R. (1990). 'Environmental fate of pesticides: a perspective', in *Progress in Pesticide Biochemistry and Toxicology* (eds D. H. Hutson and T. R. Roberts), Vol. 7, Chapter 1, John Wiley, Chichester.

Rohrbaugh, R. H., and Jurs, P. C. (1987). 'Descriptions of molecular shape applied in studies of structure/activity and structure/property relationships', *Anal. Chim. Acta*, **199**, 99–109.

Rordorf, B. F. (1986). 'Thermal properties of dioxins, furans and related compounds', *Chemosphere*, **15**, 1325–1332.

Rordorf, B. F. (1989). 'Prediction of vapour pressure, boiling points and enthalpies of fusion for twenty-nine halogenated dibenzo-*p*-dioxins and fifty-five dibenzofurans by a vapor pressure correlation method', *Chemosphere*, **18**, 783–788.

Sabljic, A. (1990). 'Topological indices and environmental chemistry', in *Practical Applications of Quantitative Structure–Activity Relationships (QSAR) in Environmental Chemistry and Toxicology* (eds W. Karcher and J. Devillers), pp. 61–82, ECSC, EEC, EAEC, Brussels and Luxembourg.

Sabljic, A., and Trinajstic, N. (1981). 'Quantitative structure–activity relationships: the role of topological indices', *Acta Pharm. Jugoslav.*, **31**, 189–214.

Saito, S., Tanoue, A., and Matsuo, M. (1992). 'Applicability of the i/o characters to a quantitative description of bioconcentration of organic chemicals in fish', *Chemosphere*, **24**, 81–87.

Schaper, K.-J. (1979). 'Simultaneous determination of electronic and lipophilic properties (pK_a, P(ion), P(neutral)) of acids and bases by nonlinear regression analysis of pH-dependent partition measurements', *J. Chem. Res.*, 357.

Schoenherr, J., and Reiderer, M. (1989). 'Foliar penetration and accumulation of organic chemicals in plant cuticles', *Rev. Environ. Contam. Toxicol.*, **108**, 1–70.

Shorter, J. (1978). 'Multiparameter extensions of the Hammett equation', in *Correlation Analysis in Chemistry* (eds N. B. Chapman and J. Shorter), Chapter 4, Plenum Press, New York.

Simamoro, P., Miller, A. H., and Yalkowsky, S. H. (1993). 'Melting point and normal boiling point correlations: application to rigid aromatic hydrocarbons', *J. Chem. Inf. Comput. Sci.*, **33**, 437–440.

Smeeks, F. C., and Jurs, P. C. (1990). 'Prediction of boiling points of alcohols from molecular structure', *Anal. Chim. Acta*, **233**, 111–119.

Stanton, D. T., Jurs, P. C., and Hicks, M. G. (1991). 'Computer-assisted prediction of normal boiling points of furans, tetrahydrofurans and thiophenes', *J. Chem. Inf. Comput. Sci.*, **31**, 301–310.

Suntio, L. R., Shiu, W. Y., Mackay, D., Seiber, J. N., and Glotfelty, D. (1988). 'Critical review of Henry's law constants for pesticides', *Rev. Environ. Contam. Toxicol.*, **103**, 1–59.

Taft, R. W., Abraham, M. H., Doherty, R. M., and Kamlet, M. J. (1985). 'The molecular properties governing solubilities of organic nonelectrolytes in water', *Nature*, **313**, 384–386.

Taft, R. W., and Topsom, R. D. (1987). 'The nature and analysis of substituent electronic effects', *Prog. Phys. Org. Chem.*, **16**, 1–84.

Takahashi, N., Mikami, N., Matsuda, T., and Miyamoto, J. (1985). 'Hydrolysis of the pyrethroid insecticide cypermethrin in aqueous media', *J. Pestic. Sci.*, **10**, 643–648.

Thomas, R. G. (1982). 'Volatilization from water', in *Handbook of Chemical Property Estimation Methods* (eds W. J. Lyman, W. F. Reehl and D. H. Rosenblatt), Chapter 15, McGraw-Hill, New York.

Tomlinson, E., and Hafkensheid, T. L. (1986). 'Aqueous solubility and partition coefficient data from high pressure liquid chromatography data', in *Partition Coefficient Determination and Estimation* (eds W. J. Dunn III, J. H. Block and R. S. Pearlman), pp. 101–141, Pergamon, Oxford.

Tomlinson, E., Davis, S. S., Parr, G. D., James, N., Farraj, N., Kinkel, J. R. M., Gaissner, D., and Wynne, H. J. (1986). 'The filter probe extractor: a versatile tool for the rapid determination of solute oil–water distribution coefficients', in *Partition Coefficient Determination and Estimation* (eds W. J. Dunn III, J. H. Block and R. S. Pearlman), pp. 83–100, Pergamon, Oxford.

Tsakanikas, P. D., and Yalkowsky, S. H. (1988). 'Estimation of melting point of flexible molecules: aliphatic hydrocarbons', *Toxicol. Environ. Chem.*, **17**, 19.

Tucker, W. A., and Nelken, L. H. (1982). 'Diffusion coefficients in air and water', in *Handbook of Chemical Property Estimation Methods* (eds W. J. Lyman, W. F. Reehl and D. H. Rosenblatt), Chapter 17, McGraw-Hill, New York.

Twu, C. H. (1983). 'Prediction of thermodynamic properties of normal paraffins using only normal boiling point', *Fluid Phase Equilib.*, **112**, 65–81.

Wauchope, R. D., Buttler, T. M., Hornsby, A. G., Augustijn-Beckers, P. W. M., and Burt, J. P. (1992). 'The SCS/ARS/CES pesticide properties database for environmental decision-making', *Rev. Environ. Contam. Toxicol.*, **123**, 1–155.

Weininger, D. (1988). 'SMILES, a chemical and information system. 1. Introduction to methodology and encoding rules', *J. Chem. Inf. Comput. Sci.*, **28**, 31–36.

Wilke, C. R., and Chang, P. (1955). 'Correlation of diffusion coefficients in dilute solution', *AIChE J.*, 264–270.

Yalkowsky, S., and Banerjee, S. (1992), *Aqueous Solubility: Methods of Estimation for Organic Compounds*, Dekker, New York.

Yalkowsky, S., and Mishra, D. (1992). 'UPPER: Unified physical property estimation relationship', *QSAR 92 conference, Duluth*.

CHAPTER 2

Recommended approaches to assess pesticide mobility in soil

M. H. Russell

Environmental Behaviour of Agrochemicals
Edited by T. R. Roberts and P. C. Kearney © 1995 John Wiley & Sons Ltd

INTRODUCTION

The many economic and food quality benefits of pesticide use have led to their increasing use over the past several decades. This trend, coupled with ever more precise analytical capabilities, has led to the occasional detection of traces of some pesticides in ground water. While most of these detections are at levels far below those thought to cause any measurable human health or environmental effect, these data have nevertheless spawned increased scrutiny of the fate and transport of pesticides within soil. In particular, greater scientific attention has been paid to new approaches for assessing the potential for individual pesticides to move through the soil and into ground water.

A wide range of experimental approaches are used for assessing pesticide mobility in soils. These methods range from laboratory studies involving only a few grams of soil to large field studies encompassing multiple locations, each consisting of many hectares. Intermediate in size are field dissipation studies and lysimeter studies in which pesticide mobility is assessed in small-scale test plots with undisturbed soil profiles. The laboratory studies are useful for isolating individual processes such as hydrolysis, sorption, biodegradation or volatilization. Field experiments reveal the integrated effects of the various individual phenomena and help assess the influences of soil type, climate and cultural practice.

In addition to these direct experimental approaches, numerous mathematical descriptions of the leaching process have been developed in the form of computer models. Environmental fate simulation models can play a significant role in the development and registration of pesticides by helping rationalize the results of laboratory and field studies, by providing estimates of likely leaching behavior in various settings, and by assessing the probability of ground water contamination from normal product use.

It is critically important to properly interpret the results of laboratory, field and modeling studies to provide a reasonable, scientifically sound assessment of the mobility of pesticides in soil. The environmental risks resulting from the leaching behavior of pesticides must be weighed against the benefits that accrue from responsible use of these products.

PURPOSE OF THE CHAPTER

The purpose of this chapter is to review and critically appraise current approaches for assessing pesticide mobility in soils, including the experimental approach of using lysimeters and the computational approach of environmental fate computer modeling.

This review leads to a recommendation of a tiered approach to assessing pesticide mobility in soils. The first step consists of laboratory studies of both mobility and the various individual mechanisms responsible for pesticide dissipation and degradation in the environment. The second tier of studies consists of field studies which evaluate the integrated effects of mobility, dissipation and degradation. In this tier, it is generally recommended that either a lysimeter study or a ground water study be conducted to experimentally evaluate the ground water contamination potential of a pesticide. The final tier consists of post-registration ground water monitoring that may be necessary to support continued product registration for those pesticides that are detected in ground water in environmentally significant concentrations. A parallel tiered approach is recommended for modeling, beginning with relatively simple screening assessments and leading to more detailed modeling.

The approach advocated in this chapter is of equal utility to the regulatory community, which has the responsibility of protecting the public, as well as the agrochemical industry itself, which is constantly seeking to better understand and improve the environmental characteristics of its products.

FACTORS AFFECTING THE MOBILITY OF PESTICIDES IN SOIL

Any assessment of the mobility of a pesticide in soil must take into account four principal factors:
- molecular properties,
- characterization of soil profile,
- climate,
- usage pattern.

Each of these factors is briefly reviewed below.

Molecular properties

The primary molecular properties responsible for the potential of a pesticide to leach into ground water are its persistence and mobility. In many ways, mobility is the easier of the two properties to assess. Persistence is much more sensitive to environmental factors and is likely to be more variable.

Mobility

The intrinsic mobility of a pesticide is inversely related to its degree of soil sorption. A pesticide having no molecular interaction with soil material would travel with infiltrating water and, in permeable soils, it would reach ground water unless it is degraded. The sorption of most neutral organic chemicals to soil correlates well with lipophilicity and soil organic carbon content. The situation is more complex for charged molecules and those that are able to interact chemically with the clays present in most soils. Nevertheless, most researchers currently choose to rank the relative intrinsic soil mobility of pesticides by using the soil/water partition coefficient, K_{oc}, which is the distribution coefficient, K_d, normalized by soil organic carbon content. Values of K_{oc} can be obtained through relatively simple laboratory experiments and can often be estimated reasonably well from chemical structure alone.

Persistence

The rate of dissipation of a pesticide in soil generally cannot be predicted with any accuracy from its chemical structure alone. Near the soil surface, the rate of field dissipation may be influenced by volatilization, photolysis and storm run-off events. Once the pesticide has passed below the surface, either due to incorporation or movement with rainwater, its rate of dissipation is more closely tied to its susceptibility to chemical and biological transformation reactions. For many compounds, hydrolysis and photochemical conversion represent significant degradative pathways. This is particularly true for foliarly applied pesticides which have a significant time of exposure to sunlight following application.

Characterization of soil profile

The physical properties of soil most often associated with susceptibility to leaching are low organic matter content, low field (moisture holding) capacity and relatively sandy texture. Soil structure can be a significant factor when it leads to the development of cracks and fissures that can act as preferential flow paths. Other factors that may play an important role for certain pesticides are soil pH (as it affects mobility or persistence) and the size and vigor of the soil microbial population when biotransformation represents an important dissipation mechanism.

Climate

Obviously, the intensity and extent of precipitation is a critical factor in determining the leaching potential of pesticides. Rainfall not only creates the driving force for soil mobility via infiltration, but it also influences degradation by altering the moisture content of the soil. Temperature plays a vital role in determining the rates of both chemical and biological transformation reactions. As a result of these climatic factors, many pesticides exhibit a greater tendency to leach in cool, moist climates than in warm, dry areas and autumn-applied pesticides have a greater tendency to leach than spring applications.

Usage pattern

The rate, timing and method of application can be important factors in determining the amount of pesticide leached. Compounds applied to the foliage when a crop is actively transpiring are less likely to leach than those which are soil-incorporated before planting (all else being equal). Canopy interception and root uptake of pesticides can also play a significant role in minimizing the amount of pesticide available to leach through the soil profile. For pesticides used in areas receiving irrigation, the timing and amounts of irrigation relative to the timing of pesticide application may prove critical.

REVIEW OF EXISTING APPROACHES TO ASSESSING PESTICIDE MOBILITY IN SOIL

A number of systematic regulatory approaches have been either proposed or implemented to assess the mobility of pesticides in soil. Historically, regulatory requirements for the testing and evaluation of pesticides (e.g. US EPA, 1975 and BBA, 1973) preceded more general approaches for the testing of all classes of chemicals such as the OECD guidelines.

Most of the approaches are similar in that they generally support stepwise or tiered testing of pesticide mobility. However, the various approaches differ significantly in the content of specific testing tiers and the specific criteria that are used to require additional testing at the next tier.

Four systematic, tiered approaches have been outlined by various non-regulatory groups: OECD, FAO, IUPAC and GIFAP.

OECD guidelines

The Organization for Economic Cooperation and Development has issued *Guidelines for Testing of Chemicals* to describe standard, internationally acceptable studies for the assessment of physical–chemical properties, degradation and accumulation behavior, and human and ecological toxicology (OECD, 1981a). The guidelines recommend the use of a phased approach in collecting and assessing environmental data.

To assess pesticide mobility and degradation, the guidelines specifically endorse measurement of vapor pressure, aqueous solubility, aqueous hydrolysis, adsorption/desorption in soil and aerobic soil degradation studies conducted in the laboratory. No further mobility or degradation studies are specifically described at present. However, OECD has initiated a pesticides program to develop international guidelines for environmental fate, ecotoxicology and toxicology studies.

FAO approach

The Food and Agriculture Organization of the United Nations has outlined *Revised Guidelines on Environmental Criteria for the Registration of*

Pesticides in a document published in 1989 (FAO, 1989). These guidelines, which are updated from an original document issued in 1985 (FAO, 1985), support the initial prediction of probable loading in various environmental compartments based on an initial group of laboratory and field tests. The physical–chemical tests include evaluation of vapor pressure, aqueous solubility and octanol/water partition coefficient. Mobility and degradation tests include measurement of adsorption/desorption characteristics, uptake by plants, degradation rates in soil and leaching.

Based upon these initial studies, the estimates of the environmental concentrations are determined and compared with acute, subchronic and chronic toxicity data to assess possible hazards to humans and the environment. Soil mobility tests that are specifically mentioned in this approach include soil thin-layer chromatography (TLC), soil column leaching studies, lysimeter studies and field leaching studies.

The FAO guidelines indicate a general sequence of studies but do not outline specific triggers. When concentrations of environmental significance are estimated for specific compartments, additional testing is recommended.

Revised FAO guidelines are in draft at present, the most recent version being Revision 3, August 1993 and these are currently favoured by the European Union as data requirements (91/414/EEC).

IUPAC approach

The IUPAC commission on agrochemicals issued a document titled 'Recommended approach to the evaluation of the environmental behavior of pesticides' (Esser *et al.*, 1988). This approach proposes flexible, stepwise testing of the environmental behavior of pesticides that will ensure evaluation of all aspects of a pesticide's environmental fate in sufficient detail. At the various stages of testing, this approach also indicates which results are of relevance in estimating exposure to the pesticide, thereby permitting risk assessment calculations to be made.

Specific tiered outlines are described for environmental mobility and behavior in soil. The first step in assessing mobility is to determine the aqueous solubility, octanol/water partition coefficient, vapor pressure and soil partition coefficient. From this information, initial assessments of soil mobility and volatility from water can be made. Next, laboratory degradation studies are performed and combined with K_{oc} data and use rate and use pattern data to assess the leaching potential of the compound when compared to suitable 'benchmark' compounds. Compounds with low leaching potentials are then further assessed using models to estimate their potential to reach surface water via run-off. Compounds with high leaching potentials are tested in either additional field tests or in lysimeter testing with undisturbed soils.

GIFAP approach

A GIFAP publication, *Enviromental Criteria for the Registration of Pesticides*, outlines guidelines on environmental regulatory criteria along with risk/benefit analysis (GIFAP, 1990a). Four specific tiers of testing are outlined in this report:

(1) initial testing in standard laboratory tests;
(2) supplementary laboratory studies;
(3) simulated field and field studies; and
(4) post-registration monitoring.

Initial laboratory testing includes measurement of vapor pressure, solubility in water and octanol–water partition coefficient and degradation in soil. In addition, adsorption/desorption studies and/or leaching studies are recommended to quantify soil mobility. Supplementary laboratory studies that could be triggered include degradation in water/sediment and leaching of major degradation products. If there are still doubts about the environmental acceptability of a chemical due to soil persistence or high mobility, then additional field studies designed to quantify degradation and/or mobility are appropriate. Finally, post-registration monitoring, combined with biological assessments, can be appropriate where there are remaining uncertainties.

Regulatory schemes of individual countries

In order to register a new pesticide, it is generally necessary to conduct an extensive series of laboratory and field studies of both degradation and mobility. The specific requirements for these studies vary by country. Typically, laboratory studies include metabolism studies in soil, evaluation of the extent of various degradative mechanisms and measurement of sorption and mobility in soil. Field testing requirements can include dissipation studies, lysimeter studies and ground water monitoring.

In many countries, the registration authorities have constructed detailed tiered schemes to assess the mobility of pesticides in soil. Outlines of the current schemes in several countries (Denmark, Germany, the Netherlands and the United States) are provided in Appendix 1, along with a brief description of the various steps.

European Union registration scheme

The European Union (EU) has a new Directive (91/414/EC) concerning the placing plant protection products on the market, which aims to harmonize the registration of agrochemicals within the EU. The data requirements have still to be finalized even though the scheme came into effect in July 1993. However, the EU scheme encourages the use of international test guidelines (EU, FAO, OECD) and will not result in the production of new test guidelines.

EXPERIMENTAL APPROACHES USED TO ASSESS PESTICIDE MOBILITY IN SOIL

Laboratory studies

Numerous laboratory studies are currently conducted to measure various physical and chemical properties of the pesticide as well as interactions between the pesticide and soil. These laboratory studies include:

- Determination of physical and chemical properties: octanol/water partition coefficient, water solubility and vapor pressure;
- Evaluation of mobility potential: soil thin-layer, thick-layer and column chromatography, soil batch adsorption and volatilization from soil;
- Measurement of degradation and dissipation: aerobic and anaerobic metabolism in soil, hydrolysis and photolysis.

Determination of physical and chemical properties

A large number of physical and chemical properties of pesticides are commonly measured to identify and describe the active ingredient of a crop protection product. Measurements of octanol/water partition coefficient, solubility in water and vapor pressure have particular value in the prediction of the environmental behavior of a pesticide. To obtain the best overall understanding of the environmental fate of a pesticide it is generally necessary to consider the combined effects of the key physical–chemical properties of a compound rather than evaluating each property individually.

Octanol–water partition coefficient: The octanol/water partition coefficient, K_{ow}, is a measure of the distribution of a substance between a lipophilic phase (specifically, n-octanol) and the aqueous phase of the test system. K_{ow} is an indicator of the bioaccumulation potential of a compound in the fatty tissue of animals and the lipophilic portions of plants as well as the adsorption potential of a compound to the organic matter in soils.

Several methods are available to determine the octanol/water partition coefficient in the laboratory. A commonly used method (OECD, 1981b; US EPA, 1982a) is to introduce the pesticide into octanol and water, to mix the biphasic system thoroughly to assure distribution equilibrium and to measure the concentration of the chemical in each of the two phases.

However, application of this technique becomes increasingly difficult at higher values of lipophilicity (log $K_{ow} > 4$).

Chromatographic methods based on high performance liquid chromatography (HPLC) (Veith and Morris, 1978; Eadsforth, 1986) and reverse phase thin-layer chromatography (RP-TLC) (Hülshoff and Perrin, 1976; Biagi *et al.*, 1979) have been described in the literature. A reverse phase HPLC method has recently been included in the OECD test guidelines (OECD, 1989). A preliminary estimate of the partition coefficient can be obtained by dividing the solubility of the test substance in *n*-octanol by its solubility in pure water (K_{ow}).

Substances which dissociate or associate in water and/or *n*-octanol do not have a constant value of K_{ow} with varying concentration due to speciation effects. Because of the multiple equilibria involved, the recommended method should not be applied to compounds which reversibly ionize or protonate, and buffer solutions instead of water should be used for such compounds.

Solubility in water: The solubility of a chemical in water can play a role in influencing envronmental distribution and transport, particularly if the solubility is very low. However, the aqueous solubilities of most pesticides are generally orders of magnitude higher than actual environmental concentrations and thus this property alone is not usually a key determinant of leaching behavior.

Several methods are reported in the literature (Mader and Grady, 1971; May *et al.*, 1978; OECD, 1981c; US EPA, 1982b; Whitehouse and Cooke, 1982) but no single method covers the entire range of water solubility. At least two methods are therefore needed. The first method (OECD, 1981c) is applicable to substances of low solubility ($<10\,\mathrm{mg\,l^{-1}}$) and is based on the elution of the pesticide with water from a microcolumn which is charged with an inert carrier material and an excess of test substance. The water solubility is determined when the mass concentration of the eluate is constant when tested at different flow rates. The second method (OECD, 1981c) tests for compounds with higher solubilities ($\geqslant 10\,\mathrm{mg\,l^{-1}}$). In this method, the test substance is dissolved in water at a temperature slightly higher than the test temperature. When saturation is achieved, the mixture is cooled and kept at the test temperature until equilibrium is achieved. Undissolved particles are removed from solution and the dissolved pesticide is measured by a suitable analytical method.

Estimates of the solubility of pesticides in water can also be made by chromatographic methods, such as HPLC (Whitehouse and Cooke, 1982).

Vapor pressure: The vapor pressure of a pesticide is important in assessing its distribution in the environment. It is the major property which can be used to calculate the volatility of a substance and to predict whether or not a pesticide is likely to enter the atmosphere in significant concentrations. Loss of a pesticide from the soil by volatilization represents a competitive process to leaching, reducing the total amount available to move through the soil profile.

The vapor pressure at a given temperature is measured in the laboratory at that temperature or calculated from an experimentally derived vapor pressure curve. Several methods are reported in the literature (Thompson and Douslin, 1971; ASTM, 1975; OECD, 1981d; US EPA, 1982c; Spencer and Cliath, 1983) which cover a range of vapor pressures from less than 10^{-3} to 10^5 Pa, but no single method covers the entire range of vapor pressures. Five methods are recommended, each covering a definite range of vapor pressure (OECD, 1981d);

(1) Dynamic method (10^3 up to 10^5 Pa, between 20 and 100 °C);
(2) Static method (10 up to 10^5 Pa, between 0 and 100 °C);
(3) Isoteniscope (from 10^2 to 10^5 Pa, between 0 and 100 °C);
(4) Vapor pressure balance (10^{-3} to 1 Pa, between 0 and 100 °C);
(5) Gas saturation method (up to 1 Pa).

In addition, estimation of ambient vapor pressures of pesticides can be made from gas chromatographic retention data (Seiber *et al.*, 1981).

Evaluation of mobility potential

Three laboratory approaches have been developed to quantify the potential of a compound to move through the soil profile under the influence of infiltrating water. Dynamic measurements of the movement of a pesticide in soil can be made using soil chromatography on flat plates or in columns. Measurement of the equilibrium partitioning of pesticides between the soil and soil solution is obtained from adsorption and desorption studies. Finally, loss of pesticides from soil and plant surfaces can be evaluated through volatility testing.

Soil chromatography: Soil chromatography studies provide comparisons of the leaching behavior of a compound in a specific soil setting under the influence of a mobile phase of water. This determination is important because the rate of leaching indicates how long a chemical is retained in the top layers of the soil where it is most subject to degradation or dissipation.

Three different methods can be used to study the leaching of pesticides in the laboratory. Soil thin-layer chromatography (soil-TLC) is an adaptation of the widely used thin-layer chromatography using soil or a soil fraction as the adsorbent and distilled water as the mobile phase (Helling and Turner, 1968). If radiolabelled, pesticides can be detected by autoradiography

(Helling and Turner, 1968), and radiochromatographic scanning (Helling, 1971). Otherwise, detection can be performed using bioassays (Chapman *et al.*, 1970; Helling *et al.*, 1971) and chromogenic spraying reagents (Helling and Thompson, 1974). Similar to normal thin-layer chromatography on commercially available carriers, frontal or relative R_f (ratio to front) values are used to characterize the mobility of the pesticides.

Soil thick-layer chromatography (ThLC), described by Gerber *et al.* (1970), can be considered a compromise between soil-TLC and dry column chromatography. A shallow metal plate (30 cm × 5 cm × 5 mm deep) is filled with air-dried soil with a particle size of ≤2 mm. The compounds to be tested are mixed with soil and the mixture applied as a streak to the starting line of plate. A cotton cloth dipping into water provides a uniform water flow on the plate which is developed using the descending technique. After leaching, the soil layer is divided into 2-cm zones which are separately analyzed by chemical, radiochemical or bioassay methods. The analyses indicate the location of the maxima of the pesticide distribution pattern and the distance of the maxima from the starting point gives the so-called 'leaching index'.

Another alternative method is packed soil column chromatography (CC) which has found widespread use for more than 30 years. A large variety of soil columns of different materials, including glass, plastic and metal, have been used (Helling, 1970; Weber, 1971; Guth, 1972; Burkhard *et al.*, 1975; US EPA, 1982d; BBA, 1986). A very useful method is the slotted tube test technique where a longitudinal slot is cut into a glass column which can then be removed after the leaching test. A bioassay can then be carried out directly on the uncovered surface to determine pesticide distribution in the soil profile (Lambert *et al.*, 1965). The mobility potential of pesticides can be assessed from soil-CC by measuring distribution maxima, penetration depths, elution curves or simply total percentages found in the column leachate. A complete analysis of column layers and column leachate is certainly the preferred method since it allows a mass balance to be determined.

Comparisons between mobility data obtained by the three methods described above show that, within certain limits, the same relative mobility was observed (Guth *et al.*, 1976). Although the column method is the most time consuming study to conduct, it is potentially more closely related to a field situation than either the TLC or ThLC methods. However, the common practice of maintaining a fixed head of water on a packed soil column results in data that are not representative of compound behavior in a field setting where soil moisture alternates between wetting and drying conditions. As a result, soil-CC results are best interpreted in comparison with standard compounds to assess likely environmental leaching behavior.

All three methods are based solely on chromatographic transport and do not consider other non-chromatographic transport mechanisms in the soil profile. Soil-CC has the following advantages over the plate methods:

(1) The water flow and thus the rate of infiltration can be easily controlled.
(2) It is straightforward to vary the application method, amount and formulation of the applied pesticide and the method of analysis.
(3) Soil-CC can be adapted for use with volatile pesticides.
(4) Whereas soil-TLC uses only a certain soil fraction, soil-CC can use natural soils and even intact cores.
(5) Soil-CC provides the ability to carry out long-term leaching/incubation studies such as 'aged residue' leaching experiments.

'Aged residue' studies are carried out to obtain information on the mobility of both the parent compound and its degradation products. For this purpose, radiolabelled pesticides are incubated in soil for a period of time sufficient to transform approximately half of the parent compound. The aged soil is then applied either to the top of soil column or to the starting zone of a soil thick-layer plate. Soil-TLC can be used but, in this case, soil extracts have to be applied to the plates and the duration of such an experiment is relatively short.

Adsorption/desorption: The immediate purpose of adsorption/desorption studies is to determine the equilibrium partitioning of pesticides between the solid and liquid phases of the soil. Because of the many different physical and chemical interactions that result between pesticides and the surfaces of the various components of the soil, this overall process is best described generically as sorption rather than adsorption. However, both terms are used in the literature to describe the interactions that occur.

Sorption of pesticides to soil influences virtually all the transport and kinetic processes that occur in soil including pesticide mobility, volatilization and degradation. Sorption data are used to assess a compound's potential for ground water contamination by providing a measure of the mobility of the compound under the influence of percolating water.

The most widely used method for carrying out soil-pesticide sorption studies is the 'batch' or slurry procedure (OECD, 1981e; US EPA, 1982d) where air- or oven-dried soil is shaken with aqueous pesticide solutions of known concentration until equilibrium is reached. According to Hance (1967), a period of 24 hours is generally sufficient to reach equilibrium conditions. The aqueous suspensions are then separated by centrifugation and the equilibrium concentration in the water phase determined either by direct analysis or after extraction with an appropriate organic solvent. The use of radiolabelled compounds with high purity provides a means of determining a mass balance and assures a higher accuracy than other analytical methods due to improved analytical recoveries.

In general, batch adsorption studies are designed for the evaluation of the sorption of nonpolar organic compounds in soil. Additional steps have to be taken to conduct sorption studies of compounds which are highly charged, have low water solubilities, are highly volatile or are unstable in the time scale of the test. Other methods used for adsorption/desorption studies include continuous-flow equilibrium (Bunhill *et al.*, 1973) and dialysis techniques (Hayes *et al.*, 1968) but these have not found wide use.

A number of reviews have been written concerning the sorptive behavior of pesticides in soil (Hamaker and Thompson, 1972; Guth *et al.*, 1977; Calvet, 1989). Sorption of a pesticide in soil depends upon three factors which interact in a complex manner: the physical–chemical properties of the pesticide, the composition of the soil and the experimental conditions under which adsorption is studied (Calvet, 1980). For any given compound, there is a continuum of sorptive mechanisms responsible for its retention on or near the surface of the soil particles, including interactions with organic matter, clays and metal oxides. In addition, the sorption process does not always involve only weak, reversible bonds. Therefore, hysteresis is commonly observed in adsorption/desorption studies. For these reasons, it is generally necesary to evaluate the sorption of a pesticide on an number of different soils to adequately understand its sorption behavior.

Because pesticide sorption typically invovles a very small fraction of the total sorptive capacity of the soil, it is usually sufficiently accurate to describe the sorption process using a linear relationship between the solid and liquid phases of the soil:

$$x/m = K_d C$$

where x/m is the sorbed concentration of pesticide on the soil (mass/mass) that is in equilibrium with a concentration C in solution. The parameter K_d is commonly referred to as the distribution coefficient or adsorption coefficient and is a constant for the given system and temperature.

A common variation of this equation is the Freundlich equation:

$$x/m = K_f C^N$$

where the Freundlich coefficient, K_f, and N are empirical constants. The constant N (or $1/n$ in some literature) generally ranges between 0.7 and 1.0, indicating that sorption data is frequently slightly nonlinear.

When sorption measurements are made over an extended range of concentrations, some degree of nonlinearity is commonly observed in the sorption isotherm. For the range of concentrations typically encountered in the soil, the nonlinearity is usually negligible. However, many experimental studies are conducted using concentrations that are orders of magnitude higher than the concentrations present in the soil. As a result, the

Freundlich equation is frequently used to obtain a better statistical fit to the sorption data than is possible with a linear equation.

A common problem encountered with fitting sorption data with Freundlich isotherms is that most environmental models assume that sorption isotherms are linear and, as a result, K_f values cannot be directly used in these models. When sorption data are used in modeling, the use of the linear sorption relationship is generally a desirable simplification since it allows simpler mathematical equations and the linear K_d value is generally sufficiently accurate to describe sorptive behavior in soil (Green and Karichoff, 1990; Koskinen and Harper, 1990). The obvious exception to this is the case where the model of interest is specifically written to accommodate the use of the Freundlich isotherm values, K_f and N. Using either Freundlich or linear isotherms, it is best to place more reliance on sorption data generated at environmentally expected concentrations, particularly if curvature is observed in the isotherm.

Adsorption constants vary from soil to soil but they can be normalized to some extent by dividing them by the proportion of soil organic carbon or organic matter. By this procedure, K_{oc} or K_{om} values are obtained, giving the amount of pesticide adsorbed by unit mass of organic carbon or organic matter:

$$K_{oc} = K_d \times 100/(\%\,OC)$$

$$K_{om} = K_d \times 100/(\%\,OM)$$

In general, this normalization works best in the root zone of the soil where the majority of the soil organic carbon is located. The coefficient of variation of K_{oc} values obtained from a wide range of different soils and sediments ranges from 10% to 140% (Bailey and White, 1970). A significant portion of this variability is due to variations in the protocols followed to conduct this study (e.g. concentrations tested, water-to-soil ratio, and selection of soils). A recent adsorption/desorption ring test involving 26 European laboratories demonstrated that the standard deviation of K values determined using radiolabelled compounds was typically less than half as high as that for non-labelled materials, primarily due to differences in analytical accuracy (ISPRA, 1990). Other key parameters that can influence the accuracy of sorption measurements include the stability and purity of the test substance, the attainment of sorption equilibrium, the concentrations chosen, the magnitude of the solution concentration change, soil/solution ratio, changes in the soil structure during the equilibration process and analysis of both the solution and adsorbed phases (Green and Yamane, 1970; Grover and Hance, 1970; Boesten, 1990; Green and Karickhoff, 1990).

Knowledge of the adsorption coefficient can provide valuable insights into the potential for a compound to move and distribute itself in the environment. Leaching through soil, volatility from soil and partitioning between various environmental compartments are all influenced by the sorptive behavior of pesticides. An increasingly important use of sorption data is in environmental fate modeling. It is vital to have appropriate values of the adsorption coefficient for the soil profiles being simulated.

Volatilization from soil and crops: Volatilization losses from soil and plants can be a major dissipative mechanism for certain pesticides which reduces the total pesticide available for leaching. The first indication of the volatilization potential of a compound is its vapor pressure. Actual volatilization losses of pesticides from plants and soil are influenced by the extent of adsorption to soil, solubility in water, soil moisture and climatic conditions (Hamaker, 1972; Spencer *et al.*, 1973).

Numerous experimental systems have been designed to evaluate the volatilization of pesticides from non-adsorbing surfaces, from soil and from plants (Phillips, 1972; Gückel *et al.*, 1974; Burkhard and Guth, 1981). Laboratory studies of volatilization losses can be conducted in either stagnant or dynamic systems, with or without plants present and in open or closed systems. In any of the closed experimental configurations, pesticide is applied to the test system and the amount that volatilizes is determined by material balance. In a dynamic system, a constant, measured stream of air or inert gas is passed over the soil surface and/or plants and is then trapped using liquid absorption methods or by adsorption on solid carriers (Russell, 1975; Seiber *et al.*, 1975; Van Dyk and Visweswariah, 1975). The extent of volatilization can then be directly determined from analysis of the trapped compound.

It is self-evident that compounds which have relatively high evaporative losses have a lower amount of parent material available for leaching through the soil profile. As a result, volatility can generally be viewed as a competitive dissipation process to leaching in soil. Both processes serve to lower the concentration of active ingredient in the zone of placement in the soil profile.

Measurement of degradation and dissipation

Aerobic/anaerobic degradation in soil: The primary purposes of aerobic and anaerobic soil studies are to determine degradation pathways of the test compound in soil and to identify important soil metabolites. These studies provide quantitative evaluations of the extent of degradation, mineraliza-tion, formation of bound residues and volatilization in a controlled laboratory setting. Since radiolabelled pesticides are typically used for these studies, it is possible to determine a mass balance at sampling interval.

Evaluation of the relative importance of microbial degradation can be determined by comparison of the results of sterile and nonsterile degradation studies and by comparison of soil studies with physical–chemical studies such as aqueous hydrolysis.

A wide variety of laboratory experimental approaches have been designed to study pesticide degradation in the soil (Guth, 1980; US EPA, 1989; BBA, 1986). These approaches include soil perfusion systems, soil biometers, gas flow-through systems and integrated systems. The first three methods can be used for aerobic as well as anaerobic soil studies and have been extensively discussed by Kaufman (1977). Anaerobic conditions can be established by waterlogging of the soil, by addition of sucrose and/or by ventilating the systems with an inert gas such as nitrogen.

The major systems include:

(1) Biometer-type systems consist of a closed system in which evolved CO_2 is trapped using a solution of sodium or potassium hydroxide. This approach was originally developed by Bartha and Pramer (1965) and is very widely used. Biometers allow monitoring of both $^{14}CO_2$ and total CO_2. Since the experimental system is closed, there is no exposure of the system to atmospheric CO_2 and a material balance can be determined. Consequently, this system is useful for evaluating the extent of mineralization of labelled pesticides by trapping the $^{14}CO_2$ produced.

This static method does have several practical limitations. Volatile products may not be adsorbed by KOH solution or, if trapped, may require additional analytical effort for separation from $^{14}CO_2$. In long-term experiments without air exchange, it is possible to generate anaerobic conditions. A modification of the basic biometer design has been developed to maintain a uniform partial pressure of oxygen to address this concern (Laskowski et al., 1983). Useful modifications of the biometer flask have been published (e.g. Anderson, 1975; Marvel et al., 1978; Loos et al., 1980).

(2) Flow-through systems have varied from relatively simple systems (Goswami and Koch, 1976; Kearney and Kontson, 1976) to the complex manifold developed by Parr and Smith (1976). These systems are widely used and have numerous practical advantages. The experiment can be conducted under more precisely defined conditions and different gaseous environments can be used. The system is usually operated under negative pressure but can also be run with positive pressure. Exposure to atmospheric CO_2 is eliminated and drying of soil is eliminated by preconditioning the incoming air with a KOH trap. By using sequential traps of ethylene glycol and KOH on the exiting air, it is possible to separately collect volatile products and CO_2. In this manner, total CO_2 production can be monitored and used as an indication of soil microbial activity. The major criticism of flow-through systems is that they are

complex in design and generally more expensive than the alternative systems.

(3) Soil perfusion systems involve pumping a pesticide solution through a column of soil and analyzing the perfusate. Although these systems are reasonably versatile and have been widely used to study soil microbiological processes, they are not recommended for the study of pesticides. The conditions of the test system are far from actual practice and the perfusion technique tends to favor the growth of bacteria.

(4) Integrated systems have been developed, for example, by Best and Weber (1974), Lichtenstein *et al*. (1974), Beall *et al*. (1976) and Harvey and Reiser (1973). These systems allow the simultaneous measurement of various processes such as degradation, respiration, volatilization, plant uptake, leaching, etc.

Although these integrated systems have tried to introduce the complexity of the environment, they are not really able to simulate actual environmental conditions. In addition, complex systems of this type typically produce highly variable results which are then difficult to interpret. It is not reasonable to monitor various processes in a complex system routinely for each compound when certain processes are not of significance. Rather the importance of each individual process should be evaluated first in a specifically designed study. After this is done, integrated systems may be of value as an additional testing step.

A very important aspect in soil degradation/metabolism studies is the microbial activity of the soils used (Guth, 1980, 1981), particularly for *in vitro* systems, because microbial metabolism plays a significant role in the transformation of many pesticides in soil. It is therefore important that soil sampled for use in transformation studies is treated as a living tissue and handled with care (Pramer and Bartha, 1972). Air-drying, prolonged storage, and freezing and thawing should be avoided since these actions will drastically alter the biochemical activity of soils by inactivating extracellular enzymes in the soil. Lay and Illnicki (1975) have made similar observations when studying the effect of soil storage on propanil degradation. They proposed that fresh samples should be used for all biotransformation studies. In addition, it is recommended that the soil moisture be kept uniform during the study.

In special cases when it is impossible to use fresh soils (e.g. when transformation studies are carried out on foreign soils), it may be necessary to amend the soil or cultivate plants to restore microbial populations prior to use (Chisaka and Kearney, 1970). It is recommended that biological activity be monitored during the study.

For anaerobic systems, it is recommended that the redox potentials or free oxygen be measured throughout the course of the experiment. Redox potentials should remain below 300 mV and the concentration of free oxygen in an anaerobic test system should generally be less than $0.1\,\mathrm{mg\,l^{-1}}$.

Hydrolysis: The purpose of conducting an aqueous hydrolysis study is to obtain quantitative estimates of the rate and extent of chemical decomposition that occurs due to chemical hydrolysis and to identify degradates formed by this degradative mechanism.

Hydrolysis experiments are normally carried out with the pure chemical dissolved in sterile aqueous buffer solutions of different pH values and at different temperatures (OECD, 1981f; US EPA, 1982e). Three pH levels, 5, 7 and 9, are generally used to encompass the wide range of natural conditions present in agriculturally important settings. The experiments can be carried out at room temperature (20–25 °C) but also at other temperatures to allow extrapolation to any temperature of interest. Sealed, sterile containers should be used to maintain sterility and to minimize volatilization.

The decrease of pesticide concentration in the aqueous solutions is followed over a specified period of time, typically 1 month. Working at higher temperatures can decrease the duration of the experiments and minimize potential problems with sterility.

The rate of hydrolytic breakdown can be described using first-order reaction kinetics. A first-order rate constant can be calculated for each temperature and at each pH. Rate constants and half-lives can be estimated for other temperatures using the Arrhenius equation.

Caution needs to be exercised in interpreting the results of sterile hydrolysis studies. In general, it is beneficial for a compound to have a measurable rate of hydrolysis. However, stability in a sterile, soil-free hydrolysis study does not necessarily indicate stability in ground water or in other aqueous environmental compartments.

Photolysis on soil surfaces: Studies of the rate of photolysis on soil surfaces are conducted to assess the rate and extent of degradation of pesticides under the influence in sunlight. This study also provides information on the degradation products that are formed due to photolysis.

Screened soil is combined with water to make a slurry which is applied to the surface of a metal plate to a typical depth of 2 mm (US EPA, 1982f). The pesticide is applied to the soil which is then placed into a closed environmental chamber and continuously irradiated with an artificial light source for 15 days, representing approximately 30 days of natural sunlight. Either radiolabelled or unlabelled compounds can be used, but the use of a radiolabel facilitates performing a material balance at the end of the study. The use of a closed environmental chamber makes it possible to sample the airspace for the presence of possible volatile degradates formed during the test.

It is evident that photodegradation occurs only on or near the soil surface where light can penetrate. As a result, this degradative mechanism only has significance for that portion of the applied pesticide that remains exposed to the degradative effects of sunlight. Interception of the pesticidal spray by plant foliage will also expose the compound to sunlight and significant photodegradation is known to occur for some foliarly applied compounds although this result can be confounded by volatilization and microbial degradation on plant surfaces.

Published information on experimental approaches for measuring photo-degradation on soil surfaces is limited (Guth, 1980, 1981). One of the key difficulties in this type of study is to obtain a light source that reasonably represents the spectrum and energy distribution of natural sunlight on a clear day. Studies conducted under natural sunlight are invariably affected by meteorological variability. Various germicidal, mercury and fluorescent lamps that have been used in numerous studies typically have inappropriate spectral energy distributions. Commerical equipment is now available that uses a reflected and filtered xenon arc lamp to obtain a reasonably accurate representation of sunlight (Burkhard and Guth, 1979; Parker and Leahey, 1988).

An important aspect of soil photolysis studies is the thickness of the soil layer. The layer thickness reported in the literature ranges from 30 μm to 15 mm. Since light does not penetrate deeper than 4 mm into soil it generally does not make sense to use a soil thickness greater than this depth. Typically, a depth of 2 mm is used.

Field studies

Field dissipation studies

Purpose of study: The purpose of a field dissipation study is to determine the extent of dissipation and distribution of pesticides and metabolites under a reasonable range of actual use conditions. This study can include the effects of groundcover, if appropriate. The major mechanisms responsible for dissipation of compounds following application include degradation, sorption, volatilization, plant uptake, leaching and wind and water transport. In a field dissipation study, relatively flat sites are chosen to minimize or eliminate run-off. All of the remaining dissipation mechanisms are then integrated with actual climatic conditions, cultural practices and soil conditions to yield the overall rate of decline of parent and metabolites with time.

Description of study: Guidelines for conducting field dissipation studies have been published by a number of regulatory agencies (e.g. US EPA, 1982f; Schinkel *et al.*, 1986). To adequately characterize the effects of environmental variability on the dissipative behavior of a specific compound, it is generally recommended that field dissipation studies be conducted at a minimum of three sites. It is appropriate to increase this number to four to six sites when the compound will be used in widely diverse geographical settings or on a large number of agronomically distinct crops.

Field dissipation studies should be conducted with the highest proposed use rate of formulated product. Either radiolabelled or unlabelled compounds can be used, with the choice depending primarily upon the use rate, available analytical methodology and local country regulations on the use of radiolabelled materials.

Field soil dissipation studies can be conducted either on bare soil or in the presence of crops, depending upon normal use. A cropped study will result in more realistic infiltration rates in the test plots due to canopy interception of precipitation and evapotranspiration of soil water. When a dissipation study is conducted with a crop present, it is necessary to document the relative interception of the plant canopy and the soil after compound application to establish the appropriate soil concentration on Day 0.

One of the most important aspects of a field dissipation study is the strategy followed to sample the plots following application. Sampling of the soil profiles in the field plots is commonly performed either by collecting a set of samples from various depth increments as a hole is progressively deepened or by using manual or hydraulic probes to obtain intact soil cores which are then divided into depth increments. Bucket augers have been successfully used to collect disturbed soil samples to depths of over 7 m. (Norris *et al.*, 1990). The collection of intact soil cores is generally restricted to depths of less than 2 m using relatively simple equipment. However, this is generally sufficient to meet the requirements of a field dissipation study. It is recommended that all holes left from sampling should be filled and marked to prevent inadvertent resampling of the same location at a later time.

Soil cores should extend at least 30 cm into the soil profile (e.g. 0–10 cm, 10–20 cm, and 20–30 cm) to adequately characterize the dissipative behavior of the pesticide in the soil. Sampling for more mobile compounds should extend deep enough to ensure that detectable soil residues are sampled. The soil cores are analyzed in discrete depth increments to allow a quantitative evaluation of the pesticide residues present in the soil profile as a function of time. To minimize the effects of field variability, compositing of spatially randomized sampling should be performed (15–20 cores composited per sampling event) (Smith *et al.*, 1987; US EPA, 1988a).

Sampling of the plots should be scheduled to provide two or three sampling events prior to dissipation of the first 50% of the compound from the soil. Later sampling points can be spaced out increasingly long increments to demonstrate the decline of parent and proximate metabolites to levels of less than 10% of that applied. It is recommended that a mass balance be calculated at each sampling event to estimate the proportion of pesticide residues present in the soil over time. This calculation is often complicated by incomplete analytical recovery of pesticide residues from soil.

At a minimum, monthly average (or cumulative, as appropriate) values of precipitation, air temperature and soil temperature should be collected. If modeling of the data is being considered, it is recommended that daily values of total precipitation, maximum and minimum air temperature and maximum and minimum soil temperature be recorded. It is also useful, but not absolutely necessary, to determine the daily evaporation potential either through recording pan evaporation data or by recording the insolation rate, wind speed and relative humidity. It may be necessary to provide supplemental irrigation for sites located in areas subject to drought conditions.

The total duration of a typical field dissipation study is 12–18 months. Based on the total amount of parent material in the sampled profile over the time period of the study, the time to 50% dissipation (DT-50) and the time to 90% dissipation (DT-90) can be determined. Movement of the pesticide residues in the soil profile can be estimated if multiple soil layers were analyzed.

Discussion of study: Results from field soil dissipation studies provide extremely useful information in assessing the persistence of an applied compound in the surface layers of the soil. As described above, the DT-50 and DT-90 values that are calculated for the parent compound reflect the integrated effects of all loss and degradation mechanisms over the time frame of the study. It is generally not possible to extract detailed information on any single dissipation mechanism such as volatilization from the soil surface from the integrated results. For many compounds, the most significant dissipation mechanisms in the field are chemical and microbial degradation.

The variability of the experimental results from location to location can indicate the relative importance of various dissipative mechanisms to the overall dissipation rate, particularly when compared with laboratory studies of individual mechanisms such as hydrolysis, aerobic soil degradation, etc.

In addition to data on the parent compound, information on the formation and decline of significant metabolites can often be obtained from field soil dissipation studies when analytical methods for degradates are available. Estimates of mobility and persistence can also be determined from the collected data to address concerns for specific degradates.

Lysimeter studies

Introduction: A lysimeter is a large column (or block) of undisturbed soil which is collected from a representative field location and enclosed in a suitable container to allow collection and analysis of the leachate which exits at the bottom of the soil column (Figures 1 and 2). This experimental method provides a link between the laboratory and the field by providing agronomic conditions that approach those in nature.

Purpose of study: Lysimeter studies are primarily designed to quantify the leaching behavior of a pesticide in a simulated field setting, integrating the influences of weather, soil and crops on mobility. The primary emphasis of these studies is the quantitation and evaluation of the traces of the applied compound and its degradates that appear in the leachate collected from the bottom of lysimeters. However, a great deal of additional information can also be determined from measurement of the distribution of the applied compound among the air, plants, soil and soil water in the soil column.

In order to reasonably approximate actual field behavior, lysimeter studies should be conducted using intact soil cores. A common collection technique is to hydraulically press a rigid steel casing with a sharpened leading edge into a selected field, excavating the surrounding soil periodically as the casing is pressed. The sampled soil column is then separated from the lower soil by means of a cutting device. The cutting plate can also be designed to be a sieve plate so that it serves as the bottom support for the

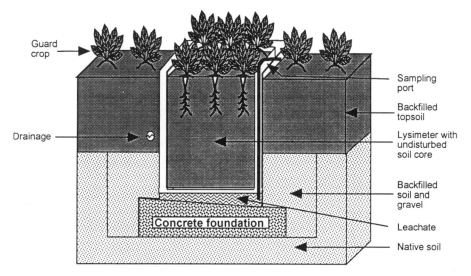

Figure 1 Typical free-standing lysimeter installation in profile (source: F. Führ, Jülich, Germany)

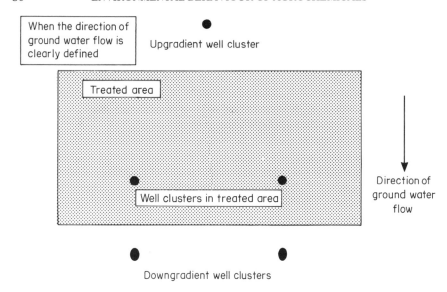

Figure 2 Typical design for locating initial monitoring well clusters where the direction of ground water flow is clearly defined (source: Jones and Norris, 1991)

lysimeter following sampling. The casing should be inserted into the soil as vertically as possible to prevent the creation of voids adjacent to the wall.

Once the soil core has been collected in a casing, it is moved to a research site where it is either placed in a buried outer shell that closely fits the casing or else is directly buried (Figure 1). When an outer shell is used, a weather-tight seal should be provided between the lysimeter casing and the shell to prevent the direct introduction of precipitation and pesticides into the sump of the outer casing and to facilitate the establishment of thermal equilibrium of the lysimeter with the soil surrounding the outer shell.

The water which percolates through the soil profile and collects in the bottom of the lysimeters is then periodically collected either by pumping of the leachate from a sump in the outer casing or by gravity drainage. The collected samples are then analyzed to provide a discrete assessment of pesticide and metabolite concentrations that have leached from the soil core during specific sampling intervals.

Lysimeter studies are typically conducted with a crop present, creating a more realistic water balance in the soil profile than the field soil dissipation study on bare soil. In addition, it is generally recommended that a guard crop be grown around the lysimeter to help ensure normal development of the crop in the lysimeter by creating a more realistic microclimate and minimizing field edge effects.

The cultivation of the lysimeter surface and its adjacent surroundings should correspond to common agricultural practice. Soil sampling of lysimeters throughout the course of the study is not recommended, particularly for regulatory studies. Caution needs to be taken to prevent disturbance of the soil deeper than the plow layer. It is recommended that netting be provided around the lysimeter installation to prevent the entry of small birds and burrowing animals that could disturb either the test crop or the soil in the lysimeters.

To minimize variability in the water balance among lysimeters due to varying plant demands for water, it is recommended that a significant number of plants be maintained in an individual lysimeter. As a result of this general guidance, small lysimeters (e.g. those with surface areas $<0.2\,m^2$) should only be used to evaluate pesticides used on densely planted crops such as cereals, whereas large lysimeters ($\geqslant0.5\,m^2$) can be used for most commonly grown field crops.

During the course of the study, the lysimeters should be exposed to ambient precipitation and, if necessary, provided with supplemental irrigation to ensure that sufficient quantities of water are added to represent typical climatic conditions for the crop being grown. Changes in the water content of the soil can be monitored by continuously weighing the lysimeter block using load cells. Appropriate daily weather data should be collected, including maximum and minimum soil and air temperatures and precipitation at a minimum. It is useful to collect a sufficient amount of climatic, soil and agronomic data to permit rationalization of the experimental results using computer modeling at the conclusion of the study.

Although not absolutely necessary, it is useful to evaluate the hydrologic characteristics of the soil core prior to application of the active ingredient through the use of tracer. Analysis of tracer data can help ensure that the soil core is free of significant cracks or cavities that could result in preferential flow of infiltrating water.

Although lysimeter studies can be conducted using either radiolabelled or unlabelled compounds, analytical screening and analysis of the leachate samples is facilitated by the use of radiolabelled material. The identification of previously unknown metabolites will normally not be possible due to the low amounts of material present in the leachate.

As currently required by national regulatory agencies, lysimeter studies are carried out for a minimum of 2 years. The data obtained from lysimeter studies consist of pesticide concentrations detected in leachate at periodic intervals throughout the study and in plant tissue and the soil at the end of the study. In addition, information is collected on the extent and timing of infiltration events for the soil, crop and climate being studied.

Discussion of study: A number of reviews of lysimeter technology have been published (e.g. Führ *et al.*, 1975, 1990; BBA, 1990a; Bergström, 1990; Stork *et al.*, 1990). A specific list of recommendations concerning the use of lysimeters to evaluate the mobility and degradation of pesticides were made at a recent workshop (Stork *et al.*, 1990) (Table 1). The general consensus of the workshop was that lysimeters provide useful data on degradation and mobility provided appropriate precautions are taken in the selection of test equipment, collection of soil cores, execution of the study and interpretation of the results.

A wide range of cross-sectional areas have been successfully used to conduct lysimeter studies. A number of research and regulatory studies have been conducted with lysimeters that have a surface area of approximately 0.5–1 m^2 and a depth of 1.0–1.5 m (Stork *et al.*, 1990). The surface can be of any shape. Small lysimeters (e.g. 30 cm diameter) have also been used to assess the leaching potential of pesticides and offer some practical advantages over the large lysimeters in certain situations, particularly if soil compaction is avoided by obtaining cores by drilling rather than pressing casings into the soil (Bergström, 1990; Persson and Bergström, 1991). In addition, the ease of collection of small lysimeters makes it possible to increase the number of soil cores in a study which allows evaluation of multiple soil types and increases the potential utility of the study.

Lysimeters generally work best with light soils. The experimental operation of a freely draining lysimeter creates a zone of saturated soil at the bottom of the lysimeter. In order for soil water to drain from the bottom of the lysimeter, the matric potential of the soil must be close to zero. For coarsely textured soils, the capillary rise from this zone of saturation is perhaps a few centimeters but can be significant for soils with high silt and clay fractions (Fetter, 1988). This effect can be reduced but not eliminated altogether by the introduction of a tightly packed fine sand layer to the

Table 1 Major conclusions of the second lysimeter workshop, November 22–23, 1990, Neustadt, Germany

1 Monolith lysimeters are a useful link between laboratory and field studies
2 Monoliths should have a surface area of at least 0.5 m^2 and be 1.0–1.3 m deep
3 Before use, lysimeters should be allowed to equilibrate over one winter season
4 Experimental conditions used in lysimeters should be close to farming practice
5 Lysimeters can be used to study processes other than leaching, but care must be taken to prevent disturbing or influencing the leaching process
6 The use of tracers to determine breakthrough curves is possible but not routinely recommended
7 Careful analysis of both parent compound and metabolites in the leachate is vital
8 The spatial variability of a field cannot be represented by a single lysimeter

SOURCE: J. A. Guth and J. T. Mani, 1991, unpublished report.

bottom of the lysimeter when it is necessary to study fine-grained soils. Alternatively, a porous ceramic plate can be installed in the bottom of the lysimeter to create a constant negative matric potential similar to that of a moist soil.

Lysimeter studies can potentially provide a more detailed assessment of pesticide mobility in the soil than is possible from a field dissipation study, which is primarily focused on evaluating the overall rate of pesticide dissipation in soil. Lysimeter measurements provide a direct assessment of compound mobility by monitoring the concentration in soil water at a fixed depth in the soil. When lysimeter studies are conducted with radiolabelled pesticides, it is possible to evaluate the mobility of both the parent compound and major metabolites.

Lysimeter studies have a number of practical limitations. The size of the lysimeter provides a natural restriction of the type of crop that can be grown. Gravity drainage from a lysimeter creates an artificial saturated zone at the bottom of each soil core that may not exist in the field setting. Finally, a lysimeter study conducted with large monoliths typically consists of only two cores, making the evaluation of the natural variability of leaching difficult.

In some circumstances, interpretation of the leaching data obtained from lysimeters is not straightforward. It is possible to obtain significant preferential flow (i.e. bypass flow of water through preferential flow paths) in lysimeters, depending upon the type of soil as well as the methodology used to collect and handle the soil cores. Preferential flow can result in detections in the leachate prior to the chromatographic movement of the bulk of the pesticide through the soil profile. A significant amount of scientific study is currently underway to evaluate the extent and significance of preferential flow in contributing to the movement of pesticides through the soil profile. It is not recommended that soils highly prone to extreme preferential flow (e.g. cracking clays) be used to conduct regulatory studies due to the difficulties in assessing the significance of the experimental results.

With appropriate selection of soils and care in collection and handling of cores, useful data are obtainable from lysimeter testing. In a comparison study of the results of parallel lysimeter and field studies, Kubiak concluded that degradation and mobility data of methabenzthiazuron were acceptably similar between the lysimeter and field data (Kubiak, 1986; Kubiak et al., 1988).

If vulnerable soils and wet climates are used for lysimeter studies, these studies can produce 'worst-case' leaching results which have a low probability of occurring naturally. It is often possible to calibrate a computer model to the experimenal results and then use the model to estimate the environmental behavior of the compound in less extreme settings. In this manner, the most critical setting is experimentally evaluated and degradation and mobility behavior in less critical settings are estimated through the use of modeling.

There are currently no definitive studies available that have correlated the concentrations measured in the leachate of lysimeters with the concentrations measured in shallow ground water using monitoring wells. In the absence of these data, some regulatory agencies have attempted to apply ground water standards to the leachate exiting the bottom of lysimeters. Such an approach is inappropriate since it ignores the continued degradation and dispersion of the trace residues that can occur both below the root zone and in the saturated zone (GIFAP, 1992).

At the present time, the evaluation of leaching of pesticides in lysimeters is an area of ongoing research to determine the effects of collection methodologies, monolith size and configuration and to determine the significance of the results obtained from these studies. Lysimeter studies are relatively expensive and there are not currently many laboratories available to conduct these studies.

Field ground water studies

Purpose of studies: The objective of a field ground water study is to evaluate the impact of an agricultural chemical on ground water quality. The general approach to conducting a field study can be separated into two parts, the first dealing with movement and dissipation in the unsaturated zone and the second focusing on saturated zone behavior (Jones and Norris, 1991). Unsaturated zone behavior is studied by collecting and analyzing soil samples at regular time intervals after a carefully controlled application. Residue movement into ground water is detected by collecting and analyzing water samples from clusters of shallow monitoring wells.

Description of studies: (1) Types of studies. Two types of ground water research studies are conducted to evaluate the potential of pesticides to reach ground water: small-scale prospective ground water studies and small-scale retrospective ground water studies (US EPA, 1988b). Both these studies are conducted using normal agronomic practices, especially for those compounds applied to the soil surface or incorporated into the soil.

Prospective studies are generally intensive investigations of a single worst-case site, 1–2 ha in size, that has not previously received any applications of the compound being evaluated (US EPA, 1988b; Norris *et al.*, 1990). These studies include sampling of ground water, soil pore-water and soil at relatively frequent time intervals to provide a detailed characterization of the leaching behavior of the compound. Prospective studies typically involve the installation of five to six well clusters with each cluster consisting of two to three wells, each installed at a different depth in the aquifer being studied to allow characterization of both horizontal and vertical movement of pesticide residues.

Small-scale retrospective studies are conducted on multiple sites that have documented prior use of the pesticide for 2 or more years. Sites of 4–8 ha in areas of high product use and moderate to high ground water vulnerability are typical chosen for retrospective monitoring. In this study design, soil sampling is only performed once at the outset of the study and installation of a minimum of three well clusters is recommended (US EPA, 1988b).

(2) Site selection. The most critical phase of a ground water study is the selection of appropriate sites for investigation. The objective of the study needs to state clearly whether the site to be selected is representative of a typical use area or a worst-case situation. For a worst-case site, the selected soil should have the appropriate physical and chemical properties to favor residue movement, e.g. a sandy soil with low organic matter content and low water holding capacity. Important hydrogeological conditions at a site include the depth to ground water and characteristics of the aquifer such as its composition and hydraulic conductivity. Finally, it is important to evaluate the influences of the climate at each site being considered. The precipitation received at a specific site should meet or exceed long-term averages to meet the objectives of a ground water study. This generally implies that each selected site should be capable of being irrigated in the event of extended dry periods.

(3) Unsaturated zone sampling. The number and location of soil cores collected during a single sampling interval is greatly influenced by the inherent variability of soil samples as well as the sampling method. Current studies of sampling variability indicate coefficients of variation ranging from 100 to 200% (Jones and Norris, 1991). As a result, it is recommended that 15–20 cores be collected at each sampling interval (Hörmann, 1973; US EPA, 1988b). To reduce analytical costs, these cores can be composited in the laboratory to a smaller number of samples.

Soil sampling intervals should be tailored to the expected dissipation rates of the agricultural chemical under study. For field studies designed to address ground water concerns, these intervals may be significantly different from those normally used in field dissipation studies. A typical sampling schedule would be to collect soil samples prior to and immediately after application, and 0.25, 0.5, 1, and 2 months and at 2-month intervals thereafter until the unsaturated zone monitoring is terminated. The earlier intervals may be eliminated when residues have relatively slow soil dissipation rates.

The depth of sample cores depends on chemical properties, soil character-istics and the climatological conditions encountered during the study. Except for samples collected immediately after application (these require only surface sampling), soil samples should be taken to the water table or to a depth sufficient to include all unsaturated zone residues. The depth increments generally used are 0.3 m increments down to 0.6 m and 0.6 m thereafter. These depth increments are significantly larger than the 0.15 m increments commony collected in field dissipation studies but are appropriate to characterize the distribution of residues over an extended distance in the soil profile.

(4) Saturated zone sampling. The number and location of monitoring wells depends on the study type, on site characteristics and on whether residues reach ground water during the study. One approach that has been used successfully in a number of studies is to install an initial grid consisting of five to six well clusters which monitor the upper 3 m of the saturated zone (Hornsby et al., 1983; Jones, 1990). A typical well cluster is composed of two to three wells screened at different depths so that the vertical position of any residue plume can be determined. For field studies conducted in areas where the water table is relatively shallow, the clusters generally consist of three wells per cluster with 0.3 m long slotted screens located just below the water table and at 1.5 and 3 m below the water table at the time of installation.

A typical layout for the initial five well clusters is illustrated in Figure 2. One well cluster is located immediately upgradient of the treatment area, two well clusters are located in the treatment area and two well clusters are located 3–30 m downgradient. The direction of ground water flow can usually be estimated satisfactorily based on surface topography. When the direction of ground water flow within the plot is not known or varies with time, well clusters may be placed on each side of the plot as illustrated in Figure 3.

If residues are found in shallow ground water during the course of the study, the well network can be expanded according to the existing monitoring data. When residues are found in only one cluster located in the treated area, it is useful to place additional clusters near the cluster containing the residues to characterize the movement and dissipation of the residues. Timely analyses of soil and ground water samples are helpful to guide the next sampling intervals, to install additional clusters and to collect deeper soil samples when necessary. Thus, it is necessary to maintain a flexible protocol to ensure the quality of the study results.

Samples should be collected from monitoring wells prior to application and at regular intervals thereafter. Because agricultural chemicals are applied over large areas and residue plumes in the saturated zone move relatively slowly (typically less than 0.2 m/day), water samples do not need to be collected in response to individual recharge events. One common schedule is to sample monthly during the first 6–12 months after application and, if necessary, every 2 or 3 months thereafter. If residues are detected in monitoring well samples, sampling is usually continued until the saturated

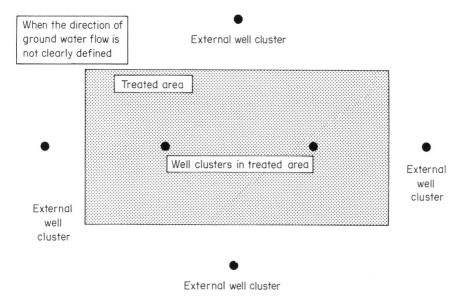

Figure 3 Alterative design for locating initial monitoring wells where the directoin of ground water flow is not clearly defined (source: Jones and Norris, 1991)

zone dissipation rate is determined or residues drop below detectable levels in all wells for two consecutive sampling events.

Care must be taken during the well installation, development and sampling to ensure that inadvertent contamination of samples does not occur. Guidance on proper techniques for well drilling, for selection of well construction materials and for sampling of wells is available in many recent publications (US EPA, 1987, 1988b; GIFAP, 1987, 1990b; Nielsen, 1991).

Discussion of studies: To date, prospective ground water studies have been performed primarily for unregistered agricultural chemicals showing the potential to leach below the root zone based on the results of laboratory and field studies of degradation and mobility. Retrospective studies have been performed primarily for agricultural chemicals which are already registered but which have been detected in ground water in concentrations of toxicological concern in the course of various ground water monitoring programs. Approximately 60% of the ground water studies that have been required by the US EPA have been small-scale retrospective studies (US EPA, 1990). However, this trend is slowly reversing in favor of the more detailed and more easily interpreted prospective ground water study, even for studies of currently registered compounds. The most important data obtained from a ground water study are dissipation rates in the unsaturated and, if residues reach ground water, the saturated zones. These rates can then be used in model simulations to predict the extent of movement in a variety of soils and aquifers under a range of weather conditions (Carsel *et al.*, 1985; Jones, 1990). If no residues reach the saturated zone, and the study was conducted in what is agreed to be 'worst-case', the experiment is used as evidence that the pesticide is unlikely to reach ground water. In this case, no further testing is necessary.

A few large-scale retrospective ground water surveys have been conducted using existing drinking water or irrigation wells. Typically, very little if any unsaturated zone sampling is performed in this type of study. Instead, survey studies are intented to provide statistical information on the magnitude and geographical extent of potential residues of pesticides in ground water.

Ground water studies are among the most expensive environmental studies performed to evaluate the environmental behavior of pesticides. In addition, these studies are relatively complex to conduct and to interpret. For this reason, it is appropriate to reserve this tier of testing for those compounds with significant mobility concerns.

The significance of any concentrations detected in ground water should be interpreted using available toxicology data for human, animal or plant exposure (WHO, 1991; GIFAP, 1987, 1990c; US EPA, 1988c) Quantitation of the frequency and extent of detected concentrations reaching shallow ground water can serve as the basis for a range of regulatory actions ranging from labelling requirements or use rate restrictions to localized restrictions or denial of registration.

CALCULATIONAL APPROACHES TO ESTIMATING PESTICIDE MOBILITY IN SOIL

Correlations between environmental properties

A number of correlations have been developed to determine the mobility potential of pesticides based on single physical–chemical properties. Some of these simple correlations can be useful for providing preliminary assessments of pesticide mobility in soil. However, it is important to determine any restrictions known to apply to the correlation to ensure proper use of the equations.

Water solubility in relation to adsorption and leaching

The physical processes governing solubility and sorption are fundamentally different. Solubility is a measure of a compound's relative tendency to self-associate as a solid phase versus association with water as a liquid phase. Sorption is a measure of the relative association of a pesticide with water in the soil solution phase versus association with the various components of the solid phase of the soil. In addition, pesticide residues are generally present in the soil at concentrations that are orders of magnitude lower than aqueous solubility limits. As a result, the solubility of a pesticide in water is of limited value in predicting the mobility of pesticides in soil.

Within groups of structurally similar pesticides, a correlation can be derived between water solubility and the extent of adsorption. This connection has been demonstrated by Bailey *et al*. (1968) for substituted ureas and s-triazines except atrazine, but other authors were unable to discover a relationship (Hance, 1965; Chapman *et al*., 1970; Guth, 1972). Although it is evident from the above results that no general relationship exists between water solubility and adsorption, the water solubility of a non-ionized pesticide can give a first, rough indication of its adsorption and consequently its potential for movement in the soil environment. Correlation equations for some classes of pesticides are listed in Lyman *et al*. (1982) and in Kenaga and Goring (1980).

Relation between octanol/water partition coefficient and mobility data

The octanol–water partition coefficient, K_{ow}, can be used as a measure of the lipophilicity of pesticides and thus as a good predictor of their bioaccumulation potential (Branson *et al*., 1975; Ellgehausen *et al*., 1980). In addition, satisfactory correlations have been found between K_{ow} and soil adsorption (Briggs, 1981) as well as leaching data (Guth, 1983). These correlations can be used to estimate the leaching potential of pesticides, particularly during the early stages of product development. Some commonly used correlations between K_{ow} and K_{oc} are presented in Lyman *et al*. (1982).

Relation between adsorption and leaching data

Rhodes *et al*. (1970) demonstrated a clear relationship between Freundlich adsorption constants and R_f values determined by soil TLC. Osgerby (1973) observed a similar correlation between R_f and K_{om} values. An empirical relationship between Freundlich adsorption constants determined by the slurry technique and penetration depths in soil columns was found by Guth (1972) after testing phenylureas, s-triazines and phosphorus acid esters. Hamaker (1975) used simple chromatographic theory to correlate adsorption constants with R_f values from soil TLC. Additional correlations are described in McCall *et al*. (1981).

Based on the correlations that have been established to date, it is clear that adsorption constants can be used to predict the leaching of pesticides on soil layers and also in soil columns and therefore could replace these leaching experiments.

Screening assessments of pesticide mobility

A number of authors have developed screening methods for determining whether a pesticide is likely to leach to ground water in detectable quantities. One approach has been to assign mobility descriptions to ranges of numerical values obtained from the various mobility studies. Descriptive classification schemes have also been developed for approaches which consider the combined influences of mobility and persistence. Finally, some approaches have attempted to provide estimates of the percentage of compound which leaches to specific depths in the soil.

Classification of mobility potential in soil based on laboratory studies

Several proposals have been published for assessing the mobility potential of pesticides in soil based on the results of the laboratory studies of pesticide mobility. Helling (1971) used R_f values from soil TLC to classify a variety of herbicides and insecticides into five mobility classes which range from low mobility in class 1 to high mobility in class 5 (Table 2). A similar classification scheme based on the results of octanol/water partition coefficients, soil column leaching relative mobility factors (RMF) and the adsorption coefficient K_{om} is presented in Table 3 (based on Guth, 1983; Guth, 1985; Guth and Hörmann, 1987). McCall has classified pesticide mobility into six descriptive classifications based upon the value of adsorption partition coefficients, K_{oc} (McCall et al., 1981) (Table 4).

Classification of persistence in soil based on laboratory and field studies

Several schemes have also been proposed for the assessment of the degradability of pesticides in soil. Goring et al. (1975) used the time for 50% disappearance of chemical from soil to classify products according to their relative persistence (Table 5). The IUPAC commission on agrochemicals (Esser et al., 1988) proposed half-lives in soils be used for assessing degradability of pesticides (Table 6). Half-lives, DT-50 and DT-90 values have also been introduced by some regulatory authorities (US EPA, 1982g; DK AEP, 1988; BBA, 1990b) as triggers in tiered persistence testing systems (Table 7).

In addition to DT-50 and DT-90 values, the evaluation criteria used by the BBA include the percentage of active ingredient or metabolite left in the soil

Table 2 Mobility classification based on soil-TLC

Class 5 (0.90–1.0)		Class 4 (0.65–0.89)		Class 3 (0.35–0.64)		Class 2 (0.10–0.34)		Class 1 (0–0.09)	
TCA	0.96	Picloram	0.84	Propachlor	0.63	Siduron	0.30	Chloroxuron	0.09
Dalapon	0.96	Fenac	0.84	Prometone	0.60	Prometryne	0.25	Diquat	0.06
Dicamba	0.96	MCPA	0.78	2,4,5-T	0.54	Propanil	0.24	C-6989	0.04
Amiben	0.91	Amitrole	0.73	Propham	0.51	Diuron	0.24	Morestan	0.02
		2,4-D	0.69	Fluometuron	0.50	Dichlobenil	0.22	Dieldrin	0.00
		Bromacil	0.69	Diphenamid	0.49	Chlorpropham	0.18	Paraquat	0.00
		Nortron	0.65	Monuron	0.48	Azinphomethyl	0.15	Trifluralin	0.00
				Atrazine	0.47			Endrin	0.00
				Simazine	0.45			DDT	0.00
				Ametryne	0.44				
				Propazine	0.41				
				Trietazine	0.36				

SOURCE: Helling, 1971.

Table 3   Mobility classification based on RMF, K_{ow} and K_{om}

Range of RMF	Range of K_{ow}	Range of K_{om}	Compounds	Mobility Class
<0.15	>10⁴	<10³	Fluorodifen, parathion	Immobile
>0.15	<10⁴	<10³	Profenophos, propiconazol, chlorbromuron, diazinon, diuron, terbutryn, terbuthylazine, prometryn, methidathion, isazophos, methoprotryn, propazine, alachlor, metolachlor, ametryn, terbumeton	Slightly mobile
	>500			
<0.8		>100		
>0.8	<500	<100	Monuron, atrazine, simazine, fluometuron	Moderately mobile
<1.3	>150	<50		
>1.3	<150	<50	Prometon, secbumeton, thiazafluron, bromacil, cyanazine, karbutilate, metalaxyl	Slightly mobile
<2.5	>50	>20		
>2.5	<50	<20	Cyromazine	Mobile
<4.0	>20	>10	Dioxacarb	
>4.0	<20	<10	Monocrotophos, phosphamidon	Very mobile

RMF, relative mobility factor = leaching distance of pesticide/leaching distance of monuron; K_{ow}, octanol/water partition coefficient; K_{om}, soil organic matter partition coefficient.
SOURCES: Guth, 1983, 1985; Guth and Hörmann, 1987.

Table 4   Mobility classification based on K_{oc}

K_{oc}	Mobility class
0–50	Very high
50–150	High
150–500	Medium
500–2000	Low
2000–5000	Slight
>5000	Immobile

SOURCE: McCall *et al*. (1981).

before next application, percent of bound residues formed and rate of application (Kloskowski *et al*., 1990; Figure 4).

Classification of ground water contamination potential based on both mobility and persistence

The mobility and persistence of a pesticide characterize two distinctly different but simultaneous kinetic processes. The mobility of a pesticide is a

Table 5   Relative persistence of pesticides in soils

Pesticide group	Pesticides	Est. time req'd for 50% to disappear from soil (months)
Non-persistent	1,3-Dichloropropene, acephate, alachlor, asulan sodium salt, azinphos methyl, bensulfuron methyl, cyanazine, diflubenzuron, fenoxycarb, fluazifop-p-butyl, fluvalinate, pirimiphos-methyl, sethoxydium, terbufos	<0.5
Slightly persistent	Aldicarb, benefin, bifenthrin, chlorothalonil, chlorsulfuron, diphenamid, esfenvalerate, ethofumesate, flucythrinate, lambda-cyhalothrin, metsulfuron methyl, molinate, propoxur, tralomethrin, triforine	0.5–1.5
Moderately persistent	Ancymidol, atrazine, benomyl/MBC, cyromazine, imazethapyr, isopropalin, metolachlor, prochloraz, trifluralin	1.5–6
Persistent	Fenarimol, prometon, tebuthiuron, thiabendazole	>6

SOURCE: Wauchope *et al.* (1992).

Table 6   Persistence ranking of pesticides in soil

Range of DT-50	Compounds	Persistence class
<1 month	Dicrotophos, dimethoate, diazinon, methidathion	Readily degradable (non-persisten)
>1 month <6 months	Atrazine, ametryn, bromacil, diuron, monuron, prometon, propazine, simazine, terbacil, trifluralin	Moderately degradable (moderately persistent)
>6 months	Aldrin, DDT, dieldrin, lindane, methoxychlor	Slowly degradable (persistent)

SOURCE: Esser *et al.* (1988).

measure of its rate of travel through the soil profile while the dissipation of a compound characterizes its rate of transformation into metabolites and other degradation products. To adequately assess the potential impact of a pesticide on ground water, it is necessary to consider both of these kinetic processes at the same time.

Table 7 Regulatory triggers for soil degradation testing

Testing step	United States	Germany	Denmark
1 → 2	None (field testing is always required)	DT-50 > 30 days and DT-90 > 100 days	Half-life > 8 months for herbicides, > 3 months for other products
2 → 3	<50% degradation until next application in the filed or total extracted residue in lab studies >50% of initial amount at next application	Residue plateau after several applications 5× as high as after single application	Exceeding above triggers in field testing ↓ No registration

Definition of testing steps: 1, laboratory testing; 2, short-term field testing (1 year); 3, long-term field testing (accumulation study, 3 years).
SOURCE: US EPA, 1982a; BBA, 1986; DK AEP, 1988.

Several relatively simple approaches to assess the combined influences of mobility and persistence have been developed. Herzel (1987) developed a graph combining mobility and persistence to illustrate the combinations of these properties that increase the probability that a compound reaches ground water (Figure 5). If suitable scales for RMF and DT-50 values are substituted for the dimensionless units used by Herzel, a useful scheme can be developed to quickly identify problem pesticides (Figure 6) (J. A. Guth, 1990, unpublished). Cohen et al. (1984) proposed defining criteria of K_{oc} less than 300–500 and a soil half-life longer than 2–3 weeks as sufficient to identify those compounds with a high potential to impact ground water via leaching (Figure 7). Additional screening methods based on individual properties have also been published (Rao et al., 1985; Arnold and Briggs, 1990).

In addition to the criteria of persistence and mobility, Jury et al. (1987) introduced a third dimension, the leaching conditions, into a leaching assessment scheme. This approach clearly demonstrates that the leaching behavior of a pesticide is determined not only by its inherent chemical and physical properties but also by the climate and hydrogeology present at its sites of use. As a result, a pesticide may adversely leach in one setting but behave acceptably in another region.

Gustafson (1989) outlined a classification scheme which calculates an index based on the soil partition coefficient, K_{oc}, and the soil half-life (Figure 8). Two values of this index were proposed to classify pesticides as probable leachers, transition leachers or improbable leachers. Classifications

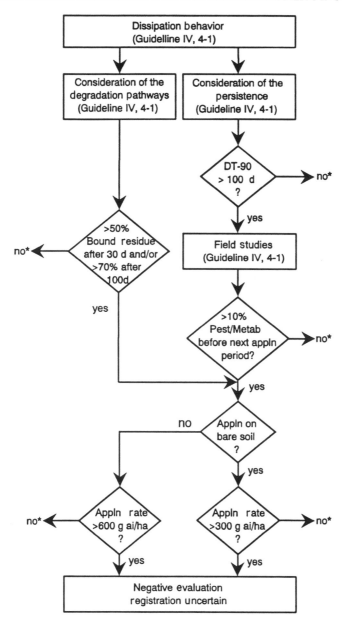

* Registration possible

Figure 4   Evaluation of data concerning dissipation in soil (based on Kloskowski *et al.*, 1990)

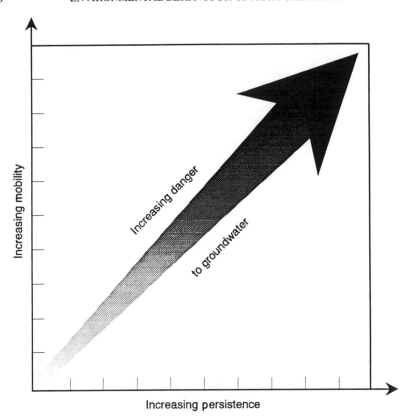

Figure 5   Schematic diagram of leaching risk (based on Herzel, 1987)

made through calculation of the indices compared well with the results of several well water monitoring programs conducted in the USA.

None of the above leaching classification schemes consider factors such as use rate, type and method of application, type of crop, etc. This is the role of the simulation model.

Computer modeling of pesticide mobility

Computer simulation modeling of pesticide mobility in soil is an attempt to integrate the effects of climate, soils and pesticide properties to provide more detailed assessments of pesticide mobility than are possible with screening assessments. This approach can provide estimates of the concentrations present in the soil, in soil solution and in ground water at any point in time and is extremely useful in providing increased understanding of the likely environmental behavior of an applied pesticide.

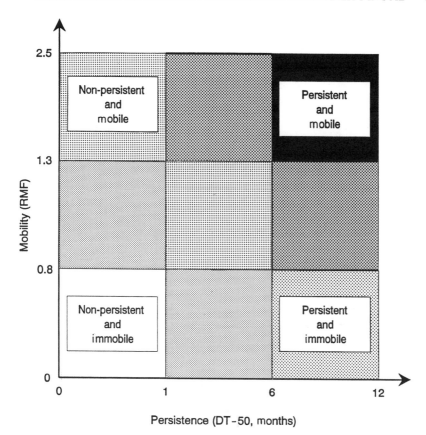

Figure 6 Classification of leaching potential of pesticides (source: J. A. Guth, unpublished)

A wide range of pesticide mobility models have been developed over the past 10–15 years and they vary dramatically in their capabilities, their documentation and their ease of use. Some models are essentially used only in research settings within a single country while others have gained wide acceptance both geographically and across various groups of end users.

Computer modeling can play an important role in the development and registration of pesticides by helping rationalize the results of field trials, providing estimates of likely leaching behavior and assessing the probability of ground water contamination.

The single most important step in simulation modeling is to define the objective of the study and to select a model that is technically appropriate to meet these objectives. Once this is done, the process of modeling consists of obtaining appropriate input data, running the computer program, and

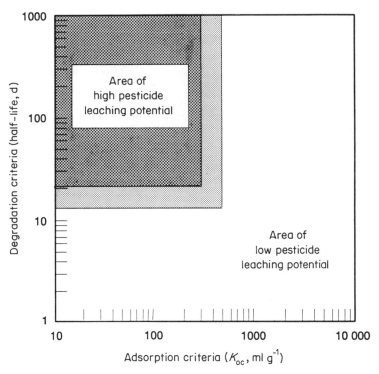

Figure 7   Criteria to define pesticides with high leaching potential (based on Cohen
et al., 1984)

interpreting the output data which is produced. This process is most
commonly performed iteratively such that the model is rerun many times
with differing input data before the overall interpretation and mobility
assessment is made.

General description of computer models of leaching

There are a very large number of pesticide leaching models (Table 8; further
information in Appendix 1). These models vary in the level of process detail
included, the way in which the processes are represented by equations, the
method of solution of the equations (and hence computing power needed)
and the extent of input data required and output produced (Addiscott and
Wagenet, 1985; Hern and Melancon, 1986). Despite many differences of
detail, the most commonly used models of pesticide mobility have much in
common in their general approach since they are all attempting to represent
the same phenomena.

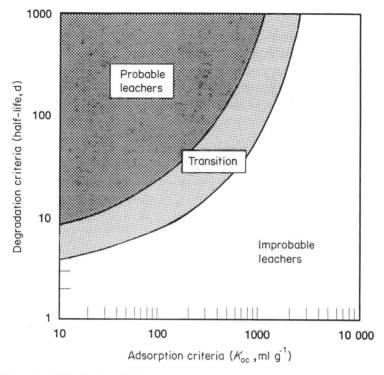

Figure 8   Relative leachability of agrichemicals (based on Gustafson, 1989)

Pesticide mobility models conceptually consist of two parts, the first handling the flows of water in the soil system (hydrology), and the second addressing the behavior of pesticides in this pattern of water movement (solute transport). The heart of all the models is the water balance submodel, in which the precipitation and irrigation are allocated to run-off (often ignored to give worst-case leaching), evapotranspiration, and storage in the soil. Flow of water in the soil is assumed to be one-dimensional and may be calculated from simple drainage rules or complex equations, based on measurable soil properties. Depending upon the algorithm used, water flow may be restricted to downward movement only or allowed to move both up and down as a function of the prevailing soil moisture and climatic conditions. Simulation is generally restricted to the unsaturated zone (above the water table) and is often restricted further than this to give a simulated soil profile depth of 1 m or less.

The solute transport submodel relies heavily on the hydrology submodel to determine movement of the pesticide in the soil profile. The transport submodel typically includes pesticide-specific properties such as adsorption,

Table 8   List of current unsaturated zone leaching models

Name of model	Authors/Institution Associated with the model	Computational requirements	Extent of use, where used
CALF	Walker and Barnes, Horticulture Research International, UK	Low	Low, Europe
PESTLA	van der Linden and Boesten, Wageningen, Netherlands	Moderate	Low, Netherlands
SESOIL-FHG	US EPA, modified by Fraunhofer Institute, Germany	Moderate	Low, Germany
PRZM-1	Carsel et al., US EPA, USA	Moderate	Most widely used model, Europe and USA
PRZM-FHG PELMO	US EPA, modified by Fraunhofer Institute, Germany	Moderate	Low, Germany
PRZM-2 (Note 1)	Dean et al., US EPA, USA	Moderate	Low, will replace PRZM-1
GLEAMS (Note 2)	Leonard et al., USDA, USA	Moderate	Moderate, USA
LEACHMP	Wagenet and Hutson, Cornell University USA	High	Moderate, Europe and USA

Note 1. PRZM-2 is a submodel of RUSTIC which is a linkage of three models: PRZM-2 (root zone), VADOFT (unsaturated zone) and SAFTMOD (saturated zoe). This model is currently under development.
Note 2. Extension of CREAMS model and currently only applicable to the root zone.
SOURCE: Based in part on Jones and Norris, 1991.

volatility and degradation as well as application rate, timing and frequency. Application may be directly to the soil surface, although some models allow for interception by a crop canopy or soil incorporation.

Of necessity, models include a number of simplifying assumptions. Pesticide adsorption is usually assumed to be completely reversible and linear with respect to concentration, while degradation is exponential (characterized by a half-life). In some models, the influences of temperature and soil-moisture content of degradation are also included. Transport in the vapor phase is included in some models as well as the formation and degradation of metabolites. A number of other processes, such as adsorption kinetics and preferential flow, are typically excluded from most currently used models.

The output from most models typically includes pesticide concentrations with depth at various time intervals after application. In addition, with some models, it is possible to obtain values of the total mass of pesticide that has moved past certain depths in the soil profile at various points in time. Most models provide an overall mass balance with time to provide information on the relative importance of various dissipative mechanisms that act upon the applied pesticide in the soil and to ensure that numerical inaccuracies produced by the code are kept within satisfactory limits.

A relatively recent development has been the development of probabilistic models. This approach to modeling typically uses a randomized calling program to vary the input parameters and then repeatedly run a deterministic model such as PRZM. The calculated results are then presented as the probability of obtaining a concentration or mass flux rather than the prediction of a single specific value. Consequently, probabilistic modeling is a very useful technique to help place the results of discrete field tests into perspective. However, this approach requires a significant amount of resources (primarily expertise, time, and input data) to obtain the desired results.

Utility of specific leaching models

The suitability of a model for a particular role or a particular pesticide mobility assessment is based on a wide range of criteria, including the theoretical basis of the model, perceived accuracy of model and accuracy of output required, nature of pesticide and crop/soil/climate to be simulated, availability of input data, output required, ease of use and quality of documentation, and quality of software programming.

The extent to which a model has been validated is an important element in model selection. Validation of pesticide mobility models has generally been limited, but certain general comments on relative model accuracy can be made based on a comparison of model content. The models that are discussed have all been used by industry and by regulators to help guide appropriate decision making concerning the mobility of pesticides in soil.

The pesticide root zone model (Carsel et al., 1984; PRZM) is a basically sound and flexible model, with the best documentation of the models reviewed. As long as care and experience are used to select appropriate input values then reasonably accurate predictions can be made in many circumstances. PRZM simulates water movement in the soil profile by a relatively simple storage and routine scheme. Due to this simplistic treatment of infiltration hydrology, PRZM works best for freely draining soil profiles where downward movement of water predominates (i.e. rainfall generally exceeds evapotranspiration). It is less appropriate for fine-textured soils and arid climates. However, this restriction does not generally represent a significant restriction since worst-case pesticide mobility scenarios are typically on sandy soils in wet climates.

Two variations of PRZM have been developed for use in Germany. The first version, identified in this paper as PRZM-FHG (PRZM-FHG, 1989) incorporates a number of changes to the source code and the model developers have chosen to release only the compiled version. The model uses a set of standardized crop and weather scenarios and provides standardized output to guide the user in making mobility assessments. The PRZM-FHG model has severely restricted the flexibility and adaptability of the original PRZM-1 model. The second variation of PRZM, called PELMO (PELMO, 1992) is a newer version of PRZM-FHG and it incorporates use of the Freundlich equation, optional non-first-order degradation kinetics, and use of the Haude formula to estimate potential evapotranspiration. In addition, the variation of soil temperature with depth is considered in determining the rate of degradation in the soil.

A relatively new model, RUSTIC (Risk of Unsaturated/Saturated Transport and Transformation of Chemical Concentrations), incorporates an enhanced version of PRZM, referred to here as PRZM-2 (Dean *et al.*, 1989). New features of PRZM-2 include the ability to handle multiple pesticides and degradates, the addition of volatilization as a dissipation mechanism and the ability of perform probabilistic Monte Carlo calculations to assess the effects of variability in the various input parameters.

For cases where the upward flow of water in soil is important, or where there is restricted drainage, LEACHMP (Wagenet and Hutson, 1989) is an appropriate model to consider. Where these conditions do not prevail, the results from LEACHMP would be expected to be similar to those obtained from the various versions of PRZM. The main limitations of LEACHMP compared to PRZM versions are its far greater computational time requirements and its somewhat lesser flexibility.

SESOIL-FHG (SESOIL-FHG, 1989) has many limitations as a model, including its theoretical basis and documentation. The use of this model for pesticide mobility calculations offers no advantages over PRZM and LEACHMP for assessing pesticide mobility and is no longer recommended for use.

PESTLA (van der Linden and Boesten, 1989) is mainly intended for use with the fixed scenarios provided with the code. When used with only the fixed scenarios, this model has extremely limited flexibility for the user and is primarily useful as a screening tool to obtain comparisons between pesticides. It is possible to modify the source code for PESTLA to alter the scenarios but both the programming language (CSMP) and the program structure make file editing an obstacle for most users.

CALF (Walker and Barnes, 1981) has been modified by numerous users with limited control over versions and content and has no manual. However, it contains several valuable features, is quick to run and has been successfully used by numerous groups, primarily in the UK.

GLEAMS (Leonard et al., 1987) is an extension of the earlier CREAMS run-off, erosion and chemistry model developed by the United States Department of Agriculture (Knisel, 1980). This model is primarily intended for use in simulating leaching processes in the root zone. Similar to PRZM, GLEAMS solves the one-dimensional convective-dispersive equation for solute transport using a simplified water balance. Soil properties and degradation rates can vary with depth.

Other existing deterministic models may offer advantages over PRZM, LEACHMP and GLEAMS for particular applications, especially when used in an informal context or where communication of model results and procedures to regulatory organizations is not of primary importance.

Two recently published examples of probabilistic leaching models are RUSTIC (Risk of Unsaturated/Saturated Transport and Transformation of Chemical Concentrations) developed by the US EPA (described above) and GRASP (Geographically Based Risk Analysis System for Pesticides) developed by DowElanco (Tillotson et al., 1991). GRASP is a program that integrates properties of the pesticide with environmental variability characteristics to provide probabilistic pesticide exposure calculations based on geographic location. The model includes a weather generator, a soils database, the PRZM transport model, a Monte Carlo shell and a sensitivity analysis module.

Discussion of computer simulation of pesticide mobility

One of the primary advantages of computer simulation with respect to laboratory and field studies is the capability of putting individual studies into a larger context. The results of computer modeling can be used to provide an improved understanding of the fate and transport of a chemical in the environment.

Another advantage of modeling is that it is a relatively inexpensive technique. Most of today's models can be run on a desktop PC or a small mainframe computer. In contrast, field studies can take many man-years of effort and expensive analytical equipment. An associated advantage is that of time. A modeling exercise may take a number of weeks or months whereas the leaching process as studied in a field or simulated-field situation can take several years.

Simulation modeling also offers the prospect of integrating the results of numerous laboratory and field studies on a compound, thus maximizing the use of data to make an overall assessment of mobility in soil. A further advantage of the modeling approach is its flexibility in addressing particular issues, depending on the amount of data available and the precise question being asked. Modeling work done in conjunction with a mobility experiment at a particular filed site or in a given lysimeter provides an example of this flexibility.

A first attempt at predicting the results from limited data with a mobility model may be useful, but is unlikely to give any more than a broad indication of the expected results. If further data such as soil profile and weather information and pesticide adsorption to that soil are subsquently determined, then skilled interpretation of these data and a revised modeling effort would give predictions of far greater reliability and accuracy. When the field measurements from our hypothetical experiment become available, then with care and skill a limited calibration of the mobility model may be performed, if necessary, to bring model results into line with the measurements. The value of this is that extrapolation can then be made to estimate likely mobility under different field conditions, e.g. with different soil properties or weather data, or for a longer time period than that of the experiment, perhaps with repeated applications of the pesticide in question. The accuracy of these extrapolated results is difficult or impossible to quantify, but will be determined by the care taken by the person using the model, and by the extent of the extrapolations and the ability of the model to handle these.

A major problem associated with modeling is assessing the accuracy and reliability of the results. One reason for this is that it is never possible to fully validate a pesticide mobility model, because of the great variability in actual field pesticide concentration distributions and the difficulties of measuring these distributions in experiments (Pennell *et al.*, 1990). Another reason is that the accuracy of model outputs is crucially determined by the accuracy of the inputs. It is important to consider the scale of the model and the acceptable degree of resolution needed to characterize pesticide mobility in soil (Rasmuson and Flühler, 1990). The availability and quality of experimental data can dramatically influence modeling results, necessitating the use of extreme care in selecting input values.

The assumptions built into the mathematics of the model and any limits of its range of applicability must also be fully understood by the model user to prevent producing results that may be more misleading than helpful. Just as with any other type of scientific study, skill and judgement are required to obtain useful results from a modeling exercise and to correctly interpret model outputs. It is crucial that adequate thought and justification be given to selecting values for the many input parameters which are not directly measured. To obtain credible modeling results, it is recommended that values for all input parameters be reviewed to ensure that they are appropriate for the compound/crop/location being considered. Simulation of pesticide mobility in fixed scenarios can provide useful comparisons between products but is rarely useful in providing estimations of actual environmental concentrations. Failure to recognize this has resulted in much inproper use of modeling in the past. Use of probabilistic modeling can provide more realistic assessments of the actual range of concentrations that are likely to be encountered in the environment.

A further limitation associated with the use of models in a formal regulatory arena is the frequent lack of a thorough model evaluation procedure prior to the adoption of a particular modeling scheme. The first step in any modeling exercise is to define the study objective. Once the objective is clear, then a model should be selected or developed, and then evaluated relative to this objective. The evaluation procedure needs to consider whether the model includes the relevant physical, biological and chemical processes, whether the level of detail is appropriate, and whether the necessary input data are available (Russell and Layton, 1991). Additionally, the chosen model should be free of programming errors, and source code should be available along with the code version number. The model should have full documentation and additional material enabling the prospective user to assess which are the key input parameters, how sensitive results are to these parameters, over what range of conditions the model is applicable, and how much model validation there has been (ASTM, 1984). This whole evaluation procedure needs to be fully documented as a point of reference.

With this foundation achieved by consensus, and with competent modelers in both industry and regulatory organizations, regulatory discussions can then focus on what model input data to use for the pesticide in question and on the interpretation of output, which is what really matters. The model evaluation procedure outlined here, if followed, makes the skilled and appropriate use of models more likely to occur, resulting in more accurate and reliable results, and better quality assessments of pesticide mobility in soil.

PROPOSED TIERED SCHEME TO ASSESS PESTICIDE MOBILITY IN SOIL

Appropriate and reproducible methods are certainly needed to determine the environmental properties of a pesticide but equally important is the assessment of the results of this testing. Such assessments should be designed to identify potential hazards and thus enable risks of adverse effects on the environment to be quantified and evaluated in relation to benefits. The nature and amount of data required to evaluate the environmental safety of a pesticide depend on the properties and use of the compound. Research resources should be focused on the identification and evaluation of significant risks and data requirements which are excessive and stifle innovation must be avoided.

A stepwise testing sequence allows an efficient selection of experiments essential to each individual risk analysis. Following each step, a preliminary assessment of risks and benefits allows decisions to be made on the need for further testing. Tests closer to practical use conditions may be required if there are doubts that benefits clearly outweigh risks.

Consistent with the previous recommendations for assessment tiers (GIFAP, 1990b), a tiered approach to assess pesticide mobility is recommended by GIFAP as a reasonable way to progressively classify the likely environmental behavior of a pesticide (Figure 9). This approach consists of the following steps:

- exploratory laboratory studies,
- verification field studies,
- surveillance field studies.

These tiers of related studies are as follows.

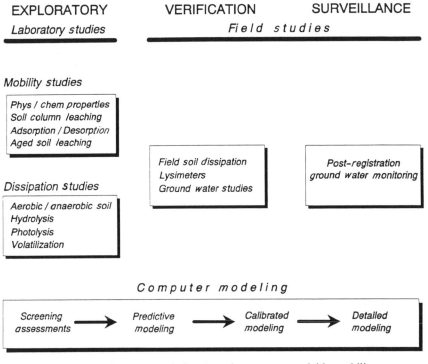

Figure 9   Recommended approach to assess pesticide mobility

Exploratory laboratory studies of pesticide mobility

This level of testing involves measurement of the basic physical, chemical and environmental properties of the test compound including evaluation of its mobility in soil and the role of major individual degradation mechanisms of its dissipation in soil. To assess the significance of the results obtained in the laboratory studies, it is useful to use the criteria that have been reviewed under 'Screening assessments'. When the various classification schemes are combined, the following stepwise testing sequence can be developed.

Octanol/water partition coefficient (K_{ow})

The octanol/water partition coefficient should be determined. If $K_{ow} < 150$, then additional mobility data are clearly needed. This determination can be made very early in the development of a compound to help anticipate the need for future testing.

Adsorption, soil column leaching, soil-TLC, aged soil leaching

It is generally not necessary to perform all of these studies to characterize the leaching behavior of a specific compound. Appropriate studies should be selected. If $K_{oc} < 500$, $R_f > 0.5$ or RMF > 1, the active ingredient or metabolite is likely to be moderately to highly mobile in the soil. Additional factors now need to be considered, such as the dissipation of the compound, the proposed use pattern (rate, method and timing) and the significance of metabolites.

Hydrolysis, photolysis, volatility

The results of these studies can serve to clarify the respective importance of non-biological dissipation of the compound in the soil. For compounds which are foliarly applied, these routes of dissipation can reduce the fraction of the applied pesticide that is available to move through the soil profile. Compounds with significant rates of hydrolysis of volatilization are less likely to persist in the soil environment and generally have a reduced potential to leach through the soil at concentrations of concern.

Aerobic soil, anaerobic soil

The primary purpose of these studies is to determine degradation pathways and to identify important soil metabolites. The degradation rates obtained from these studies provide an initial estimate of the degradation rates to be expected in a field setting. By comparing the results of these studies with the rate of hydrolysis and photolysis, the relative importance of microbial degradation can be determined.

Screening assessments

It is useful to evaluate the combined influences of compound mobility and dissipation through the use of a screening assessment such as the method of Gustafson (1989) or van der Linden and Boesten (1989). These assessments can provide a first estimate of the fraction of the applied dose that may move below the root zone in a fixed 'worst-case' scenario. In interpreting the results of this screening assessment, it is also useful to consider the proposed use rate of the compound as well as the toxicology of the parent molecule and its major metabolites. Screening assessments such as those described here should serve primarily to classify the general behavior of a compound and to indicate the potential need for additional testing. These methods should not serve as cut-off or 'guillotine' criteria upon which regulatory decisions are based.

Depending upon the results of these initial exploratory studies, supplementary laboratory testing may be needed to further clarify the environmental distribution and degradation that is likely to occur from pesticide use. When the results of exploratory testing are not conclusive in demonstrating acceptable levels of mobility in soil, it is necessary to conduct some type of field testing to verify the actual mobility behavior. Field testing of pesticide mobility is appropriate when:

(1) high mobility is directly observed in laboratory mobility studies; and
(2) a screening assessment of the combined influences of persistence and mobility indicates a reasonable potential for leaching of an environmentally significant fraction of the applied dose below the root zone.

Verification field studies of pesticide mobility

Three types of studies can be used to verify the mobility behavior of pesticides in a field setting: field soil dissipation studies, lysimeters using intact soil cores and prospective ground water monitoring studies.

Field soil dissipation studies

The primary purpose of this study is to characterize the overall rate of dissipation of pesticide from a field setting following application. Detailed information on the leaching behavior is generally not available since soil sampling is generally limited in depth and soil water samples are not collected. As a result, a field soil dissipation study can only serve to verify the leaching behavior of those compounds which are classified as having a low to moderate potential of significant leaching. For low use rate compounds or compounds with high mobility but rapid degradation, this approach can be sufficient to verify the absence of significant concentrations moving out the root zone.

The single most significant result from a set of field soil dissipation studies is the relative magnitude and variability of the calculated rate of dissipation (e.g. the half-life if dissipation if first-order). DT-50 values of less than 30 days indicate that the compound should be regarded as non-persistent. Compounds with DT-50 values of 30–100 days are moderately persistent while compounds with slower dissipation rates are generally regarded as persistent.

Lysimeters

These studies are appropriate for compounds with moderate to high mobility. The primary difficulty with a lysimeter study is interpreting the environmental significance of the low level traces of the applied compound that may show up in the leachate over the time course of the study. As discussed in the review of lysimeters, there are relatively few studies available which relate the results of lysimeter testing to concentrations of pesticide residues found in ground water monitoring. In the absence of these data, overly restrictive drinking water standards are being applied to leachate concentrations by various regulatory agencies.

The results of lysimeter testing should be evaluated by using criteria that clearly have environmental significance. It is inappropriate to establish arbitrary limits on the total cumulative percentage of parent and metabolites in leachate without providing some means of interpreting the impact of these concentrations on shallow ground water. To help provide a logical connection between lysimeter results and ground water concentrations, an appropriate set of ground water monitoring studies could be conducted in parallel with a set of lysimeter studies. Computer modeling could be used to help provide a rational link between the results observed in the two studies. Once this connection between the results of the lysimeter study and likely concentrations in ground water has been established, it will be possible to draw more definitive conclusions concerning the significance of lysimeter results.

Until appropriate research is published to allow appropriate interpretation of the results of lysimeter testing, it is likely that regulators will continue to apply arbitrary numerical standards that are thought to be conservative. A number of the current regulatory standards are listed in Appendix 1.

Prospective ground water studies

When the results of laboratory studies clearly indicate a high potential for pesticide mobility in soil, it is appropriate to conduct a detailed prospective ground water monitoring study to support product registration. This study provides a detailed investigation into the concentrations present in the soil and in shallow ground water following normal product application in a worst-case setting. Using suction lysimeters, it is also possible to obtain samples of the soil pore water when the soil moisture content is near field capacity.

It is generally reasonably straightforward to interpret the results of individual prospective ground water monitoring studies. Since sites used for these investigations have not had prior applications of the pesticide, spurious detections due to unknown prior use of the compound can be eliminated. As reviewed under 'Field ground water studies', care should be taken to ensure that soil and water samples are properly collected to avoid inadvertent contamination. It is important to consider the 'worst-case' conditions under which these studies are conducted when extrpolating the study results to other less vulnerable settings. Simulation modeling can be used to estimate the likely behavior of pesticides in more typical settings.

It is the recommendation of GIFAP that the results of ground water monitoring be compared against toxicologically based standards (GIFAP, 1987, 1990c). This recommendation is consistent with other widely accepted water quality standards that have been established for both organic and non-organic materials known to be present in water (US EPA, 1988c). If the results of prospective ground water monitoring are judged to exceed levels of toxicological concern, numerous options are available to address the problem including:

- selection of appropriate application rates and timing to fit seasonal and local conditions (subject to efficacy considerations);
- restrictions of irrigation practices;
- establishment of well setbacks from treated fields;
- restriction of product use to geographical areas with low to moderate vulnerability;
- encouragement of alternative chemical use in well defined geographical areas having especially vulnerable aquifers;
- specification of measures to prevent point source contamination:
     requirements for anti-siphoning devices,
     requirements for mixing/loading/rinsate pads,
     container storage and disposal restrictions;
- influence of formulation.

Post-registration field studies of pesticide mobility

Following registration of a product, it may be necessary to perform additional ground water monitoring of various geographical areas in which the pesticide is labelled for use. The objective of this testing is not to further clarify actual mobility mechanisms but to determine potential human and ecological exposure to pesticide residues in ground water as a result of normal use of the compound in typical settings. Post-registration ground water monitoring studies can involve the monitoring of existing shallow drinking water wells as well as the installation of new monitoring wells.

The results of ground water monitoring studies typically consist of data which describe the frequency and extent of detections. Large scale monitoring studies which have been conducted in the USA (Monsanto, 1990; US EPA, 1990) have involved a considerable degree of statistical design. As a result, the study results can be used to estimate the probable extent of exposure to various concentrations over wide geographical areas.

Other tiered schemes to assess pesticide mobility

In addition to the various non-regulatory proposals outlined under 'Review of existing approaches to assessing pesticide mobility' regulators in a number of countries have instituted tiered approaches to assess pesticide mobility. A brief overview of three of these systems is presented in Appendix 1.

The tiered regulatory approach taken by Germany includes consideration of the percentages of active ingredient and/or metabolites found in column leachates as well as DT-50 values as decision criteria for further tests in lysimeters (Figure 10). Herzel also used percentages in leachates and DT-50 values in establishing a classification system (Herzel, 1987). The Danish Environmental Protection Agency has proposed a system based no the percent eluted from soil columns (DK AEP, 1987). In the Danish classification scheme, compounds identified as mobile are then evaluated further in lysimeters.

It is obvious that the various systems for classifying pesticides according to their mobility in soil need to be harmonized in order to ensure appropriate levels of testing.

ASSESSMENT OF RISK AND BENEFIT

Although the primary purpose of this chapter is to describe approaches to assessing the risk resulting from mobility of pesticides, it is also important to take into account the numerous benefits of a pesticide when determining the acceptability of a pesticide. Although it is clear that no human activity has a zero risk, it is in the best interests of everyone to minimize the risk and maximize the benefit.

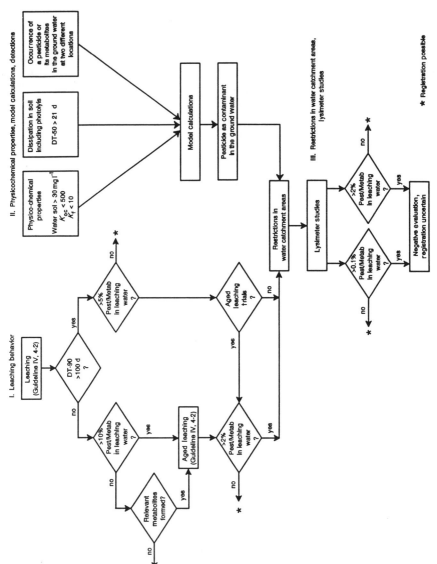

Figure 10 Evaluation of data concerning leaching potential (source: Kloskowski *et al.*, 1990)

Benefits from the use of pesticides can be described in three primary categories (GIFAP, 1990a):

- ecological benefits,
- economic benefits,
- social benefits.

Ecological benefits

The ecological effects of a specific pesticide should be compared with the known impacts of alternative pest control methods. Many alternative pest control practices can have a serious impact on the environment. As an example, excessive tillage to obtain weed control can greatly increase soil erosion by exposing loose soil to the weathering effects of the wind and water. The use of herbicides can reduce erosion by an order of magnitude by leaving the stubble of dead weeds intact in undisturbed soil. In addition, repeated cultivation also greatly reduces the number of some beneficial organisms, such as earthworms, in the soil surface and can lead to localized soil compaction problems. The use of herbicides in place of cultivation can have a positive impact on both these situations by leaving the soil undisturbed and requiring fewer trips over the field to maintain weed control.

In the past, ever-increasing demands were placed on prime areas of wildlife habitat, such as wetlands and forested areas, to serve as new production areas to meet mankind's increasing requirements. The increases in yield that are obtainable through improved agricultural methods which include pesticides have made is possible to stop and even reverse this trend in some countries.

Economic benefits

Pesticides have become a major factor in improving the efficiency and productivity of agriculture. Pesticides are estimated to be responsible for 20% of the gain in agricultural productivity since 1940. In addition, there would be an overall reduction of 25% in crop output if pesticides could not be used.

Efforts are currently in progress to harmonize the pesticide registration requirements among the EEC member countries (Lynch, 1992).

To follow any type of stepwise or tiered testing sequence, appropriate criteria (or 'trigger' values) must be established to decide when additional testing steps should be conducted. Such trigger values should not be used as simple cut-off or 'guillotine' criteria to cancel registration of a pesticide. Instead, they should establish reasonable levels of concern that can be addressed by additional testing to clarify the appropriate regulatory options that can be used to bring the environmental risks into balance with the benefits resulting from use of the product.

For a more complete evaluation of the environmental behavior of a pesticide, a number of factors need to be considered, including the application rate, the number of applications per year, type of formulation, method and timing of application, the crops treated, and typical crop rotations. In addition, geographic use areas and typical soil and climatic conditions are important considerations in evaluating the leaching potential of a pesticide.

By preventing yield losses, pesticides have substantially reduced the costs of producing food. The use of pesticides has also helped improve the quality of food. As an example of the improved quantity and quality benefits obtainable with pesticides, consider some statistics concerning the production of Cox's orange pippin apples in Germany. When pesticides were used, the yield was 162 dt/ha and 85% of the total production was marketable. When insecticides, fungicides and acaracides were not used, the average yield declined to 95 dt/ha and only 35% of this yield was marketable. In summary, over four times as many marketable apples were obtained through the use of pesticides (Schuhmann, 1976).

Pimentel *et al.* (1978) performed a risk/benefit analysis concerning pesticide use in the United States. These authors determined that 34%, 12% and 2% of the total crop land (excluding pastures) is treated with herbicides, insecticides and fungicides, respectively. Their analysis compared the economic costs and benefits of these levels of pesticide use with an alternative scenario using the best available mechanical, cultural and biological techniques for pest control. Using conservative assumptions, the general conclusion was that every $1 spent on pesticides (including application costs) returned $4 on the investment. Overall, it was concluded that pesticides prevent the loss of almost $9 billion (US) of crops per year.

Social benefits

Catastrophic crop losses due to pests and disease have inflicted suffering on humanity throughout history. The Irish potato famine, caused by potato blight, is an example from recent history where crop failures have had a devastating impact on rural communities. The use of pesticides helps to reduce the risk of such catastrophes recurring. Current threats to crop production include the locust plagues in Africa.

Another social benefit of pesticides is the role that these compounds have in controlling the vectors responsible for disease in humans and livestock. The use of insecticides to control the insects that spread malaria has saved tens of millions of human lives and freed hundreds of millions from the continued threat of this disease.

The use of pesticides has also contributed to a dramatic increase in the efficiency of agricultural food production. While agricultural outputs have continued to rise, the number of people employed in agriculture has steadily decreased, enabling many individuals to fulfill other socioeconomic roles. For those remaining in agriculture, the use of pesticides has eliminated much of the menial work of manual or mechanical weed control.

Finally, it is important to recognize that continued pesticide use is essential to ensure a reliable supply of inexpensive, high quality food, free from natural toxins such as aflatoxins and ergotamines. Pesticides therefore play a vital role in the improvement and maintenance of public health.

CURRENT RESEARCH AREAS IN PESTICIDE MOBILITY

Based on the foregoing discussions, it is clear that a sound understanding exists of the major factors that influence pesticide behavior in soil. When combined with the nature of the soil environment and climatic conditions, knowledge of the physical, chemical and environmental properties of pesticides enables appropriate experimental and computational evaluations of pesticide mobility to be performed.

Additional information from ongoing pesticide mobility research will improve both our current understanding and our current capabilities in conducting laboratory and field experiments and in performing computer modeling. These current research areas include:

- the influence of simultaneous kinetics of sorption and diffusion on mobility;
- techniques to assess the spatial variability of soils;
- assessment of the extent of water movement and degradation of pesticide residues in subsoils;
- the role of preferential flow in determining leaching behavior in soils.

In each of these areas, both experimental and computational approaches are currently being developed. In addition to these known areas of research, several other areas also need to be explored:

- development of appropriate guidance concerning sizes and protocols to conduct lysimeter studies;
- establishment of appropriate, environmentally significant criteria to evaluate the results of lysimeter testing;
- continued appraisals of computer models and their potential applications in regulatory settings.

Progress in these areas will help ensure that effective and environmentally compatible crop protection products continue to be available to enhance the world's production of food, feed and fiber.

APPENDIX 1: REGULATORY SCHEMES FOR SOIL MOBILITY IN DENMARK, THE NETHERLANDS AND THE UNITED STATES

Danish regulatory scheme

The Danish National Agency for Environmental Protection accepts soil mobility and persistence tests that are conducted according to US EPA, OECD and/or German guidelines (DK AEP, 1988). Initial testing includes studies of degradation in aerobic and anaerobic soils, field dissipation studies in at least three soils and soil column leaching studies. Based upon the results of these tests, additional field tests and/or lysimeter tests can be required.

To determine the need for higher levels of mobility testing, the Danish National Agency of Environmental Protection have proposed a trigger system based on percentages eluted from soil columns which use three standard soils. Depending upon the soil type, if more than 2–5% of the active ingredient is found in the leachate from soil column leaching studies, then lysimeter or field testing is proposed. In these field studies, if more than 5 g/ha of active ingredient or metabolites are detected in leachate collected below 1 m depth, then it is proposed that registration of the compound be denied.

German regulatory scheme

Kloskowski *et al*. (1990) has outlined a tiered regulatory scheme used by the BBA to address questions of pesticide mobility (Figure 10). Initial studies of persistence and mobility are conducted in a set of standard soils. Persistence and mobility criteria are then applied to the results of the studies to determine if lysimeter studies should be conducted. If a compound has a DT-90 of more than 100 days and more than 5% of the active ingredient and metabolites are found in leachate collected from soil column studies, then additional testing is required.

The results of lysimeter studies are interpreted using two criteria. The first standard is that total cumulative percentage of the applied dose in the lysimeter leachate should not exceed 0.1%. The second criteria is that no more than 2% of the parent compound and its metabolites should have moved below a soil depth of 60 cm after a 2-year period.

The German regulatory system also has an option to consider the results of computer modeling. The PELMO computer model can be used with a set of four fixed soil and climatic scenarios to determine whether it is likely that normal product use will result in pesticide residues of concern below the root zone. Compounds which are not predicted to present a threat to groundwater are then exempted from lysimeter testing.

Dutch regulatory scheme

The regulatory scheme used in the Netherlands uses DT-50 and K_{om} values, determined from degradation and sorption studies respectively, to estimate the percentage of parent compound that leaches to a depth of 1–2 m. The model PESTLA has been run for a set of fixed scenarios to create plots of the percentage leached to 1–2 m as a function of DT-50 and K_{om}. If less than 0.001% of the compound leaches to this depth, then a negligible risk due to leaching is assumed. Pesticides with higher leaching percentages are required to be tested in field dissipation and/or lysimeter studies. The results of these studies are interpreted in light of typical Dutch ground water conditions to determine the significance of the leaching behavior of the compound. A leaching fraction of greater than 10% or greater than 1 ppb is regarded as presenting a high risk to ground water.

United States regulatory scheme

According to guidelines established the US EPA, initial laboratory studies of pesticide mobility may include adsorption/desorption, soil-TLC and soil column leaching (US EPA, 1982g,h). Currently, adsorption/desorption studies are clearly preferred by the EPA. In addition to these laboratory studies, field soil dissipation studies are a requirement for registration in the USA.

Various schemes have been proposed by the EPA to determine when further testing is required. The current approach involves consideration of the persistence and mobility data as well as use patterns, use rate and toxicology. When it is determined that a pesticide has a significant potential for mobility in soil and its use patterns, use rate and toxicology indicate that concentrations in shallow ground water could be at levels of concern, ground water monitoring studies are requested. As of the autumn of 1990, monitoring studies have been required for 36 pesticides (US EPA, 1990). Ground water monitoring may be required either as a condition of registration or for registered pesticides that have been detected in ground water.

If significant detections are found in ground water monitoring studies, various regulatory options are considered including:

- restrictions on certain soils and in certain climates;
- restrictions in certain geographical areas;
- establishment of well setbacks;
- classification of pesticide as 'Restricted Use';
- requirement of state management plans;
- cancellation or ban.

APPENDIX 2: CURRENT COMPUTER MODELS OF PESTICIDE MOBILITY

LEACHMP

LEACHMP was developed at Cornell University in the USA (Wagenet and Hutson, 1989). It forms parts of LEACHM, a suite of leaching models which also includes nitrogen and inorganic ion models. Water flow is by a solution of Richards equation and pesticide movement is by a coupled solution of the convection-dispersion equation. This complexity results in fairly extensive soil data input requirements and the model is computationally intensive. Metabolites are considered, as is a vapor phase. Input parameters can be adjusted to simulate specific situations, but the facilities for doing this are somewhat more limited than those in PRZM. The manual is well written.

PESTLA

PESTLA is a pesticide mobility model developed in The Netherlands (van der Linden and Boesten, 1989). The general concept of the water flow and pesticide movement equations is similar to that in LEACHMP. The most significant process differences from LEACHMP are that pesticide adsorption is represented by the Freundlich equation, degradation is affected by temperature and soil moisture content, and there is no vapor phase. A single set of soil and climate data is built into the model to give a 'standard scenario'. For this scenario, the two parameters which can be varied are the pesticide half-life and adsorption coefficient, which combine to give a graph of maximum leachate concentration. Use of PESTLA in this manner effectively reduces this model to a screening or early assessment model. While the 'standard scenario' is well documented, it is not possible to change other input parameters or to adapt the model since the source code is not generally available. Reports of results for this model refer to four standard scenarios: spring and autumn pesticide application, each on two soil types.

Of these, only the spring application to the lighter soil is documented or provided with the software.

PRZM versions

The original version of PRZM is designated in this paper as PRZM-1 (Carsel *et al.*, 1984). Water flow is by the water storage routing method (e.g. 'tipping bucket' method) which alternates between the field capacity and wilting point of the soil. Water redistribution can be instantaneous (daily) or time-dependent. Pesticide movement is controlled by water flow, with numerical dispersion to simulate actual dispersion with depth. Degradation is simulated as a first-order process and only parent pesticide can be simulated. A vapor phase is not included. Extensive facilities for parameter input allow for detailed specification of the simulated system (e.g. crop characteristics, varying pesticide adsorption and degradation with depth, etc.), and this task is aided by an excellent manual.

PRZM-1 has been adapted for use in Germany by the Fraunhofer Institut in Schmallenberg and the most current version is designated as PELMO in this paper (PELMO, 1992). This version provides the capability of simulating degradation as a function of both soil moisture and temperature. This has certain advantages, but it also makes it very difficult to use field dissipation data as input for the model—the user is thus pushed towards using laboratory degradation data. PELMO also can directly use the Freundlich parameters K_f and N to account for nonlinear isotherms. Parameter alteration has been restricted and 'standard scenarios' for crop, climate and soil have been provided. Source code is not available and documentation is limited.

A new version of PRZM, hereafter called PRZM-2, has been developed by the EPA as a submodel of RUSTIC, which also includes deep subsoil and saturated zone submodels (Dean *et al.*, 1989). Significant enhancements include the simulation of metabolites, an improved treatment of dispersion, and the inclusion of a vapor phase which allows volatile compounds to be simulated. Documentation has been further improved.

SESOIL-FHG

When PRZM was initially modified in Germany to produce PRZM-FHG, a second model was also adapted. That model was SESOIL (Bonazountas and Wagner, 1984), originally developed by the US EPA Office of Toxic Substances for simple assessment of leaching from waste dumps and industrial chemical spills. SESOIL was not primarily written for chemicals which adsorb and degrade, as do all pesticides, and the water flow submodel cannot cope with soils with horizons having differing properties. The basis of the model is quite unlike any other model in this review. The documentation is very poor and the software has evolved over the years with little or no coordination or documentation of changes made or version numbers. Watson and Brown (1985) provided a good review of the model and its limitations. It is not known which version was taken and adapted by the Fraunhofer Institut to produce the version called SESOIL-FHG, but while several beneficial changes were made (SESOIL-FHG, 1989), the basic form of the original version remains. As with PRZM-FHG, standard crop, climate and soil scenarios are provided with SESOIL-FHG, but most of the standard soils have horizons with very different hydrologic properties and the fact that SESOIL cannot handle this has been ignored. SESOIL-FHG is slower to run than PRZM-1 or PRZM-FHG but faster than LEACHMP or PESTLA.

Other models

Other models which are used for pesticide mobility prediction in soil include CALF (Walker and Barnes, 1981) and GLEAMS (Leonard *et al.*, 1987). Both of these models have been used in both regulatory and non-regulatory settings.

ACKNOWLEDGEMENTS

This chapter was prepared by a GIFAP (International Group of National Associations of Manufacturers of Agrochemical Products) group of experts to provide an agrochemical industry recommended approach. The group comprised Mark Russell (DuPont, USA; Chairman), Philippe Adrian (Rhône-Poulenc, France), David Gustafson (Rhône-Poulenc, USA), Johann Guth (Ciba, Switzerland), Helmuth Morgenthaler (BASF, Germany), Terry Roberts (Hazleton, UK), Konrad Scholtz (Bayer, Germany) and Kim Travis (Zeneca, UK).

REFERENCES

Addiscott, T. M., and Wagenet, R. J. (1985). 'Concepts of solute leaching in soils: a review of modeling approaches', *J. Soil Sci.*, **36** 411–424.
Anderson, J. P. E. (1975). 'Einfluss von Temperatur und Feuchte auf Verdampfung, Abbau und Festlegung von Diallat im Böden', *Z. Pflkrankh. Pflschutz, Sonderheft*, **VII**, 141–146.

Arnold, D. J., and Briggs, G. G. (1990). 'Fate of pesticides in soil: predictive and practical aspects', in *Progress in Pesticide Biochemistry and Toxicology* (eds D. H. Hutson and T. R. Roberts), Vol 7, John Wiley & Sons Ltd, Chichester.

ASTM, (1975). *Standard Test Method for Vapor Pressure*, D2879-75, American Society of Testing Materials (ASTM), Philadelphia.

ASTM (1984). *Standard Practice for Evaluating Fate Models of Chemicals*, E978-84, American Society of Testing Materials (ASTM), Philadelphia.

Bailey, G. W., and White, J. L. (1970). 'Factors influencing the adsorption, desorption, and movement of pesticides in soil', *Residue Rev.*, **32**, 29–92.

Bailey, G. W., White, J. L., and Rothberg, T. (1968). 'Adsorption of organic herbicides by montmorillonite: role of pH and chemical character of adsorbate', *Soil Sci. Soc. Am. Proc.*, **32**, 222–234.

Bartha, R., and Pramer, D. (1965). 'Features of a flask and method for measuring the persistence and biological effects of pesticides in soil', *Soil Sci.*, **100**, 68–70.

BBA (1973). Prüfung des Versickerungsverhaltens von Pflanzenbehandlungsmitteln, Merkblatt Nr. 36, 1 Auflage, Februar.

BBA (1986). *Guidelines for the Official Testing of Plant Protection Products, Part IV, 4–2, Seepage Behavior of Plant Proctection Products*, Biological Research Centre for Agriculture and Forestry, Braunschweig, Germany.

BBA (1990a). 'Lysimeter tests for the translocation of plant protection products into the subsoil', in *Guidelines for the Official Testing of Plant Protection Products*, Part IV, 4–3, Biological Research Centre for Agriculture and Forestry, Braunschweig, Germany.

BBA (1990b). *Guidelines for the Official Testing of Plant Protection Products*, Biological Research Centre for Agriculture and Forestry, Braunschweig, Germany.

Beall, M. L., Nash, R. G., and Kearney, P. C. (1976). 'Agroecosystem. A laboratory model ecosystem to simulate agricultural field conditions for monitoring pesticides', *Proceedings of the EPA Conference on Modelling and Simulation*, pp. 790–793, EPA.

Bergström, L. (1990). 'Use of lysimeters to estimate leaching of pesticides in agricultural soils', *Environ. Pollut.*, **67**, 325–347.

Best, J. A., and Weber, J. B. (1974). 'Disappearance of s-triazines as affected by soil pH using a balance-sheet approach', *Weed Sci.*, **22**, 364–373.

Biagi, G. L., et al. (1979). 'R values of naphthols and acetophenones in structure-activity studies', *J. Chromatogr.*, **177**, 34–49.

Boesten, J. J. T. I. (1990). 'Influence of solid/liquid ratio on the experimental error on sorption coefficients in pesticide/soil systems', *Pestic. Sci.*, **30**(1), 31–41.

Bonazountas, M., and Wagner, J. (1984). SESOIL: A seasonal soil compartment model. A. D. Little and Co., Cambridge, MA.

Branson, D. R., Neely, W. B., and Blau, G. E. (1975). In *Structure-Activity Correlations in Studies of Toxicity and Bioconcentration with Aquatic Organisms* (eds G. D. Veith and D. E. Konasewich), pp. 99–117, Proceedings of a symposium held in Burlington, Ontario.

Briggs, G. G. (1981). 'Theoretical and experimental relationships between soil adsorption, octanol–water partition coefficients, water solubilities, bioconcentration factors, and the parachor', *J. Agric. Food Chem.*, **29**, 1050–1059.

Bunhill, S., et al. (1973). 'Continuous flow methods for studying adsorption of herbicides by soil dispersions and soil columns', *Proc. Eur. Weed Res. Counc. Symp. Herbicides—Soil*, pp. 70–79.

Burkhard, N., Eberle, D. O., and Guth, J. A. (1975). 'Model system for studying the environmental behavior of pesticides', in *Environmental Quality and Safety* (eds F. Coulston and F. Korte), Suppl. vol. III, 203–213, IUPAC, Stuttgart, Thieme.

Burkhard, N., and Guth, J. A. (1979). 'Photolysis of organophorphorus insecticides on soil surfaces', *Pestic. Sci.*, **10**, 313–319.

Burkhard, N., and Guth, J. A. (1981). 'Rate of volatilization of pesticides from soil surfaces: comparison of calculated results with those determined in a laboratory model system', *Pestic. Sci.* **12**, 37–44.

Calvet, R. (1980). 'Adsorption–desorption phenomena', in *Interactions between Herbicides and the Soil* (ed. R. J. Hance), Academic Press, London.

Calvet, R. (1989). 'Evaluation of adsorption coefficients and the prediction of the mobilities of pesticides in soils', in *Methodological Aspects of the Study of Pesticide Behavior in Soil* (ed. P. Jamet), INRA, Paris.

Carsel, R. F., Mulkey, L. A., Lorber, M. N., and Baskin, L. B. (1985). *Ecol. Modeling*, **30**, 49–69.

Carsel, R. F., Smith, C. N., Mulkey, L. A., Dean, J. D., and Jowise, P. (1984). *Users Manual for the Pesticide Root Zone Model (PRZM)*, Release 1, ERL, US EPA, PB85-158913, US EPA, Athens, GA.

Chapman, T., Gabbot, P. A., and Osgerby, J. M. (1970). 'Techniques for measuring the relative movement of herbicides in soil under leaching conditions', *Pestic. Sci.*, **1**, 56–58.

Chisaka, H., and Kearney, P. C. (1970). 'Metabolism of propanil in soils', *J. Agric. Food Chem.*, **18**, 854–858.

Cohen, S. Z., Creeger, S. M., Carsel, R. F., and Enfield, C. G. (1984). 'Potential pesticide contamination of groundwater from agricultural uses', in *Treatment and Disposal of Pesticide Wastes*, pp. 297–325, ACS Symp. Ser. 259, American Chemical Society, Washington, DC.

Dean, J. D., Huyakorn, P. S., Donigian, A. S., Voos, K. A., Schanz, R. W., Meeks, Y. J., and Carsel, R. F. (1989). *Risk of Unsaturated/Saturated Transport and Transformation of Chemical Concentrations (RUSTIC)*, Volumes 1 and 2, EPA/600/3-89/048a and EPA/600/3-89/048b, US EPA, Athens, GA.

DK AEP (1987). *Statutory Order from the Ministry of the Environment No. 791, December 10, 1987, on Chemical Pesticides*. Ministry of the Environment, Denmark, National Agency of Environmental Protection, March, 1988.

DK AEP (1988). *Application concerning the Active Ingredient*, National Agency of Environmental Protection, Copenhagen.

Eadsforth, C. V. (1986). 'Application of reverse-phase HPLC for the determination of partition coefficients', *Pestic. Sci.*, **17**, 311–325.

Ellgehausen, H., Guth, J. A., and Esser, H. O. (1980). 'Factors determining the bioaccumulation potential of pesticides in the individual compartments of aquatic food chains', *Ecotoxicol. Environ. Safety*, **4**, 134–137.

Esser, H. O., Hemingway, R. J., Klein, W., Sharp, D. B., Vonk, J. W., and Holland, P. T. (1988). 'Recommended approach to the evaluation of the environmental behavior of pesticides', IUPAC Reports on Pesticides (24), *Pure Appl. Chem.*, **60**, 901–932.

FAO (1985). *Guidelines on Environmental Criteria for the Registration of Pesticides*, Food and Agriculture Organization of the United Nations, Rome.

FAO (1989). *Revised Guidelines on Environmental Criteria for the Registration of Pesticides*, Food and Agriculture Organization of the United Nations, Rome.

Fetter, C. W. (1988). *Applied Hydrogeology*, p. 91, Merrill, Columbus, Ohio.

Führ, F., Cheng, H. H., and Mittelstaedt, W. (1975). 'Pesticide, balance and metabolism studies with standardized lysimeters', *Landwirtschaftliche Forschung, Sonderheft*, **32**(1), 272–278.

Führ, F., Steffens, W., Mittelstaedt, W., and Brumhard, B. (1990). 'Pesticides in the soil—lysimeter experiments with ^{14}C-labelled active ingredients', *Annual Report 1989*, Research Centre, Jülich, Germany.

Gerber, H. R., Ziegler, P., and Dubach, P. (1970). 'Leaching as a tool in the evaluation of herbicides', *Proc. 10th Br. Weed Control Conf.*, 118–125.

GIFAP (1987). *GIFAP Position Paper on Toxicological Evaluation of Pesticide Residues in Drinking Water*, GIFAP, Brussels.

GIFAP (1990a). *Environmental Criteria for the Registration of Pesticides*, Technical Monograph No. 3, GIFAP, Brussels.

GIFAP (1990b). *Water Quality Monitoring: Site Selection and Sampling Procedures for Pesticide Analysis*, Technical Monograph No. 13, GIFAP, Brussels.

GIFAP (1990c). *GIFAP Position Paper on Pesticide Residues in Water*, GIFAP, Brussels.

GIFAP (1992). *Fact Sheet: Degradation of Pesticides in Ground Water*, GIFAP, Brussels.

Goring, *et al.* (1975) 'Principles of pesticide degradation in soil', in *Environmental Dynamics of Pesticides* (eds R. Haque and V. H. Freed), pp. 135–172, Plenum Press, NY.

Goswami, K. P., and Koch, B. L. (1976). 'A simple apparatus for measuring degradation of ^{14}C-labelled pesticides in soil', *Soil Biol. Biochem.*, **8**, 527–528.

Green, R. E., and Karickhoff, S. W. (1990). 'Sorption estimates for modeling', in *Pesticides in the Soil Environment: Processes, Impacts and Modeling* (ed. H. H. Cheng), Soil Sci. Soc. Am. Book Series, No. 2, Madison, Wisconsin.

Green, R. E., and Yamane, V. K. (1970). 'Precision in pesticide adsorption measurements', *Soil Sci. Soc. Am. Proc.*, **34**, 353–354.

Grover, R., and Hance, R. J. (1970). 'Effect of ratio of soil to water on adsorption of linuron and atrazine', *Soil Sci.*, **109**, 136–138.

Gückel, W., Rittig, F. R., and Synnatschke, G. (1974). 'A method for determining the volatility of active ingredients used in plant protection. II. Applications to formulated products', *Pestic. Sci.*, **5**, 393–400.

Gustafson, D. I. (1989). 'Groundwater ubiquity score: a simple method for assessing pesticide leachability', *Environ. Toxic. Chem.*, **8**(4), 339–357.

Guth, J. A. (1972). 'Adsorptions- und Enwaschverhalten von Pflanzenschutzmitteln im Böden', *Schr. Reihe Ver. Wass-Böden-Lufthyg.*, Berlin-Dahlem, Heft **37**, 143–154.

Guth, J. A. (1980). 'The study of transformations', in *Interactions between Herbicides and the Soil* (ed. R. J. Hance), pp, 123–157, Academic Press, London.

Guth, J. A. (1981). 'Experimental approaches to studying the fate of pesticides in soil', in *Progress in Pesticide Biochemistry and Toxicology* (eds D. H. Hutson and T. R. Roberts), Vol. 1, pp. 85–114, John Wiley & Sons Ltd, Chichester.

Guth, J. A. (1983). 'Untersuchungen zum Verhalten von Pflanzenschutzmitteln 1 im Böden', *Bull. Bödenkundliche ges. Schweiz (BDS)*, **7**, 26–33.

Guth, J. A. (1985). 'Adsorption/desorption', in *Joint International Symposium, Physicochemical Properties and their Role in Enviromental Hazard Assessment*, July 1–3, 1985, Canterbury, UK.

Guth, J. A., Burkhard, N., and Eberle, D. O. (1976). 'Experimental models for studying the persistence of pesticides in soils', *Proceedings of BCPC Symposium: Persistence of Insecticides and Herbicides*, pp. 137–157, BCPC, Surrey, UK.

Guth, J. A., Gerber, H. R., and Schlaepfer, T. (1977). 'Effect of adsorption, movement and persistence on the biological availability of soil-applied pesticides', *Proc. Br. Crop Prot. Conf.*, **3**, 961–971.

Guth, J. A., and Hörmann, W. D. (1987). 'Problematik und Relevanz von Pflanzenschutzmitteln-Spuren im Grund(Trink-)Wasser', *Schr.-Reihe Verein WaBoLu*, **68**, 91–106.

Hamaker, J. W. (1972). 'Diffusion and volatilization', in *Organic Chemicals in the Soil Environment* (eds C. A. I. Goring and J. W. Hamaker), Vol. 1, pp. 341–397, Marcel Dekker, New York.

Hamaker, J. W. (1975). 'The interpretation of soil leaching experiments', in *Environmental Dynamics of Pesticides* (eds R. Haque and V. H. Freed), pp. 135–172, Plenum Press, NY.

Hamaker, J. W., and Thompson, J. M. (1972). 'Adsorption in organic chemicals', in *Organic Chemicals in the Soil Environment* (eds C. A. I. Goring and J. W. Hamaker), Marcel Dekker, New York.

Hance, R. J. (1965). 'The adsorption of urea and some of its derivatives by a variety of soils', *Weed Res.*, **5**, 98–107.

Hance, R. J. (1967). 'The speed of attainment of sorption equilibria in some systems involving herbicides', *Weed Res.*, **7**, 29–36.

Harvey, J., and Reiser, R. W. (1973). 'Metabolism of methomyl in tobacco, corn and cabbage', *J. Agric. Food Chem.*, **21**, 775–783.

Hayes, M. H. B., Stacey, M., and Thompson, J. M. (1968). 'Adsorption of s-triazine herbicides by soil organic matter preparations', in *Isotopes and Radiation in Soil Organic Matter Studies*, p. 75, International Atomic Energy, Agency, Vienna.

Helling, C. S. (1970). 'Movement of s-triazine herbicides in soils', *Residue Rev.*, **32**, 175–210.

Helling, C. S. (1971). 'Pesticide mobility in soils', *Soil Sci. Soc. Am. Proc.*, **35**, 743–748.

Helling, C. S., Kaufman, D. D., and Dieter, C. T. (1971). 'Algae bioassay detection of pesticide mobility in soils', *Weed Sci.*, **19**, 685–690.

Helling, C. S., and Thompson, S. M. (1974). 'Azide and ethylenethiourea mobility in soils', *Soil Sci. Soc. Am. Proc.*, **38**, 80–95.

Helling, C. S., and Turner, B. C. (1968). 'Pesticide mobility: determination by soil thin-layer chromatography', *Science*, **162**, 562–563.

Hern, S. C., and Malancon, S. M. (eds) (1986). *Vadose Zone Modeling of Organic Pollutants*, Lewis Publishers Inc., Chelsea, MI.

Herzel, V. F. (1987). 'Classification of plant protection agents from the point of view of drinking water protection'. *Nachrichtenbl. Dt. Pflanzenschutzd.*, **39**, 97–104.

Hörmann, W. D. (1973). *Proc. Eur. Weed Res. Count. Symp.*, pp. 129–140.

Hornsby, A. G., Rao, P. S. C., Wheeler, W. B., Nkedi-Kizza, P., and Jones, R. L. (1983). In *Proceedings of the Conference on Characterization and Monitoring of the Vadose (Unsaturated) Zone, Las Vegas, NV, Dec. 8–10, 1983* (eds D. Nielson and M. Curl), pp. 936–958, National Water Well Association, Dublin, Ohio.

Hülshoff, A., and Perrin, J. H. (1976). 'A reversed-phase thin layer chromatographic method for the determination of relative partition coefficient of very lipophilic compounds', *J. Chromatogr.*, **120**, 65.

ISPRA (1990). 'Adsorption/desorption of chemicals on soil' Meeting minutes and draft report, Joint Research Centre, Ispra (Verese), Italy.

Jones, R. L. (1990). 'Pesticides in ground water: conduct of field research studies', in *Progress in Pesticide Biochemistry and Toxicology* (eds D. H. Hutson and T. R. Roberts), Vol. 7, pp. 27–46, John Wiley & Sons Ltd, Chichester.

Jones, R. L. and Hanks, R. J. (1990). 'Review of unsaturated zone leaching models from a user's perspective', in *Proceedings of the International Symposium on Water Quality Modeling of Agricultural Non-point Sources* (ed. D. G. DeCoursey), Part 1, USDA-ARS, ARS-81, USDA, Tifton, GA.

Jones, R. L., Hunt, T. W., Norris, F. A., and Harden, C. F. (1989). 'Field research studies on the movement and degradation of thiodicarb and its metabolite methomyl', *J. Contam. Hydrol.*, **4**, 359–371.

Jones, R. L., and Norris, F. A. (1991). In *Groundwater Residue Sampling Design* (eds R. G. Nash and A. R. Leslie), ACS Symposium Series 465, pp. 165–181, American Chemical Society, Washington, DC.

Jury, W. A., Focht, D. D., and Farmer, W. J. (1987). 'Evaluation of pesticide groundwater pollution potential from standard indices of soil-chemical adsorption and biodegradation', *J. Environ. Qual.*, **16**, 422–428.

Kaufman, D. D. (1977). 'Approaches to investigating soil degradation and dissipation of pesticides', *Symposium on Terrestrial Microcosms and Environmental Chemistry, Corvallis, Oregon, June 13–17*.

Kearney, P. C., and Kontson, A. (1976). 'A simple system to simultaneously measure volatilization and metabolism of pesticide from soils', *J. Agric. Food Chem.*, **24**, 424–426.

Kenaga, E. E., and Goring, C. A. I. (1980). 'Relationship between water solubility, soil sorption, octanol–water partitioning and concentration of chemicals in the biota', in *Aquatic Toxicology* (eds J. G. Eaton, et al.), pp. 78–115, ASTM STP 707, Philadelphia.

KFA (1990).

Kloskowski, R., Nolting, H.-G., and Schinkel, K. (1990). Behavior and leaching potential of pesticides in soil: Criteria for the evaluation of data used for registration in the Federal Republic of Germany, Federal Biological Research Center for Agriculture and Forestry, Braunschweig, FRG, presented at IUPAC 1990, Hamburg, FRG.

Knisel, W. G. (1980). *A Field-scale Model for Chemical, Runoff, and Erosion from Agricultural Management Systems*, Conserv. Rep. No. 26, US Dept. Agric., Sci. Educ. Adm., USDA, Tifton, GA.

Knisel, W. G. (ed.) (1980). CREAMS: A field-scale model for chemicals, runoff, and erosion from agricultural management systems. USDA-SEA Conserv. Res. Rep. 16. U.S. Gov. Print. Office, Washington, D.C.

Koskinen, W. C., and Harper, S. S. (1990). 'The retention processes: mechanisms', in *Pesticides in the Soil Environment: Processes, Impacts and Modeling* (ed. H. H. Cheng), Soil Sci. Soc. Am. Book Series, No. 2, Madison, Wisconsin.

Kubiak, R. (1986). *Comparative Studies to evaluate the Transferability of Results from Standardized Laboratory Experiments and Ecosystem Compartments to the Real Field Situation*, Ber. Kernforschungsanlage Jülich, Jül-2055, Germany, 203 pp.

Kubiak, R., Führ, F., Mittelstaedt, W., Hansper, M., and Steffens, W. (1988). 'Transferability of lysimeter results to actual field situations', *Weed Sci.*, **36**, 514–518.

Lambert, S. M., Porter, P. E., and Schieferstein, R. H. (1965). 'Movement and sorption of chemicals applied to the soil', *Weeds*, **13**, 185–190.

Laskowski, D. A., Swann, R. L., McCall, P. J., and Bidlack, H. D. (1983). 'Soil Degradation Studies', *Residue Reviews*, **85**, 139–147.

Lay, M. M., and Illnicki, R. D. (1975). 'Effect of soil storage on propanil degradation', *Weed Res.*, **15**, 63–66.

Leonard, R. A., Knisel, W. G., and Still, D. A. (1987). 'GLEAMS: Groundwater Loading Effects of Agricultural Management Systems', *Trans. ASAE*, **30**, 1403–1418.

Lichtenstein, E. P., Fuhremann, T. S., and Schulz, K. R. (1974). 'Translocation and metabolism of [^{14}C]-phorate as affected by percolating water in a model soil-plant ecosystem', *J. Agric. Food Chem.*, **22**, 991–996.

Loos, M. A., Kontson, A., and Kearney, P. (1980). 'Inexpensive soil flask for [^{14}C]-pesticide degradation studies', *Soil Biol. Biochem.*, **12**, 583–585.

Lyman, W. J., Reehl, W. F., and Rosenblatt, D. H. (1982). *Handbook of Chemical Property Estimation Methods*, McGraw-Hill, New York.

Lynch, M. R. (1992). *Development of Uniform Principles in Relation to the Authorization of Plant Protection Products*, Commission of the European Communities, Brussels.

Mader, W. J., and Grady, L. T. (1971). 'Determination of solubility', in *Physical Methods of Chemistry* (eds A. Weissberger and B. W. Rossiter), Vol. 1, Part V, Chapter V, Wiley Interscience, Chichester.

Marvel, J. T., Brightwell, B. B., Malik, J.M., Sutherland, M. L., and Rueppel, M. L. (1978). 'A simple apparatus and quantitative method for determining the persistence of pesticides in soil', *J. Agric. Food Chem.*, **22**, 684–688.

May, W. E., Wasik, S. P., and Freeman, D. H. (1978). 'Determination of solubility behavior of some polycyclic aromatic hydrocarbons in water', *Anal. Chem.*, **50**(7), 997.

McCall, P. J., Laskowski, D. A., Swann, R. L., and Dishburger, H. J. (1981). 'Measurement of sorption coefficients of organic chemicals and their use, in environmental fate analysis', in *Test Protocols for Environmental Fate and Movement of Toxicants*, Proceedings of AOAC Symposium, AOAC, Washington DC.

Monsanto Technical Bulletin (1990). The National Alachlor Well Water Survey (NAWWS): Data Summary (July, 1990), Monsanto Company, St. Louis, MO, USA.

Nielsen, D. M. (1991). *Practical Handbook of Ground-Water Monitoring*, Lewis, Chelsea, Michigan, USA.

Nolting, H.-G. (1990). 'Drinking water contamination by pesticides—fact and fiction', Conference presentation at Fortbildungszentrum Gesunheits- und Umweltschutz Berlin e.V.

Norris, F. A., Jones, R. L., Kirkland, S. D., and Marquardt, T. E. (1990). 'Techniques for collecting soil samples in field residue studies', ACS Symposium Series, **465**, pp 356–394, ACS, Washington DC.

OECD (1981a). *Guidelines for Testing of Chemicals*, OECD, Paris. ISBN 92-64-12221-4.

OECD (1981b). *Guidelines for Testing of Chemicals, 107, Partition Coefficient (n-Octanol/Water)*, OECD, Paris. ISBN 92-64-12221-4.

OECD (1981c). *Guidelines for Testing of Chemicals, 105, Water Solubility*, OECD, Paris. ISBN 92-64-12221-4.

OECD (1981d). *Guidelines for Testing of Chemicals, 104, Vapor Pressure Curve*, OECD, Paris. ISBN 92-64-12221-4.

OECD (1981e). *Guidelines for Testing of Chemicals, 106, Adsorption/Desorption*, OECD, Paris. ISBN 92-64-12221-4.

OECD (1981f). *Guidelines for Testing of Chemicals, 111, Hydrolysis as a Function of pH*, OECD, Paris. ISBN 92-64-12221-4.

OECD (1989). *Guidelines for Testing of Chemicals, 117, Partition Coefficient (n-Octanol/Water), HPLC Method*, OECD, Paris. ISBN 92-64-12221-4.

Osgerby, J. M. (1973). 'Processes affecting herbicide action in soil', *Pestic. Sci.*, **4**, 247–258.

Parker, S., and Leahy, P. (1988). 'Development of a method to investigate the photodegradation of pesticides', in *Pests and Diseases*, pp. 663–668, BCPC, Surrey, UK.

Parr, J. F., and Smith, S. (1976). 'A multipurpose manifold assembly: use in evaluation microbiological effects of pesticides', *Soil Sci.*, **107**, 271–276.

PELMO (1992). 'PELMO Documentation', M. Klein, Fraunhofer-Institut fur Umweltchemie und Okotoxikologie, Schmallenberg.

Pennell, K. D., Hornsby, A. G., Jessup, R. E., and Rao, P. S. C. (1990). 'Evaluation of five simulation models for predicting aldicarb and bromide behavior under field conditions', *Water Resources Research*, **26**(11), 2679–2693.

Persson, L., and Bergström, L. (1991). 'Drilling method for collection of undisturbed soil monoliths', *Soil Sci. Soc. Am. J.*, **55**, 285–287.

Phillips, F. T. (1972). 'Persistence of organochlorine insecticides on different substrates under different environmental conditions. I. The rates of loss of dieldrin and aldrin by volatilization from glass surfaces', *Pestic. Sci.*, **2**, 225–266.

Pimental, D., Krummel, J., Gallahan, D., Hough, J., Merill, A., Schreiner, I., Vittum, P., Koziol, F., Back, E., Yen, D., and Fiance, S. (1978). *BioScience*, Vol. 28, No. 12, p. 772.

Pramer, D., and Bartha, D. (1972). 'Preparation and processing of soil samples for biodegradation studies', *Environ. Lett.*, **2**, 217–224.

PRZM-FHG (1989). *Simulation Model using PRZM for Scenarios in Germany*, R&D Project UBA-10602065, Fraunhofer Institut für Umweltchemie und Ökotoxicologie Schmallenberg/Grafschaft, Germany.

Rao, P. S. C., Hornsby, A. G., and Jessup, R. E. (1985). 'Indices for ranking the potential for pesticide contamination of groundwater', *Proc. Soil Crop Sci. Soc. Fla*, **44**, 1–8.

Rasmuson, A., and Flühler, H. (1990). 'Flow and transport modeling approaches: philosophy, complexity and relationship to measurements', in *Field-Scale Water and Solute Flux in Soils*, Birkhäuser Verlag, Basel.

Rhodes, R. C., Belasco, I. J., and Pease, H. L. (1970). 'Determination of mobility and adsorption of agrichemicals in soils', *J. Agric. Food Chem.*, **18**, 524–528.

Russell, J. M. (1975). 'Analysis of air pollutants using sampling tubes and gas chromatography', *Environ. Sci. Technol.*, **9**, 1175–1178.

Russell, M. H., and Layton, R. J. (1991) 'Models and Modeling in a Regulatory Setting: Considerations, Applications and Problems', *Weed Technol.*, **6**, pp. 673–676.

Schinkel, K., Nolting, H.-G., and Lundehn, J.-P. (1986). 'Persistence of plant protection products in the soil—degradation, transformation and metabolism', in *Guidelines for the Official Testing of Plant Protection Products*, Part IV, 4–1, Federal Biological Research Centre for Agriculture and Forestry, Braunschweig, Germany.

Schuhmann, G. (1976). 'The economic impact of pesticides on advanced countries', in *Pesticides and Human Welfare*, Gunn, D.L., and Stevens, J. G. R., eds, Oxford University Press, Oxford, UK.

Seiber, J. N., *et al.* (1975). 'Determination of pesticides and their transformation products in air', in *Environmental Dynamics of Pesticides* (eds R. Haque and V. H. Freed), pp. 17–43, Plenum Press, NY.

Seiber, J. N., Woodrow, J. E., Sanders, P. F. (1981). 'Estimation of ambient vapor pressures of pesticides from gas chromatographic retention data', abstract, *82nd Am. Chem. Soc. Mtg., New York*.

SESOIL-FHG (1989). *Simulation Model using SESOIL for Scenarios in Germany*, R&D Project UBA-10602065, Fraunhofer Institut für Umweltchemie und Ökotoxicologie Schmallenberg/Grafschaft, Germany.

Smith, C. N., Parrish, R. S., and Carsel, R. F. (1987). 'Estimating sample requirements for field evaluations of pesticide leaching', *Environ. Toxic. Chem.*, **6**, 343–357.

Spencer, W. F., and Cliath, M. M. (1983). 'Measurement of pesticide vapor pressures', *Residue Rev.*, **85**, 57–71.

Spencer, W. F., Farmer, W. J., and Cliath, M. M. (1973). 'Pesticide volatilization', *Residue Rev.*, **49**, 1–45.

Stork, A., Traub-Eberhard, U., and Führ, F. (eds) (1990). *Lysimeter Workshop.* Langzeitverhalten von Pflanzenshutzmitteln in Böden, KFA, Jülich, Germany.

Thompson, G. W., and Douslin, D. R. (1971). 'Vapour pressure', in *Physical Methods of Chemistry* (eds A. Weissberger and B. W. Rossiter), Vol. 1, Part V, Chapter II, Wiley Interscience, Chichester.

Tillotson, P., Fontaine, D., and Martin, E. (1991). *A User's Guide to GRASP: a Geographically-based Risk Analysis System for Pesticides*, Dow-Elanco, USA.

US EPA (1975). *Federal Register*, **40**, No. 123, June 25.

US EPA (1982a). *Pesticide Assessment Guidelines, Subdivision D, Product Chemistry §63-11, Octanol/Water Partition Coefficient*. NTIS PB83-153890, Springfield, VA.

US EPA (1982b). *Pesticide Assessment Guidelines, Subdivision D, Product Chemistry, §63-8, Solubility*, NTIS PB83-153890, Springfield, VA.

US EPA (1982c). *Pesticide Assessment Guidelines, Subdivision D, Product Chemistry, §63-9, Vapor Pressure*, NTIS PB83-153890, Springfield, VA.

US EPA (1982d). *Pesticide Assessment Guidelines, Subdivision N, Chemistry: Environmental Fate, §163-1, Leaching and Adsorption/Desorption*, NTIS PB83-153973, Springfield, VA.

US EPA (1982e). *Pesticide Assessment Guidelines, Subdivision N, Chemistry: Environmental Fate, §161-1, Hydrolysis*, NTIS PB83-153973, Springfield, VA.

US EPA (1982f). *Pesticide Assessment Guidelines, Subdivision N, Chemistry: Environmental Fate, §164-1, Field Dissipation Studies for Terrestrial Uses*, NTIS PB83-153973, Springfield, VA.

US EPA (1982g). *Pesticide Assessment Guidelines, Subdivision N, Chemistry: Environmental Fate* (eds R. K. Hitch *et al.*), NTIS PB83-153973, Springfield, VA.

US EPA (1982h). *Pesticide Assessment Guidelines, Subdivision D, Product Chemistry* (eds G. J. Bensh *et al.*), PB83-153890.

US EPA (1987). *Handbook-Ground Water*. EPA/625/6-87/016.

US EPA (1988a). *Standard Evaluation Procedure, Terrestrial Field Dissipation Studies*. NTIS PB90-208935, Springfield, VA.

US EPA (1988b). *Guidance for Ground-water Monitoring Studies*. Draft document.

US EPA (1988c). *Health Advisories for 50 Pesticides*, NTIS PB-245931, Springfield, VA.

US EPA (1990). Poster presentation, in *Monitoring Studies as a Tool for Estimating Groundwater Exposure* (eds E. Behl *et al.*), IUPAC, Hamburg, Germany.

van der Linden, A. M. A., and Boesten, J. J. T. I. (1989). *Calculation of the Rate of Leaching and Accumulation of Pesticide as a Function of their Adsorption and Degradation rate in Plough Layer Soil*, IOB-Staring Centrum, RIVM, The Netherlands.

Van Dyk, L. P., and Visweswariah, K. (1975). 'Pesticides in air: sampling methods', *Residue Rev.*, **55**, 91–134.

Veith, G. D., and Morris, R. T. (1978). *A Rapid Method for Estimating Log P for Organic Chemicals*, EPA-600/3-78-049, NTIS PB-284386, Springfield, VA.

Wagenet, R. J., and Hutson, J. L. (1989). *Leaching and Estimation Model— LEACHM*, Cornell University, New York, USA.

Walker, A., and Barnes, A. (1981). 'Simulation of herbicide persistence in soil: a revised computer model', *Pestic. Sci.* **12**, 123–132.

Watson, D. B., and Brown, S. M. (1985). Testing and evaluation of the SESOIL model. U. S. Environmental Protection Agency, EPA, August 1985, Athens, GA.

Wauchope, R. D., Buttler, T. M., Hornsby, A. G., Augustijn Beckers, P. W. M., and Burt, J. P. (1992). 'The SCS/ARS/CES Pesticide Properties Database for Environmental Decision-Making', *Rev. Environ. Contam., Toxicol.*, **123**, 1–164.

Weber, J. B. (1971). 'Model soil system, herbicide leaching, and sorption', in *Research Methods in Weed Science* (ed. R. E. Wilkinson), pp. 145–160.

Whitehouse, B. G., and Cooke, R. C. (1982). 'Estimating the aqueous solubility of aromatic hydrocarbons by high performance liquid chromatography', *Chemosphere*, **11**, 689–699.

WHO (1991). *Revision of the WHO Guidelines for Drinking-water Quality*, EUR/ICP/CWS 031, WHO, Copenhagen.

CHAPTER 3

Soils and pesticide mobility

C. D. Brown, A. D. Carter and J. M. Hollis

Environmental Behaviour of Agrochemicals
Edited by T. R. Roberts and P. C. Kearney   © 1995 John Wiley & Sons Ltd

INTRODUCTION

The presence of pesticide residues in surface and ground water sources has focused much needed attention on fate and behaviour in the environment and particularly on the soil properties and processes which influence pesticide mobility. There is increasing evidence to suggest that the specific environmental and agricultural conditions which prevail during or shortly after pesticide application are critical in determining pesticide fate and behaviour and hence mobility. There is also a much greater awareness of the inextricable link between pesticide applications, the soil hydrological cycle and the aquatic environment. Within this wider context, the chapter will discuss the soil factors which influence pesticide mobility, experimental and computational methods for assessing pesticide mobility and possibilities for extrapolating results to larger scales. The reader is referred to the previous chapter for specific details on the techniques available to assess pesticide mobility in soils.

Figure 1 illustrates the link between all pesticide usage (agricultural, forestry and non-cropped land) and the aquatic environment. Where soil layers are present at sites of application there is the potential for retention, degradation, dilution and attenuation of a pesticide and this is the case for most pesticides used in agricultural or forestry situations. Transport to ground or surface waters can be more rapid when agricultural drainage systems are installed or when water moves over slowly permeable surfaces as run-off. Where the soil layer is absent or hard surfaces are sprayed (typical situations in urban and amenity usage) there is less potential for pesticide retention and degradation and there may be rapid infiltration of a chemical into subsoils, artificial drainage systems or underlying aquifers. Pesticides in soil may also arise from disposal through soakaways, spillages or even direct discharge, and some of the most serious cases of soil and water contamination occur as a result of pesticide misuse, accidental spillage or inadequate handling/storage conditions (Ritter, 1990; Harris *et al.*, 1991). In order that appropriate control or legislative procedures can be taken, it is important to be able to identify and distinguish between these point-source contamination events and those which arise as a result of diffuse movement through the agricultural environment. Diffuse pollution is not easy to monitor, legislate against or control. Many studies are now taking place in order to identify and characterize the soil processes and factors which influence pesticide mobility so that the contamination risk from diffuse sources can be minimized.

Figure 1 Pesticide movement in the aquatic environment

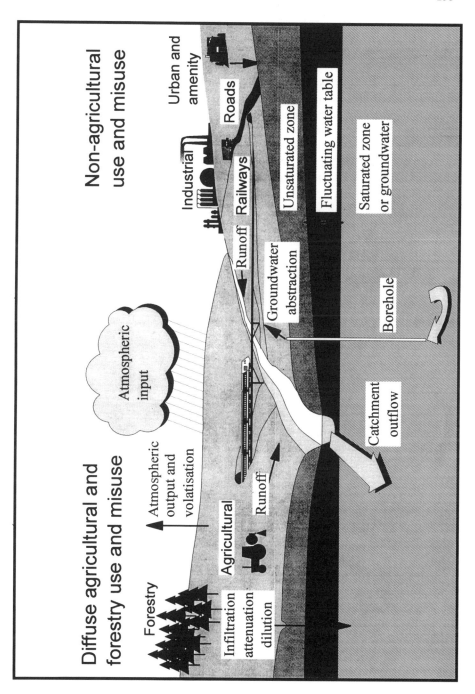

SOIL FACTORS WHICH DETERMINE PESTICIDE MOBILITY

Mineral soils are composed of varying proportions of sand, silt, clay and organic material, water and air (Baver, 1956). Pesticides may exist in any of these phases within a soil but their partitioning and relative concentrations will vary according to many environmental factors. In their undisturbed field state, soil particles aggregate to produce a range of soil structures which determine the porosity and density characteristics of a soil. Soil porosity and density largely determine the water holding or retention capacity of the soil as well as the rate of water movement within soil profiles. There is a considerable body of literature documenting the chemical characteristics of soil which influence pesticide mobility (e.g. Cheng, 1990; see Chapter 2, in this volume), but the same is not true for the physical and hydrological properties of the soil. Hence, this section will consider those properties of a soil which are less well understood or documented in relation to pesticide mobility. Recent research shows that the pathways and rates of transport of pesticide both in the dissolved phase and sorbed to sediment suspended in solution can be determined by the physical and hydrological characteristics of a soil (Johnson *et al.*, 1993). The two most important properties of the soil are the water retention and hydraulic conductivity characteristics. These properties can determine the soil water regime of a particular soil.

Water retention

At zero pore water pressure the soil is saturated with water, but as the soil dries the relative volumes of soil, air and water change, large pores spaces fill with air and water is held in finer pores and more tightly to particle surfaces. The drying out process is associated with an increase in soil pore water pressure which has a negative value or for convenience is sometimes expressed as a suction and does not require the negative. This pressure is measured in units of Pascals (the SI unit), bars, cm of water or pF (the logarithm of the negative pore water pressure when expressed in cm of water or millibars). The negative pore water pressure and the soil water content can be plotted to give a curve called the soil water release characteristic which is different for each soil type and soil horizon (Hall *et al.*, 1977). Figure 2 illustrates average water retention curves for the ranges of UK soil textures. The shape and position of each curve is strongly influenced by soil texture, structure, organic carbon content and density (Reeve and Carter, 1991). The soil pore water pressure is equivalent to the amount of energy which a plant or microorganism needs to exert before it can obtain water from the soil. The greater the 'suction' the more stress is imposed on the organism.

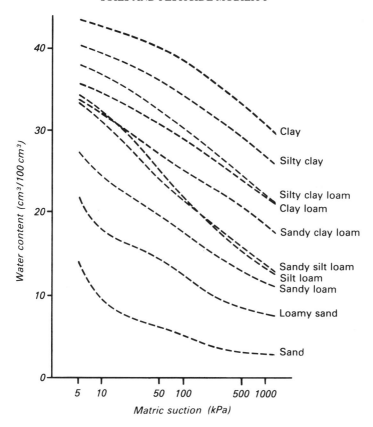

Figure 2   Release characteristics of different particle-size classes over the available water range

Laboratory degradation studies have previously been carried out for regulatory submission using a range of soil water contents or 'suctions', making comparison of degradation data between soil types extremely difficult. Figure 3 illustrates the water release characteristics for a clay loam and a sandy loam with organic carbon contents of 5.5 and 4.8%, respectively. The water content/suction requirements for pesticide degradation studies of several different regulatory organizations are marked and it can clearly be seen that study conditions are not comparable for the two soil types. In Europe, the Harmonisation Directive (European Community, 1991) has been able to address this unscientific situation and stipulates a soil water suction range of 10–32 kPa (0.1–0.32 bar, pF 2.0–2.5) for all degradation studies.

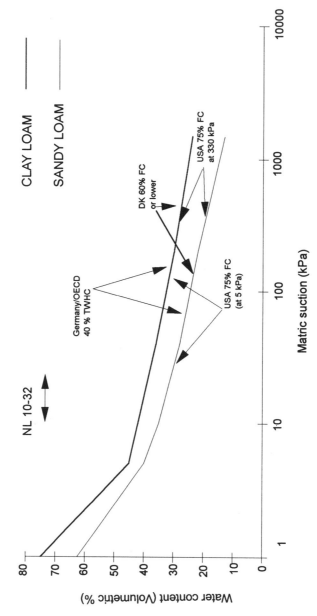

Figure 3   Different regulatory requirements for soil degradation studies

Hydraulic conductivity

When the soil is saturated, all pores are, in theory, water filled and contribute to any drainage or flow. The saturated hydraulic conductivity is usually measured in $cm\,day^{-1}$ or $m\,s^{-1}$ and varies according to soil type. Hydraulic conductivity determines the rate that water (and thus soluble pesticide) moves either vertically or laterally within a soil profile. In a sandy soil flow is very rapid in the saturated state, but quickly diminishes as soil suction increases, i.e. when the larger pores have drained (Figure 4). In a clayey soil matrix, flow is slower in the saturated state, but greater in the unsaturated state because there is a greater number of finer soil pores contributing to flow than in the sandy soil at the same suction. The conductivity of clay soil can be complicated by large cracks, interpedal voids and other macropores, particularly where they connect with mole channels or drains (Jarvis and Leeds-Harrison, 1987). The hydraulic conductivity of these bypass routes can be extremely rapid with bypass water reaching outflows within minutes of the commencement of a rainfall event. Since the majority of applied pesticide remains in the top few centimetres of the soil surface it is important to be aware of any hydrological pathway which connects the soil surface with a drainage outflow (Johnson et al., 1993). The soil matrix need not be saturated for water to flow in these bypass systems. That pesticide which is held in the soil solution within larger soil pores (but against gravity) is most likely to be leached and transported down the profile

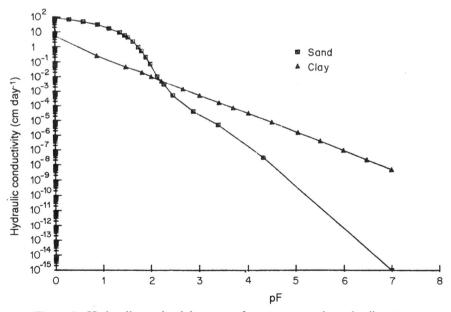

Figure 4 Hydraulic conductivity curves for two contrasting subsoil textures

following rainfall. Leaching still occurs when a soil is unsaturated but it is a much slower process since the hydraulic conductivity of an unsaturated soil at field capacity can be 2–5 orders of magnitude lower than the saturated soil.

Soil water regimes

Where soil horizons contain relatively few coarse pores, downward percolation of excess water is impeded, causing periodic saturation in the overlying soil layer. Such impeding layers are known as slowly permeable horizons and any excess water must be redistributed horizontally. If there are insufficient laterally interconnected voids, movement will be relatively slow and drainage restricted. In some soils, such as those developed in river alluvium, the local ground water table may rise and fall within the soil profile during the seasons. Evidence of soil wetness is given by characteristic pale or greyish colours on structure or pore faces contrasting with ochreous mottles and/or ferri-manganiferous concretions within the soil structures. This mottling represents a temporary anaerobic condition in the soil layer whereby iron and manganese deposits become reduced and then mobilized within the soil water fraction. The pale or greyish colours are iron-depleted surfaces and highlight the main conducting channels within the slowly permeable horizon. Such soils typically have water tables which exhibit seasonal fluctuations. These complex oxidation–reduction processes occur simultaneously within the same soil layer and as such will have a major influence on the degradation of a pesticide.

Rates of degradation of pesticides are generally slower in subsoils than in topsoils, in part because of smaller active microbial populations. However, Lewis and Dyson (1994) show that this is not necessarily always true and consider that specific subsoil soil physical conditions may promote the biological activity of microbial populations and hence increase rates of degradation. A knowledge of the duration and depth of waterlogging in a soil profile is essential to the understanding of pesticide availability, degradation and mobility in soils. Nicholls et al. (1987) report an important example of this in which they noted damage to a succeeding sugar beet crop grown on a site previously treated with the sulphonylurea herbicide, chlorsulfuron. The authors postulate that persistence of chlorsulfuron in the waterlogged subsoil at the site was followed by re-entry of the pesticide into the rooting zone with rising water table levels, thus creating the possibility for damage to the following crop. Clearly, the characteristics of the soil water regime at a site are an important consideration not only in site selection for leaching, dissipation and accumulation studies but also for interpretation of experimental data and extrapolation of results to the wider environment.

POTENTIAL PATHWAYS IN SOIL

Water as a solvent can hold or carry pesticides in solution and can also transport soil particles to which pesticides have sorbed. An understanding of soil water regimes is thus essential when determining pesticide mobility. Figure 5 summarizes the potential flow pathways in soil.

Infiltration

When water is applied to the soil as rainfall or irrigation it will infiltrate the soil surface but under certain circumstances may pond or move as surface run-off across a slope. The rate of infiltration of a particular soil surface will determine the amount of water which dissipates through the soil profile and how much, if any, runs off. The infiltration rate is determined by surface properties and conditions such as permeability, structural composition of the surface tilth, presence of slaking or capping, the antecedent moisture content, the presence of a shallow water table. It is generally expected that sandy soils will be more permeable than clayey ones and that water will percolate readily through the profile. However, Williamson and Carter

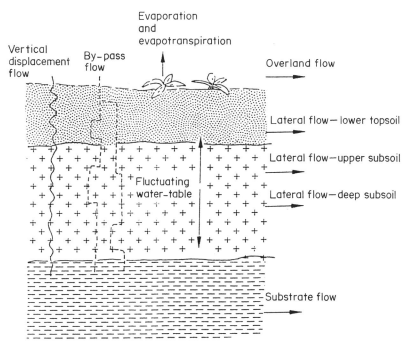

Figure 5   Pathways of water movement in soil

(1993) found that heavy rainfall 10 days after pesticide application generated greater volumes of run-off and contained greater quantities of pesticide than at a clayey site within the same field. The surface of the sandy soil was badly slaked following the breakdown of soil aggregates during wetting and a 5 mm crust virtually sealed the sandy soil surface.

Surface run-off

Surface run-off or overland flow occurs when the surface storage capacity is exceeded and flow is generated. It can be a critical process, since water is in direct contact with the immediate soil surface which contains the largest portion of applied pesticide and this is particularly true where flow and/or sediments are transported directly into surface water sources. Burgoa and Wauchope discuss this process and pathway further in Chapter 5.

Evapotranspiration

Evapotranspiration of water from the soil profile and crop, if present, initiates drying of the soil and the net flux of water eventually becomes upwards as more water is used by crops or evaporated from the soil surface than is supplied in the form of natural or artificial precipitation. Typical evapotranspiration rates for a growing crop in the UK may be 1–3 mm of water per day (Jones and Thomasson, 1985) and during the summer period a soil moisture deficit (SMD) develops with the result that leaching of soluble pesticide does not normally occur. Agroclimatic conditions vary considerably and knowledge of basic information on soil temperature and deficit situations are fundamental to understanding the potential for pesticide mobility in soils.

Storage and field capacity

When evapotranspiration rates decrease, the net flux of water in the soil is usually downwards and internal drainage becomes the dominant process. As a soil profile begins to wet, the SMD begins to decrease, the soil water suction decreases, more water is held in each layer and the hydraulic conductivity of the soil increases. When all but the very coarsest of pores are water filled, the soil is often considered to be at field capacity, although it is difficult to attribute a particular suction to this state. It has however, for various purposes, become convenient to define such a suction, and most soils are generally considered to have field capacity values between 3 and 10 kPa (0.03–0.10 bar, pF 1.47–2.0). Soils in Northern Europe are usually at or near a field capacity status from late autumn to early spring. Pesticides applied during this period are generally more likely to be involved in downward and/or lateral fluxes of internal drainage.

Leaching and bypass flow

The redistribution or movement of water is partly determined by the prevailing moisture and hydraulic gradients within the soil profile. Water can move both laterally and vertically through pores, cracks and fissures, but the size and number of water-filled voids at any one time depends on the drainage characteristics of the soil and the pore water pressure. The retention process is important as it determines how much water will flow through the profile and out of the rooting zone and thus controls the amount of leaching of pesticide which takes place. Recent research by Walker *et al.* (1995) has shown that total alachlor residues in soil declined over time but the aqueous extractable residues declined more rapidly, indicating an increased strength of adsorption with residence time in soil (Figure 6a). The data in Figure 6b are calculated percentages of the total residues that were readily extractable with water. These declined from 5–7% initially to approximately 0.5% after 65 days, with little difference between the results of two contrasting soil textures.

The movement of water between individual soil particles within the main soil matrix is the dominant flow process in sandy or light loamy soils. However, there is increasing evidence to suggest that there can be preferential movement or fingering of water in sandy soils (as opposed to bypass flow), particularly in those soils which exhibit hydrophobic character- istics (Flury *et al.*, 1994). As the clay content of a soil increases, there is a tendency for aggregation of soil particles into structural units called peds. Ped sizes and shapes vary according to soil type (Hodgson, 1976) and good structural development is essential for a well drained and aerated soil. Water and solutes frequently move between structural units, interacting with ped surfaces and diffusing in and out of the ped matrix (Addiscott and Whitmore, 1991). Strong vertical structural development is characteristic in many clay soils, particularly those with the potential to shrink and swell.

Bypass flow through cracks and fissures can be an important process (Nicholls *et al.*, 1993). In England and Wales, 33% of soils have the potential to crack, but bypass flow through macropores may be a dominant process in a further 31% of soils (Hollis and Carter, 1990). Where bypass movement takes place, there is potential to transport pesticide from surface layers to greater depths within the soil profile than conventional theory dictates. Where these macropores or cracks connect with artificial drainage systems or underlying aquifers contamination of water sources can occur. The use of herbicides on hard surfaces in non-agricultural situations can create a comparable situation as run-off rapidly enters drains and water flows to ditches or soakaways, possibly causing contamination. Sketches of contrasting soil profiles in Figure 7 emphasize the importance of soil structure and anthropogenic controls in determining pesticide mobility in soils.

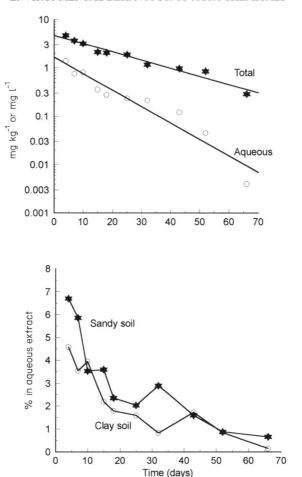

Figure 6   Decline in total and aqueous extractable residues of alachlor in soil. (a) Clay loam. (b) Changes in aqueous availability of alachlor.

Artificial drainage

Land drainage design has as its objective the removal of excess water from soil or land surface. Since the mid-eighteenth century, large-scale attempts have been made to improve subsurface water control by the installation of various types of underdrainage systems. In the UK, early stone drains were followed by baked clay horseshoe tiles and later still by round clay pipes. Various other forms of early drainage control are described in detail by Trafford (1970). Despite their age many of these old systems are still effective and are responsible for draining many slowly permeable soils or

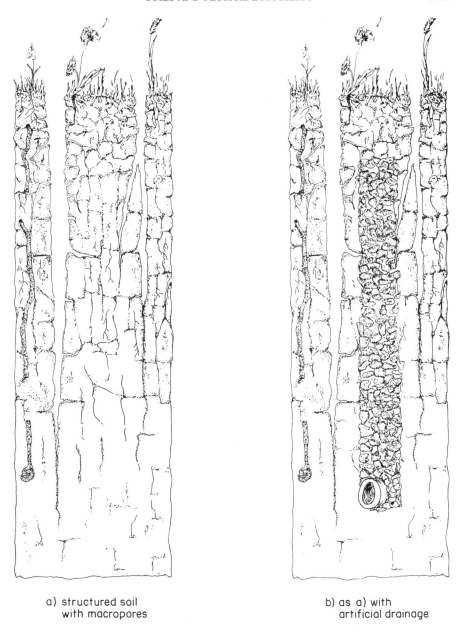

a) structured soil
   with macropores

b) as a) with
   artificial drainage

Figure 7 Pathways for water movement in clay soils

those with shallow water tables, which would not otherwise be cultivated. Artificial drainage has been shown to be responsible for the transport of significant quantities of dissolved pesticide (Johnson *et al.*, 1993; Brown *et al.*, 1995), particularly when rainfall and subsequent drainage occur shortly after pesticide application. This hydrological pathway is responsible for many of the fluctuations of pesticide concentrations in surface water and it is thus useful to describe the drainage process and its influence on soil in further detail.

The design of a drainage scheme is influenced by many considerations, including rainfall volume, cropping practice, soils and slope. Drain depth and spacing are used to control the depth of the water table and the rate of drainage and the size, type and gradient of pipe (e.g. slotted plastic pipe, clay tiles) determine how and when water is transported from the soil. Most clay subsoils have saturated hydraulic conductivities of $0.1 \, \mathrm{m \, day^{-1}}$ or less, so that effective drainage is mainly confined to the surface horizon. In such soils, pipe drainage is usually ineffective unless the subsoil properties are modified to increase physically the number and size of cracks and fissures. This can be done by moling or by subsoiling the site. Mole drains are unlined channels which convey water readily through the soil to the pipe drainage system. They are formed by a mole plough consisting of a steel shaft with a circular steel foot or bullet at the base trailing a cylindrical expander behind. Typical dimensions for the bullet and expander are 75 and 100 mm diameter, respectively. Mole drains are normally drawn at 45–60 cm depth, 2–3 m spacing, and at right angles to the primary drainage treatment. Ideally, a stable circular drainage channel is created with shattering and fissuring in overlying layers. Moles fill and empty at very low suctions and provide a direct route for water flow, hence creating the possibility for drainage to occur before the entire soil profile is saturated. Subsoiling is another option in heavy soils and is carried out to break up soil layers, usually below the plough depth, to make the soil less dense and allow water and air entry. Small tines up to giant rippers can be used depending on the requirement, but the objective of subsoiling is always to speed horizontal movement of water to drains. Permeable fill is commonly placed in the trench above a drainage pipe to form a permanent connector for moling or subsoiling treatments. The fill may consist of crushed gravel, hard crushed stone or synthetic material, and serves to ensure that flows of water have easy access to the underlying drain pipe. Ragg *et al.* (1984) describe drainage requirements and design for different soil types.

Drainage treatments of some form are commonplace in most slowly permeable soils or in areas where shallow water tables limit agricultural production. In the UK, for example, it is estimated that approximately 40% of lowland, arable soils are drained.

RELEVANCE OF MEASUREMENTS

Monitoring the presence and characterizing the mobility of pesticides in soils is fraught with many theoretical and practical problems. The previous chapter provides a detailed discussion of the recommended experimental approaches to assess pesticide mobility in soil. This short section discusses some of the further limitations/advantages of the techniques and then describes their relevance to the objective of determining fate and behaviour of pesticides in undisturbed agricultural soils within the context of the wider environment.

Laboratory leaching columns

Laboratory leaching studies on pesticide compounds are normally carried out on stored, dried, disturbed and sieved soil samples, repacked into laboratory containers. The advantage of such studies is that different soil types and leaching volumes can be chosen to simulate a range of soil climatic environments, whilst repacking the columns eliminates much of the natural soil variability which would otherwise necessitate a far larger number of replicated measurements.

The structural condition of the soils used in these column leaching tests has, however, little relevance to most agricultural soils. Disturbed, sieved and repacked soil tends to have a granular structure, causing water to flow mainly around individual soil particles. This type of structure occurs naturally in relatively loose sandy soils. In most other cases, soil particles are aggregated into compact structures of varying shape, density and coherence, and water tends to flow preferentially along fine fissures or pores between aggregates. The more compact and dense the aggregate, the greater the likelihood that flow will bypass parts of the soil matrix and that the soil moisture release characteristic will differ significantly from that of granular soils. Hollis and Carter (1990) state that the artificial structure of soil packed into a laboratory leaching column is only comparable with about 7% of agricultural situations in the UK.

Figure 8 compares the water retention properties of a sieved, repacked soil and the same soil in its natural state. The disturbed soil is wetter at low suctions and is likely to have a greater saturated hydraulic conductivity than the undisturbed soil. The shape of the water retention curve is critical in determining pesticide availability for movement and this can be seen by comparing specific points on the water release characteristic which are relevant to degradation studies. The rate of water flow through the disturbed soil is likely to be greater than in undisturbed soil, but the area of soil surface interacting with the pesticide is also likely to be greater, leading

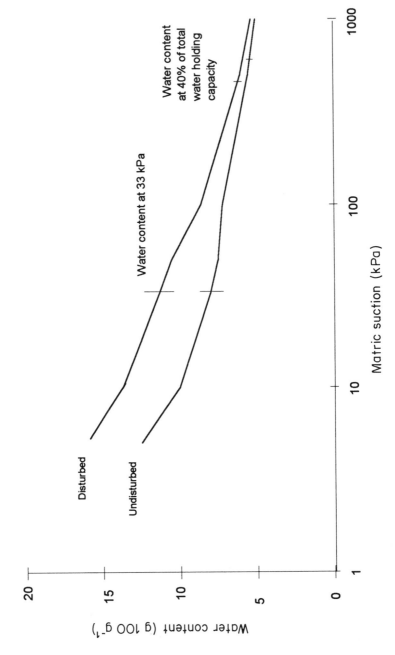

Figure 8   Water release characteristics of disturbed and undisturbed sandy loam soil

to an increase in the number of sorption sites in the disturbed sample. The effect of repacking columns with disturbed soil may thus be either to under- or overestimate pesticide mobility, depending upon the strength of sorption of the given chemical to the soil in question.

An additional consideration in interpreting results from laboratory leaching columns is that the volume of water used to generate leachate is usually far in excess of any natural precipitation event. This can frequently initiate vertical saturated flow within the column, a situation which is not likely to occur naturally. The results of small leaching column experiments thus cannot and should not be extrapolated directly to the field situation. However, this type of study has uses in determining the potential mobility of an active ingredient and its metabolites as a basis for relative comparison between pesticides. Pesticides or metabolites which do leach would then trigger further, more detailed, investigations into their mobility in soils.

Lysimeters

Increasingly, soil lysimeters are being used for pesticide leaching studies (Führ and Hance, 1992), enabling work to be carried out on undisturbed field soil whilst still retaining the ability to control environmental conditions and use radiolabelled chemicals. The use of lysimeters is not, however, a new technique and there are a number of reviews on their use for studying soil hydrology or solute movement through soil (e.g. Lawes et al., 1881; Millar and Turk, 1943; Kitching and Day, 1979). A major limitation to the use of lysimeters is the interpretation and extrapolation of resulting data and it would thus be of considerable benefit to evaluate previous uses of these instruments, particularly the soil hydrological data which have been generated. Comparison of data with the many 'in field' experiments at locations with similar soil would ultimately enable a wider, more reliable, interpretation of results from lysimeters on pesticide mobility in soil.

Although lysimeters can give a relatively accurate simulation of flow in both granular and structured soils, leaching study results are only applicable to situations where vertical flow directly to a groundwater table or aquifer predominates. Only 30–40% of UK soils conform to this flow pattern (Hollis and Carter, 1990). On the remaining land, soils have slowly permeable or impermeable layers within about 1.2 m depth which give rise to lateral flow, often under temporarily waterlogged conditions. Lysimeter studies cannot simulate leaching in these conditions and run-off plots or field studies are more appropriate. Sandy, arable soils with low organic carbon contents are frequently selected for lysimeter studies as representing a 'worst case' for leaching. However, it is by no means certain that this scenario genuinely gives a worst case as bypass flow in soils with well developed structures may contribute greater quantities of pesticide to leachate outflow.

This phenomenon was noted by Yon (1992) in a comparative study using a sandy soil and a medium loam over clay. Where preferential bypass channels terminate in relatively free draining substrates leaching is likely to be greater than in granular substrates. Interpretation of data from soils which demonstrate bypass flow should be carried out with great care since *in situ* these bypass channels may not have connectivity and solutes would normally slowly dissipate through surrounding voids and into the soil matrix. Soils with potential for bypass flow to a permeable substrate can be difficult to identify but it is estimated that they may constitute about 2–3% of cropped land in the UK.

Field studies

Environmental fate studies carried out within individual fields or small catchments represent the ideal monitoring situation, but experimental conditions are difficult to characterize or control. In most cases it is not possible to use radiolabelled chemicals and the analysis of the applied active ingredient or its metabolites may be difficult. Field studies are also very difficult and expensive to monitor comprehensively (Simmons, 1991). Many different monitoring techniques are available to investigate pesticide movement in soil and examples of these are discussed by Nash and Leslie (1991). The most difficult problem associated with monitoring is to ensure that the installation procedure, the equipment itself, or its operation does not influence soil processes or pesticide concentrations which prevail in the soil profile. This is particularly difficult in structured soil, but can also create problems in soils with high dry bulk densities. Table 1 compares the bulk density data from a sandy subsoil with the same soil after it had been excavated and then repacked around a soil water suction sampler (Williamson and Carter, 1993). Examination of the site hydrological and analytical data revealed that there had been preferential movement of a herbicide in the repacked soil around the sampler. The same problems of disturbance are applicable to all types of installations.

Soil coring to determine the depth of leaching of an agrochemical is not necessarily the most appropriate method to determine whether a pesticide or its metabolites have the potential to leach and contaminate water. The limit of analytical detection is often at least one, but can be several orders of magnitude higher for pesticides in soil compared to water samples. Many of the field soil leaching studies carried out with potentially mobile products show there to be no leaching in the soil profile, yet surface or ground water analyses are contradictory. Field studies need to take into account the small but environmentally significant concentrations which do leach from soil profiles and monitoring of the soil water phase may be more appropriate.

Continuous monitoring of all parameters at field experimental sites is in

Table 1 Influence of soil disturbance on soil physical properties

		Depth (cm)							
		47–52	57–62	67–72	77–82	87–92	97–102	107–112	117–122
Coarse porosity (%)	Repacked	17.52	16.67	17.41	18.29	18.97	15.63	15.86	14.25
	Undisturbed	14.37	12.89	14.47	12.75	15.45	13.67	11.95	8.53
Bulk density (g cm^{-3})	Repacked	1.41	1.46	1.44	1.51	1.55	1.59	1.52	1.56
	Undisturbed	1.57	1.50	1.66	1.70	1.69	1.67	1.64	1.69

Values for repacked and disturbed samples are means of 2 and 3 measurements respectively.

many cases technically impossible and where technology is available it is extremely expensive to provide adequate human and analytical resources. Fogg *et al.* (1994) compare the mobility of a seed dressing fungicide using field, lysimeter and predictive modelling techniques in accordance with different regulatory requirements and conclude that a simplified field study in a 'realistic worst-case' situation provided the data that were most easily interpreted to assess the potential for the fungicide to leach. Carefully selected, representative study areas and monitoring sites are essential if the resulting data are to be applicable to a range of situations where a product might be used and to allow successful extrapolation to other agroclimatic conditions. The use of predictive modelling offers the opportunity to determine worst- or typical-case scenarios for leaching/run-off of a given pesticide under a given usage and this can facilitate selection of the most suitable site for an experiment. In the UK, for example, the use of soil hydrological classifications (Boorman and Hollis, 1990) enable relevant, representative sites and sampling methodology to be selected and monitoring strategies to be more effectively targeted. The same principles can be applied to catchment monitoring which is an increasingly popular retrospective/post-registration monitoring tool employed particularly by water protection agencies and national regulatory bodies (e.g. Kreuger, 1991; Breach and Porter, 1993; USEPA, 1994).

EVIDENCE OF PESTICIDE LEACHING

Experimental determination using the techniques described above can give an indication of pesticide mobility under specific conditions, but such data must be interpreted within their wider environmental context. Relatively simple measurements to derive sorption coefficients and half-lives for pesticides in soil can be coupled with relationships between pesticide physicochemical characteristics to give a good idea of the *potential* environmental mobility of an active ingredient (see Chapter 2). Such approaches can be very useful in a regulatory context as they give an indication of the likelihood of leaching of a given active ingredient compared to existing ones. However, they are only useful if comprehensive monitoring data are available to calibrate the assessments and validate the comparisons. The presence of a number of pesticides in surface and ground water sources has been recognized for some years (e.g. IUPAC, 1987; Lees and McVeigh, 1988; Battaglia, 1991). This recognition, coupled with tight regulation of water quality standards, has prompted widespread monitoring of water supplies for contamination using standardized methodologies and has resulted in improved, more detailed data sets which are now becoming available.

Results of ground water monitoring in the USA by government at all levels, the pesticide industry and private institutions have been summarized for the period 1971–1991 by the US EPA (1992). Monitoring of 69 000 wells in 42 states for 258 parent compounds and 45 metabolites is presented, although the authors warn that 70% of analyses and the same proportion of positive detections are from the states of New York, Florida and California. The report lists 117 parent compounds and 16 metabolites whcih have been positively detected in wells, but 69% of detections for parent compounds are accounted for by just five compounds. The report is particularly valuable in listing pesticides which have been analysed for, but not detected, as these data are rarely documented. Consideration of the properties of such pesticides should play an important part in risk management for the future. Similarly, although metabolites are considered in the regulatory process, they are frequently omitted from monitoring strategies. Selected data for the analysis of parent compounds and their metabolites in the USA between 1971 and 1991 are summarized in Table 2 (US EPA, 1992). It is likely that monitoring of pesticide metabolites will become more important as the environmental fate and behaviour of these compounds are better understood.

Analysis of water supplies in England and Wales by individual water companies during 1990 and 1991 have been reported by DWI (1991, 1992). Forty-five different pesticides were detected at least once at concentrations $>0.1 \, \mu g \, l^{-1}$ and an initial examination of the physicochemical properties of these pesticides shows a very large range. Whilst information on properties is not available for all of these compounds, representative values for half-life in soil ($t_{1/2}$) range from 1 to 400 d and for the organic carbon partition coefficient (K_{oc}) range from 1 to 100 000 ml g^{-1} (Howard, 1991; Worthing and Hance, 1991; Wauchope et al., 1992). If frequency of detection is also considered, however, then more coherent patterns in data are observed. Figure 9 separates pesticides detected in water supplies by drinking water companies in 1990 and 1991 (DWI, 1991, 1992) according to the frequency of detection and also includes pesticides analysed for, but not detected by the National Rivers Authority (NRA) in 1991 (A. Ferguson, personal communication). Eight pesticides were detected an average of more than 50 times per year, and form a relatively tight grouping which can be defined in terms of $t_{1/2}$ and K_{oc} as:

$$\log t_{1/2} = 1.5 + (0.44 \times \log K_{oc}) \qquad [r = 0.68].$$

Gustafson (1989) proposed that risk of contamination of water supplies could be assessed by a ground water ubiquity score (GUS) which is defined by:

$$\text{GUS} = \log t_{1/2} \times (4 - \log K_{oc}),$$

Table 2 Summary of analyses for selected pesticides and their metabolites in the USA between 1971 and 1991

Parent compound	Analyses	Detections	Metabolite	Analyses	Detections
Aldicarb	43 786	3002	Aldicarb sulphone	37 652	5070
			Aldicarb sulphoxide	37 593	4991
Carbofuran	28 020	4127	3-hydroxycarbofuran	22 314	42
			3-ketocarbufuran	839	3
Atrazine	26 909	1512	Des-ethyl atrazine	689	27
			Des-isopropyl atrazine	689	27
DCPA	2033	5	DCPA acid metabolites	118	59
Heptachlor	3241	55	Heptachlor epoxide	3115	32

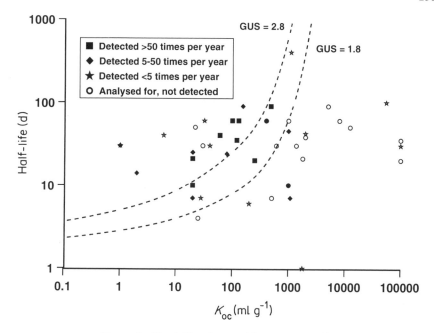

Figure 9   Pesticides detected in water supplies

and suggested that values for GUS of 1.8 and 2.8 bracket the region in which transition occurs from compounds likely to leach to those not likely to leach. Whilst the data presented in Figure 8 are for both surface and ground waters, these values for GUS still define with a reasonable degree of accuracy the transitional region between pesticides with frequent detections and those for which detection in water supplies is relatively uncommon.

The usage pattern of a pesticide will also be an important factor in determining the risk of contamination of water sources. Clearly, the greater the amount of a given pesticide used, the greater the risk to waters. Certain uses will have a greater risk associated with them than others. As discussed above, soil acts as a filter and buffer to the applied pesticide and offers the opportunity for sorption and microbial degradation. Where interaction with the soil layer is absent or reduced, a pesticide may have direct access to a water course via the unsaturated zone or directly into a surface water source via a drainage system. Semi-point, non-agricultural usage of pesticides (e.g. along railway tracks or road kerbs) potentially poses a considerable risk for contamination of water sources. Thus atrazine and simazine were the only pesticides detected in public supply boreholes during monitoring of a 16 000 ha catchment in the UK, despite a survey of agricultural usage showing use of simazine to be very limited and that of atrazine to be negligible (Clark *et al.*, 1991). The authors concluded that the observed contamination could only have arisen from non-agricultural applications.

Table 3 shows the usage patterns for the eight pesticides detected most frequently in water sources in England and Wales. With the exception of propyzamide, all of the pesticides make up a major fraction of the total application of pesticides to arable crops and/or for non-agricultural uses. Correspondingly, the eight pesticides listed in Table 3 accounted for 23% of all the analyses carried out by water companies in 1991, but accounted for 97% of all the positive detections (DWI, 1991).

Simple assessments of the risk of pesticide leaching based upon inherent physicochemical characteristics of an active ingredient can give only a very generalized indication of the likelihood of a pesticide leaching out of the soil. The agricultural ecosystem is heterogeneous in nature and pesticide fate is actually determined by the interaction of a complex range of factors for which direct measurement is often impractical. It is inevitable therefore that attempts are made to simulate the effects of such complex interactions by the use of mathematical models.

USE OF MODELLING TO PREDICT PESTICIDE MOBILITY

During the 1980s a large number of mathematical models that attempt to simulate pesticide fate in the soil were developed with a range of applications and levels of detail. A number of detailed reviews of modelling approaches have been published (e.g. Addiscott and Wagenet, 1985; Wagenet and Rao, 1991). Jury and Ghodrati (1989) divided models into one of three types—screening, management, and simulation (see Table 4). However, they warn that in a number of cases 'the normal close connection between theoretical inquiry and experimental validation has been severed'. It is vital that modelling should develop alongside our understanding of soil processes in order that maximum benefit is gained from both data and modelling alike. Only by careful comparison and evaluation of the results from theoretical simulations with measured experimental data can the full potential of both modelling and field or laboratory monitoring be realized. Given the range of models for pesticide fate in soil which are now available, there is a clear need for evaluation of the capabilities of models in a range of scenarios. Consequently, although new approaches are still being developed, most current work on modelling pesticide fate in soil is focusing upon validating and evaluating existing models, identifying processes which are either not described or are poorly simulated and amending source codes to include improved model routines.

Table 3 Usage patterns and mean analyses/detections in 1990 and 1991 for those pesticides most frequently detected in drinking water sources in England and Wales Davis *et al.* (1991) and DoE (1991)

Pesticide	Mean number of analyses per year	Mean number of detections above the MAC per year	Proportion of total weight of pesticides used on arable crops in 1990 (%)[†]	Proportion of total weight of pesticides used for non-agricultural applications in 1990 (%)[‡]
Atrazine	30 197	7966	–	22
Simazine	28 000	3347	0.3	14
Diuron	17 972*	2406*	–	12
Isoproturon	16 908	2353	9.9	–
Mecoprop	19 529	1122	5.0	8
Chlorotoluron	16 908	560	2.2	–
2,4-D	12 641	77	–	9
Propyzamide	3908	73	0.3	–
Total for all pesticides (1990)	830 000	22 548	31 459 tons	550 tons

*Data for 1991 only.
[†]Davis *et al.* (1991).
[‡]DoE (1991).
– No information available (<0.2 and 4% of total for arable crops and non-agricultural usage, respectively.

Table 4 Classification of a number of models developed to simulate the movement of pesticides through soil Jury *et al.* (1983)

Model	Reference
Screening	
Behaviour Assessment Model (BAM)	Jury *et al.* (1983)
Management	
Pesticide Root Zone Model (PRZM)	Dean *et al.* (1989)
Ground water Loading Effects of Agricultural Management Systems (GLEAMS)	Leonard *et al.* (1987)
Chemical Movement in Layered Soils (CMLS)	Nofziger *et al.* (1988)
CALculates Flow (CALF, now renamed VARLEACH)	Walker (1987)
MOUSE	Steenhuis *et al.* (1987)
Simulation	
Leaching Estimation and CHemistry—Pesticides (LEACHP)	Hutson and Wagenet (1992)
MACRO	Jarvis (1994)

Model evaluation

Pennell *et al.* (1990) evaluated five simulation models using a data set for a conservative tracer (bromide) and a relatively weakly sorbed pesticide (aldicarb) on a deep, sandy, well-drained soil with a large value for saturated hydraulic conductivity. Most of the models which are currently available would be expected to perform best under such a scenario, and differences in predictive ability were relatively small for the leaching models, CMLS, PRZM and LEACHP with predictions for depth to centre of solute mass differing from measured values by 30–70%. MOUSE and GLEAMS were found to perform less well than the other three models, although this was related to constraints in input and output for MOUSE and to the overall bias of GLEAMS which does not simulate below the root zone (Knisel, 1980). In making recommendations for use of the five models tested, Pennell *et al.* conclude that consideration of the model applicability in terms of available input data, desired output, ease of model use, and soil and pesticide to be modelled may be more important at present than critical evaluation of model performance.

Bergström and Jarvis (1994) reported an evaluation of seven models for pesticide leaching with a view to use for registration purposes. Leaching of dichlorprop and bentazone through undisturbed soil monoliths containing five different soils with a range of textures provided a comprehensive test of the models. Again, similarities between results from the models outweighed the differences despite a range of approaches ranging from simple functional models with capacity flow mechanisms (e.g. CALF) to comprehensive mechanistic models with solution of the Richards' equation (e.g. MACRO). Based on half-lives determined in the laboratory, all models overpredicted degradation of dichlorprop in the field, emphasizing the need for input data relevant to the scenario being modelled. Similarly, the models generally underpredicted leaching, particularly for dichlorprop, and this was concluded to result from the failure of the models to account for preferential flow. This process, which can have a significant impact upon solute movement, can be defined as the rapid transport of water and solutes through a small portion of the soil volume which is receiving input over its entire boundary (Guth and Mani, 1991). Two of the models tested, MACRO and PLM, have the ability to describe preferential flow and this improved model predictions, although selection of the input parameters was not related to measured properties.

Evaluation of models is, in the first instance, constrained by the availability of data sets with suitable measurements of input parameters for models and reliable results against which to test the model. Monitoring activity is often biased towards sand- and clay-rich soils as these are seen to represent extremes of leaching potential and preferential flow, respectively, and hence few data are available for less extreme soil types. Existing data sets often contain some, but not all, of the parameters and results required for model evaluation.

A number of methods have been proposed for evaluating simulation results against measured data. The simplest and most commonly used assessment is to plot simulated and measured values on the same graph against a parameter such as time or soil depth, and to use qualitative analysis to compare the two sets of values. Such analysis is subjective, but provides a simple insight into the performance of the model and allows visual comparisons between simulations from a number of models. A similar graphical assessment can be made by plotting simulated data against measured data as paired values. The assessment can be made objective by carrying out a linear regression according to the equation: $y = a + bx$. The goodness-of-fit can be evaluated using the product-moment correlation coefficient (r), but is more comprehensively assessed by testing the null hypothesis that $a = 0$ and $b = 1$ using separate t-tests or simultaneously by F-test (Rose, 1991).

The difference between simulated and measured values is a simple test for goodness-of-fit of a model (Addiscott and Whitmore, 1987) and is best expressed as a percentage of the measured value for pesticide fate models where measured concentrations can range through several orders of magnitude over time. The calculated difference can be used either as a single mean or as a distribution to reject or approve the use of a model in a given situation. Mueller *et al.* (1992) has suggested that simulated results for pesticide concentrations should fall within 50–200% of the measured data in order for the model to be acceptable, but any range can be used provided it is specified beforehand. The percentage difference between simulated and measured data can be readily compared to potential errors in the measured data which can be relatively large for pesticide residues. For example, the spatial variability of pesticide concentrations and a number of key processes in field soils is reviewed by Rao and Wagenet (1985). Taylor *et al.* (1971) reported a 50-fold variation in concentrations of dieldrin in soil because of the application method, but this is an extreme example; Walker and Brown (1985) found coefficients of variation of 0.25 or less in half-lives for simazine and metribuzin across a small field plot. Errors are also associated with the analysis of pesticide residues and Hance (1990) reviewed data from circulation of water samples between laboratories which show errors of $\pm 100\%$ at concentrations of $10\text{--}50\ \mu g\,l^{-1}$.

The difference between simulated and measured data also gives the residual for statistical analysis of an experiment and this can be used as the basis for more complex, objective approaches to assessing goodness-of-fit of a model. Greenwood *et al.* (1985) compared the sum of squares of the residuals with the sum of squares of the data about the mean in order to express as a percentage the proportion of variation in the data which is accounted for by the model. This method is not efficient when dealing with replicated data, but is very useful for evaluating simulations of field experiments where bulking of samples and cost implications frequently preclude replication of data. Where a limited number of replicate data are available (e.g. in lysimeter experiments), Whitmore (1991) suggests a method for partitioning the sum of squares of the residuals into components for the differences between the simulation and the mean of replicate measurements (the 'lack of fit') and for the variance within each set of replicate measurements (the 'pure error'). If the lack of fit component is not significantly larger than the pure error component, then there are no grounds for rejecting the model from these data. Whitmore's method is a useful one as it can indicate bias of the simulation, highlight outlying data which may critically affect the assessment process, and take account of effects of experimental error in initial measurements on the simulation of subsequent measurements. Pennell *et al.* (1990) propose the use of a normalized root mean square error function to quantify the differences between predicted and measured data, but emphasize that the use of both qualitative and quantitative analysis will give the most complete assessment of model performance.

There is clearly a great deal of work to be done in order to evaluate fully the range of models for pesticide fate in soil which is currently available and it is essential that the models should be tested under a range of conditions. A useful utility has been incorporated into the MACRO model (Jarvis, 1991) to evaluate automatically the goodness-of-fit of the model simulation either to unreplicated measured data by simple linear regression or to replicated data according to the method of Whitmore (1991). Another useful development is the incorporation of the simple capacity flow routine of Addiscott (1977) into Version 3.0 of LEACH (Hutson and Wagenet, 1992). This allows direct comparison of the effect on simulation results of changing from a physically based solution of the Richards' equation to an empirical flow routine whilst keeping all other input parameters the same.

Modelling preferential flow

One of the processes which is not currently well described by mathematical modelling is that of preferential flow. This process has been shown to occur in a wide range of soils via water movement through cracks, fissures, and biopores in structured soils (Beven and Germann, 1982) and via fingering flow and funnel flow in coarser-textured soils (White *et al.*, 1977; Kung, 1990). Preferential flow can greatly reduce interaction between the surface of soil particles and soil water and associated solutes. It is thus a particularly important process when considering pesticide mobility in soils because the extent of interaction with the soil is a more important factor for pesticides than for more mobile solutes such as nitrate. Thus whilst most models will predict total water loss from the soil profile fairly accurately as the difference between precipitation and evapotranspiration, pesticide movement may be greatly underestimated under certain conditions due to a failure to describe preferential flow mechanisms (e.g. Bergström and Jarvis, 1994).

Few models incorporate descriptions of preferential flow, but two models illustrate the current state of modelling this process. PLM (Hall, 1994) incorporates a development of the simple capacity flow routine developed by Addiscott (1977) to describe solute movement in structured soils. Soil water is partitioned into mobile and immobile phases with only water held at matric potentials less than $-5\,kPa$ displaced when water flows and with solute allowed to diffuse between the two phases. In order to describe preferential flow through macropores or fissures, the mobile phase is further divided empirically into 'fast' and 'slow' pore spaces with water only moving through the 'fast' pores when the 'slow' and immobile pore spaces are filled by particularly intense or prolonged rain. This means that calibration of the model is required before reasonable simulations can be made. MACRO (Jarvis, 1994) is a mechanistic model for non-steady state transport of water and solute in macroporous soil in which flow is separated into domains for microporous and macroporous flow with approximate physically-based interaction terms between the two. The model can be run in one or both domains depending upon the scale of soil structure and upon transient boundary conditions such as rainfall intensity and the initial moisture content of the soil. Evaluation of this model against breakthrough curves for chloride showed that diffusive exchange of solute between the two flow domains was the main factor causing variability in leaching (Jarvis *et al.*, 1991). The importance of this exchange could be expected to be even greater for pesticides as they are more strongly sorbed to soil. There is an urgent need for more experimental information on preferential flow processes and at present the prediction of Beven and Germann (1982) that difficulties in obtaining experimental data rather than theoretical development will be the greater obstacle in this field still holds true. However, modelling as an iterative process provides a framework around which experimental research can be structured and results interpreted. An important contribution is the identification of areas in which understanding is weak such as the effect of pore dimensions on preferential flow and the importance to solute movement of diffusive processes between regions with different flow patterns.

EXTRAPOLATION OF MODEL RESULTS

Use of a mathematical model produces a pattern of predicted concentrations of a pesticide in soil and water at a given point which relates directly to the chosen input parameters. In the case of modelling a laboratory column experiment or a lysimeter, this output may be sufficient, but extrapolation of the data is required in order to render the results applicable to the wider environment. It is impractical to perform repeat simulations with all the possible scenarios of interest and a number of alternative methods for extrapolation are currently in use:

Worst-case scenario

At present, it is common to select 'realistic worst-case' input parameters for the model and to avoid direct extrapolation by assuming that predicted concentrations are likely to be the largest expected under normal conditions (e.g. BBA, 1993). The risk to health posed by pesticide contamination of water sources is still poorly quantified (Morgan, 1992) and such an approach to extrapolation incorporates a margin of safety. However, it is important that the input parameters selected for modelling be closely examined to verify a genuine worst-case scenario. Also, it is difficult to define just what is meant by 'worst case' since it is possible to simulate contamination of water by most pesticides if the selected parameters for modelling are extreme enough.

Benchmark compounds

A relatively simple method of extrapolation of model results involves comparing simulated concentrations of a pesticide in a number of scenarios against predicted concentrations of one or more 'benchmark compounds'. Benchmark compounds are generally selected because they are widely used and monitored and have similar properties and/or usage patterns to the pesticide under investigation. The known impact of the benchmark on the wider environment can be used to extrapolate the expected impact of the pesticide of interest.

Stochastic modelling

The use of stochastic approaches to modelling allows extrapolation of modelling results beyond a single spatial point either to a localized scale such as a whole field or to encompass a range of locations which can be grouped together because of their similar properties. Stochastic modelling arises from a recognition of the problems caused by spatial and temporal variation in parameters to be used for modelling and results in a statistical description of the uncertainty associated with the output. Addiscott and Wagenet (1985) divide stochastic models into those which focus on variability in processes and take no account of mechanism and those which make allowance for variability in existing mechanistic models. An example of the former is the convective transfer function model of Jury (1982) which describes the probability of a solute applied at the surface arriving at a given depth following a given amount of applied water. Knighton and Wagenet (1987) used a Markov process to obtain a statistical distribution of solute concentration within discrete layers of the soil profile. A Monte Carlo processor is incorporated into PRZM-2 which links the root zone model PRZM with the vadose zone model VADOFT (Mullins *et al.*, 1993). The model can be run in Monte Carlo mode with a specified number of runs and varying parameters and will then produce a statistical analysis of the output distribution and the uncertainty associated with it.

Hutson (1992) warns against the use of such stochastic simulations where data bases are inadequate to merit them because the complexity of the treatment may mask weaknesses in knowledge of input parameters. The need for large data sets is thus a major limitation to the use of stochastic approaches to modelling (IUPAC, 1987). In an attempt to satisfy this need, a database system for the USA has been developed and named DBAPE (Database Analysis and Parameter Estimation system, Carsel *et al.*, 1991). DBAPE holds attribute data relating to sand, clay and organic matter content, bulk density, 'available' moisture, soil series hydrological group (Soil Conservation Service, 1972) and crop production potential for the 8880 soil series in the US characterized as being agricultural soils. In addition, spatial data detailing the area occupied by each soil series on a state and county basis is included. Using the DBAPE system it is possible to search the databases according to user-specified attribute data, retrieve all the series that qualify for the specified attributes and then use a parameter estimator module to estimate the moisture retention and moisture content/conductivity curves for the series identified. A similar system, known as SEISMIC (Spatial Environmental Information System for Modelling the Impact of Chemicals), has been developed for England and Wales (Hollis *et al.*, 1993). It holds spatial data, at 5 × 5 km resolution, on the distribution of soil series, crop type and climate characteristics within the two countries, as well as soil parameter data on the particle-size fractions, organic carbon contents, pH, bulk density, water release and hydraulic conductivity characteristics of all the 417 soil series recognized on the 1:250 000 scale regional soil maps of the two countries. For each soil series, parameter data sets are held for each of four land uses: arable, ley (short-term) grassland, permanent grassland, and 'other'. Using such integrated database systems, a comprehensive range of soil parameter data can be accessed and placed within its regional or national context for stochastic modelling purposes.

LINKING PESTICIDE MOBILITY IN SOIL TO ENVIRONMENTAL FATE

Currently, a great deal of interest is focused upon the potential for using models in a regulatory context. It should be remembered, however, that models of pesticide fate within the soil simulate only one aspect of the broader environmental fate of a compound. Whether or not use of that compound will lead to actual environmental contamination depends on complex factors characterizing usage, climate, soil and aquifer. These factors have been reviewed by many authors (e.g. Wagenet and Rao, 1985; Hance, 1987; Nicholls, 1988) and can be grouped into five categories which are outlined below and in Table 5:

(1) the extent of usage, surface impact and physicochemical properties of the pesticide;
(2) climate and weather parameters determining soil temperatures and leachate volumes;
(3) soil parameters determining sorption, degradation and leachate concentrations of the pesticide;
(4) unsaturated or vadose zone parameters determining sorption, degradation and leachate concentrations of the pesticide;
(5) aquifer characteristics determining degradation and dilution of the pesticide in the saturated zone.

Table 5 Main factors affecting the environmental fate of pesticides

Category	Factor
Pesticide	Target crop or land use
	Application date
	Application rate
	Half-life in soil
	Half-life in water
	K_{oc}
	Volatility
Climate/weather	Temperature
	Rainfall
	Evapotranspiration
Soil	Run-off characteristics
	Organic carbon content
	Bulk density
	pH
	Hydraulic characteristics
Vadose zone	Minimum depth to saturated zone
	Hydraulic characteristics
Saturated zone	Flow characteristics
	Mixing characteristics

Most of the factors listed in Table 5 vary spatially and/or temporally across the landscape and this will have an effect upon the environmental fate of the compound. Any realistic assessment of the risk to the environment posed by pesticide mobility thus needs to integrate seasonally dynamic factors relating to pesticide usage, land and crop management and weather with intrinsic but spatially variable factors relating to soil and hydrogeological characteristics. Such an integrated approach is best achieved within the framework of a geographical information system (GIS) where each compartment of the environment is simulated as a set of data layers comprising both spatial and parameter information which can then be linked and interacted by the use of models. As discussed by Hollis and Brown (1993), two main problems need to be addressed when using such an approach. First, what type of model to use in a particular situation and second, how to cope with spatial and temporal variation of environmental parameters.

Using models to link environmental compartments

Theoretically, any type of model can be used to generate an integrated environmental risk assessment, provided that it incorporates some sort of data from each of the five broad categories listed in Table 5. However, no single model which is currently available incorporates all the factors set out in Table 5. In fact, most of the models developed for predicting pesticide fate concentrate on a given compartment such as soil (e.g. LEACHP) or the vadose zone (e.g. VADOFT) and it should thus be possible to link such models within a GIS. Unfortunately, it is difficult to link complex mechanistic models that simulate processes in each environmental compartment as the processes and dimensions of flow in each compartment are different and outputs and inputs must be carefully harmonized. The RUSTIC system (Dean *et al.*, 1989) attempted to link PRZM with VADOFT and a saturated zone model called SAFTMOD, but there were problems with its use and the latest version, PRZM-2 simply links an updated version of PRZM with VADOFT (Mullins *et al.*, 1993). A more realistic option may be to combine models of differing complexity according to the perceived needs of a specific situation. For example, the most significant attenuation of a pesticide used in agriculture will occur through interaction with the crop and the root zone in the soil with much reduced potential for degradation and sorption of the pesticide once it moves out of the root zone. In such a case, it is probably justified to use a relatively complex, mechanistic model to predict leachate concentrations moving out of the soil zone and then adjust these concentrations using a much simpler model based on attenuation potential in each of the vadose and saturated zones. By contrast, attenuation in the soil zone may be lost or considerably reduced for a pesticide applied along a road or railway cutting or disposed of in a soakaway. Here, the use of more complex models for the vadose and saturated zones may be the best approach.

INCORPORATING SPATIAL AND TEMPORAL VARIABILITY

The spatial and temporal variability of important environmental parameters poses a further problem when attempting a realistic integrated approach to environmental risk assessment. However, within a GIS, each of the broad categories listed in Table 5 can be treated as an environmental data layer. For each layer, it is then possible to define a limited number of groupings within which, for modelling purposes, the characteristics are relatively homogeneous. This process of simplification and grouping together is particularly suited to the use of simpler models based on a small number of parameters. For example, Jury *et al.* (1987) describe a screening model to predict the likelihood of ground water contamination by pesticides which can be reduced to a linear inequality between K_{oc} and $t_{1/2}$ and depends only upon soil properties (bulk density, organic carbon fraction and volumetric water content), microbial population distribution, and leaching rate. Where more complex models with larger numbers of parameters are to be used, care must be taken in grouping categories into a number of classes. For each environmental compartment, classes must be as homogeneous as possible with respect to all the parameters on which the model depends. Different models may use different parameters for simulating environmental processes, so that it is possible that a set of homogeneous classes created for use with one model will not be homogeneous with respect to the use of an alternative model. Nevertheless, there are some basic principles which can be applied to the creation of homogeneous groupings for integrated environmental risk assessment.

Pesticide characteristics

Pesticides are frequently grouped according to their physicochemical properties such as K_{oc} and half-life, but some account of variation in usage must also be taken. Such variation can be simulated both from data relating to the spatial distribution of its target land use (e.g. a specific crop or a railway line, road or soakaway), and from land use-specific data relating to average timing and rate of application rate along with a factor describing interception by vegetation. The spatial distribution of different land uses can be derived directly from remotely sensed images or from a combination of more generalized cropping statistics giving the proportion of each crop present within specified areas together with maps showing settlements and communications networks.

Weather

Temporal variation in weather can be characterized from data sets comprising monthly, daily or even hourly values for each necessary parameter, depending on the requirements of the model. If measured data are not available then a number of techniques have now been developed for simulating representative data sets using statistical 'weather generator' models (Hutchinson, 1986, 1991; Richardson and Wright, 1984). Temporal data sets do not take into account spatial variation in weather characteristics. However, because temperature and excess moisture are the principal *climatic* factors which determine pesticide leaching potential, spatial variation in weather can be simulated by assigning representative weather data sets to different 'climatic zones' delineated according to the spatial variation of these parameters. For example, spatial data sets comprising long-term mean values of accumulated temperature and either excess winter rain (the volume of rainfall during significant periods of zero soil moisture deficit) or annual effective rainfall can be generated using regression algorithms which relate each parameter to altitude, latitude and longitude. Each algorithm is derived from statistical analysis of measured data from a number of sites within the area of interest (Jones and Thomasson, 1985). These spatial data sets can then be used to identify different climatic zones and an example based on accumulated temperature within the EC is shown in Figure 10.

An alternative approach to characterizing spatial variation in weather at a much broader scale is to use the classification of soil moisture and temperature regimes incorporated into the USDA soil taxonomy system (Soil Survey Staff, 1975). Different soil temperature regimes are allocated on the basis of ranges in long-term average annual, summer and winter soil temperatures, whereas soil moisture regimes are defined using the long-term average annual duration of dry, moist or wet soil conditions within specified depths. The system is used to characterize soils throughout the world and any areas with the same soil temperature and moisture regimes are considered to have similar pedoclimates. To compare this system with the more specific climatic characterization using accumulated temperature, the distribution of USDA-equivalent soil temperature regimes within the EC is shown in Figure 11.

Soil

Most soil maps show the distribution of soil types (classes), with each type having a defined range of specified characteristics. The differentiation of such soil classes can be made at various levels of detail ranging from a few very broadly defined groupings to a large number of vary narrowly and precisely defined classes, often called soil series. At the most detailed level

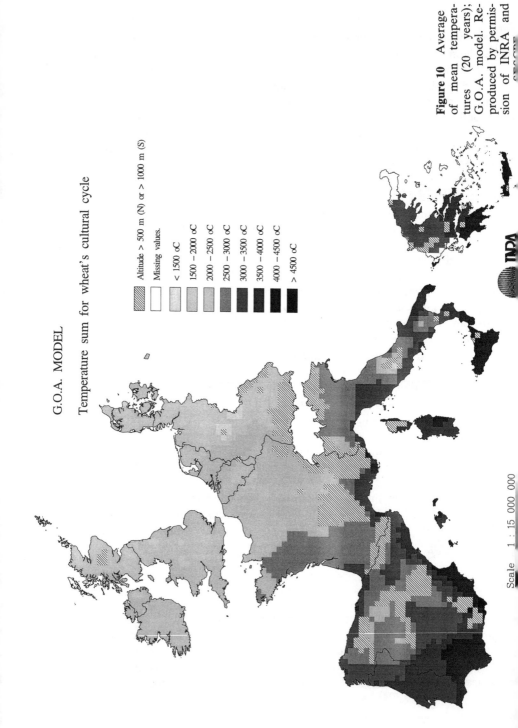

G.O.A. MODEL

Temperature sum for wheat's cultural cycle

Altitude > 500 m (N) or > 1000 m (S)

Missing values.

< 1500 oC

1500 – 2000 oC

2000 – 2500 oC

2500 – 3000 oC

3000 – 3500 oC

3500 – 4000 oC

4000 – 4500 oC

> 4500 oC

Figure 10 Average of mean temperatures (20 years); G.O.A. model. Reproduced by permission of INRA and

Scale   1 : 15 000 000

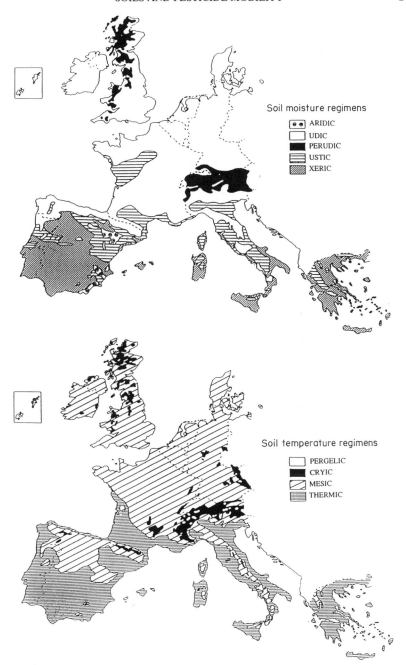

Figure 11 Distribution of USDA-equivalent soil temperature and moisture regimens in Europe

(scales of 1:50 000 and less), soil maps usually show the distribution of discrete areas of individual soil types with very similar textural, structural and morphological profiles. In such cases, each map unit characterizes areas where the soil chemical exchange and hydraulic properties can be treated as homogeneous for modelling purposes, especially if they are under a similar land use. At less detailed map scales, however, individual map units often comprise broader soil groupings in which characteristics such as texture, organic carbon content and soil depth may vary significantly. In such cases, it is necessary to estimate the proportion of each more detailed soil type within a given map unit, with each being characterized in terms of an average value for its basic properties such as particle size distribution and organic carbon content. Sometimes, other important data such as bulk density and hydraulic properties may be available (e.g. Ragg et al., 1984), but often this is not the case and a number of techniques have therefore been developed to derive such parameters from the basic data (Van Genuchten et al., 1992). Management models such as GLEAMS and PRZM-2 also require detailed data on soil run-off curve characteristics and soil erosion factors in relation to soil types and management practices. Although the accompanying manuals give detailed guidance on how to estimate these characteristics, they are based on conditions in the USA and considerable care is needed when extrapolating them to other areas.

Where the environmental vulnerability of large areas needs to be assessed, then many detailed soil types are likely to be present. In these cases, it is often possible to simplify the data required by focusing solely on those parameters required by the model to be used. As discussed by Bouma (1989a, b), differences between soil types distinguished on traditional pedological soil maps often do not reflect differences in a specific soil function, and it may be valid to combine either soil types or soil horizons into a small number of groupings within which the parameters required by a particular model do not vary significantly. Thus, on the 1:250 000 scale soil maps of England, Wales and Scotland, 974 main soil types are recognized, but these can be grouped into a small number of discrete hydrological classes, based on conceptual groupings that describe the dominant pathways of water movement through the soil and, where appropriate, substrate (Figure 12). Within each of the groupings, subdivisions may be made according to flow rates, flow mechanisms and/or water storage (Boorman and Hollis, 1990; Boorman et al., 1991). The result is 29 classes which have each been calibrated against a large database of streamflow characteristics and shown to have a characteristic hydrological response to rainfall. Similarly, in the USA, over 14 000 soil series have been recognized, but for the purposes of differentiating their infiltration and run-off potential they have been grouped into four hydrological soil groups, A, B, C, and D (Soil Conservation Service, 1972). It is also possible to group soils into a limited

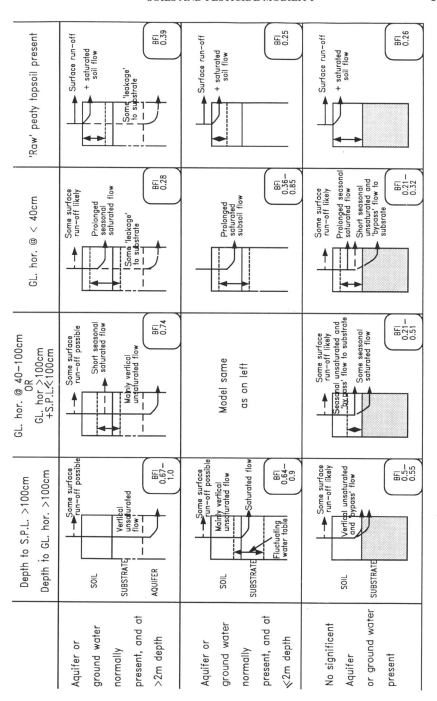

Figure 12 The HOST framework

number of classes based on the potential to sorb pesticide as determined by their organic matter content, clay content and clay mineralogy. Such a system has been incorporated into the classification of soil cover used as one of the factors for assessing the overall vulnerability of aquifers proposed by the National Rivers Authority in England and Wales as part of its policy and practice for the protection of ground water (NRA, 1992).

Aquifer

The spatial variation of aquifer characteristics in both the vadose and saturated zones is difficult to quantify. Civita (1993) presents a comprehensive review of ground water vulnerability maps which identifies two broad methodologies: the identification of homogeneous zones based on an assessment of hydrogeological complex and setting; and assessments based on 'point scoring' or parametric systems. Using the former methodology, the Institute of Geological Sciences has published small-scale hydrogeological maps for Great Britain based on the broad flow characteristics and production potential of geological formations (IGS, 1977). In France, a series of more detailed hydrogeological maps have been produced based on a combination of saturated and unsaturated flow characteristics and average depth to ground water (BRGM, 1975–1979). Point scoring or parametric systems assign a range of numerical values to the potential range of each significant hydrogeological factor and then combine them mathematically to derive an overall index value for characterizing the aquifer. Examples are the DRASTIC system (Aller *et al.*, 1987), developed for the United States EPA as a nationally consistent method of evaluation, and the GOD system (Foster, 1987) based on Groundwater occurrence, Overall aquifer class and Depth to ground water. Such systems are useful for assessing and comparing the combination of characteristics at different points, but may be difficult to extrapolate spatially. In order to overcome some of these problems, a computerized development of the DRASTIC system approach, provisionally named SINTACS, is being produced for aquifer vulnerability mapping at medium to small scales in Italy (Civita, 1993).

Integrated assessments of environmental fate

A number of attempts have been made to combine spatial extrapolation through simplification and grouping with an integrated approach to modelling to produce spatial vulnerability assessments (e.g. Khan and Liang, 1989; Pestemer *et al.*, 1993). Hollis (1991) has described a fully integrated system which uses simple models to combine geo-referenced data sets of climate, target land use, soil and vadose zone characteristics within a GIS framework and this system is described below for illustration. Currently, models used in integrated assessments tend to be relatively simple, but where data sets of sufficient quality exist, the use of more complex mechanistic models is no longer precluded by considerations such as computing requirements.

VULNERABILITY ASSESSMENTS FOR PESTICIDE MOVEMENT TO WATERS

An integrated system using simple attenuation factor models to produce maps of surface water and aquifer vulnerability for England and Wales which are specific to a pesticide and a target land use has been described in detail by Hollis (1991) and by Hollis and Brown (1993). The system is not a vehicle for predicting absolute concentrations of pesticides reaching water, but rather a scientific methodology for estimating the relative likelihood of contamination of water sources by a pesticide in a wide range of scenarios. A summary of the vulnerability assessment process and data handling is given in Figure 13.

The system is initiated by specifying a pesticide of interest and its target land use. The areas to which this combination apply are calculated using either land suitability models (Thomasson and Jones, 1991) or by using detailed cropping statistics for England and Wales resolved to a 5 km grid basis. Vulnerability assessments are then based upon integrating the HOST classification for soil and the substrate hydrology described above with a broad classification of soil sorption potential, to produce soil vulnerability classes that are homogeneous with respect to the factors used by the model. Separate classes are needed for aquifers and surface waters because the mechanisms and rates of water movement to each type of water source are different. Factors taken into account include depth to an aquifer or a seasonally saturated layer, the presence or absence of bypass flow for aquifers and run-off and stream base flow indices for surface waters. Nine and five soil/substrate vulnerability classes have been defined for aquifers and surface waters, respectively, and the proportions of each within 5 km^2 blocs has been calculated for England and Wales.

Duration of the field capacity period and the amount of excess winter rain are used as generalized climatological parameters and these have been calculated as annual averages at 5 km grid intersects within England and Wales (Jones and Thomasson, 1985). For ground water vulnerability assessments, these two indices are used to calculate daily soil water fluxes which are then used to calculate travel times to aquifers on a 5 km grid basis. For surface water vulnerability assessments, average daily soil water fluxes

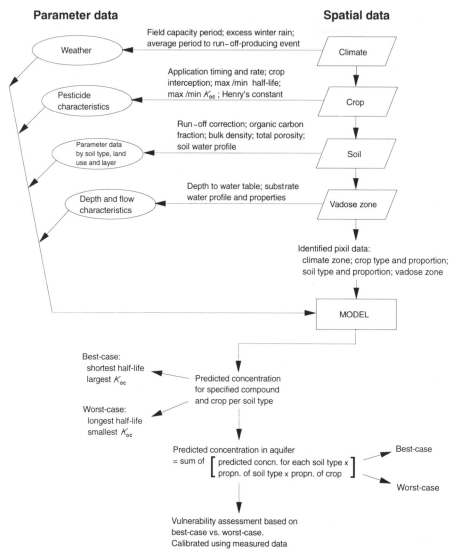

Parameter data   **Spatial data**

Figure 13   A summary of the vulnerability assessment process and data handling

are less important than individual rainfall events because soil vulnerability classes are based on parameters that predict the proportion of rain from an average event that reaches a water course relatively rapidly. The critical factor for surface water vulnerability assessments is thus the length of time between application of a pesticide and an event large enough to produce run-off. This is borne out by field experiments where the first run-off-producing event after application is observed to be disproportionately important in determining total losses of a pesticide in run-off (Wauchope, 1978). Different soil classes will require different amounts of rainfall to trigger run-off and a series of standard events have been defined ranging from 5 mm for those soils most prone to run-off to 20 mm for soils where run-off is relatively uncommon. England and Wales have been divided up into five classes of field capacity period and real meteorological data have been used to calculate the average return period for events likely to trigger run-off for each combination of soil and field capacity period classes.

Thus the vulnerability assessment system draws on spatially geo-referenced data sets for climate, land use, soil and aquifers, detailing the location of homogeneous units and the proportion of each homogeneous unit within the area of interest on a 5 km grid. Each unit is characterized by a number of representative parameters which are passed across to a simple mathematical model and used to predict the likelihood of a given pesticide contaminating water sources at concentrations $>0.1\,\mu\text{g}\,\text{l}^{-1}$ with a given combination of soil/climate parameters. The equations used to model aquifer vulnerability are based upon work by Rao *et al.* (1985) and Leonard and Knisel (1988). Daily soil water fluxes are used to calculate travel times for the pesticide through each soil horizon, which will also be a function of factors such as K_{oc} of the pesticide, organic matter content, bulk density and depth of each soil horizon. An overall travel time for pesticide to reach ground water can then be calculated and an attenuation factor which accounts for degradation of the pesticide during travel is used to predict the concentration of pesticide in leachate reaching the ground water. For surface water vulnerability assessments, the depth (d mm) to which pesticide has penetrated between application and the time of the first event producing run-off is calculated using a similar method to that described above. An attenuation factor is used to calculate the amount of pesticide remaining in the soil at the time of the event and this pesticide is evenly spread throughout the top d mm of soil. The rainfall which produces run-off infiltrates the soil and mixes with the mobile water fraction in the top 1 mm of soil, thus diluting it. The mobile water is defined as that fraction of the soil water which is readily available for run-off and is calculated by the difference between soil water content at 0.05 and 2 bar pressure. A fraction of the mobile water in the top 1 mm of soil is then assumed either to run off directly to surface waters or to pass rapidly through the drain system to surface waters. This fraction, which represents the amount of the total rainfall which moves rapidly to surface waters, is determined by the standard percentage run-off value for the soil vulnerability class, and the final concentration of pesticide in this run-off water is calculated using a simple dilution factor.

Vulnerability assessments are completed by using the above models to predict concentration of pesticide entering water sources using 'best-' and 'worst-case' pesticide properties based on reported ranges in measured K_{oc} and soil half-life. Where the models predict that concentrations of pesticide entering waters will be $<0.1\ \mu\mathrm{g\,l^{-1}}$ using worst-case parameters, the vulnerability assessment is *low*. Where models predict that using best-case parameters, concentrations will be $>0.1\ \mu\mathrm{g\,l^{-1}}$, the vulnerability assessment is *high*. All other results will be between these two assessments. Specific meanings of each vulnerability assessment are defined as follows:

- *Low*: the pesticide is not likely to contaminate water sources unless it is misapplied.
- *Low/Mod*: contamination of water sources is unlikely but could occur under conditions where the pesticide is most mobile and/or persistent.
- *Moderate*: occasional contamination of water sources is possible, particularly under conditions where the pesticide is more mobile and/or persistent.
- *Mod/High*: occasional contamination of water sources is likely to occur except under conditions where the pesticide is least mobile and/or persistent.
- *High*: frequent contamination of water sources is probable.

Illustrative results of the vulnerability assessment exercise are given in Tables 6 and 7 for a hypothetical pesticide which is applied to winter cereals in autumn and is non-volatile, persistent, and relatively immobile. The properties of the example pesticide were selected as follows:

Application rate:	$0.5\ \mathrm{kg\,ha^{-1}}$	Vapour pressure:	$1.0\times10^{-7}\ \mathrm{mbar}$
Best-case half-life:	60 d	Worst-case half-life:	200 d
Best-case K_{oc}:	$2000\ \mathrm{ml\,g^{-1}}$	Worst-case K_{oc}:	$500\ \mathrm{ml\,g^{-1}}$

Table 6 Aquifer vulnerability assessment for test pesticide 1

Soil vulnerability class	Excess winter rain (mm)					
	<175	175–200	201–300	301–400	401–750	>750
a1	Low	Low	Low	Low/Mod	Moderate	Moderate
a2	Low	Low	Low	Low	Low	Low/Mod
a3	Low	Low	Low	Low	Low/Mod	Low/Mod
a4	Low	Low	Low	Low	Low/Mod	Moderate
a5	Low	Low	Low	Low	Low	Low/Mod
a6	Low	Low	Low	Low	Low	Low
a7	Low	Low	Low	Low	Low	Low
a8	Low	Low	Low/Mod	Moderate	Moderate	Moderate
a9	Low	Low	Low	Low	Low	Low

Table 7 Surface water vulnerability assessment for test pesticide 1

Soil vulnerability class	Field capacity period (d)			
	<125	125–175	176–225	>225
s1	Moderate	Moderate	Moderate	Moderate
s2	Moderate	Moderate/High	Moderate/High	Moderate/High
s3	Low/Moderate	Low/Moderate	Moderate/High	Moderate/High
s4	Low/Moderate	Low/Moderate	Low/Moderate	Moderate
s5	Low	Low/Moderate	Low/Moderate	Moderate

Tables 6 and 7 give vulnerability assessments for test pesticide 1 with all possible combinations of soil vulnerability and climate classes. It is thus possible to overlay the data sets using a GIS to produce maps of surface and ground water vulnerability for England and Wales (see Figures 14 and 15).

CONCLUSIONS

Concerns over pesticide residues in water sources have focused attention on the specific soil properties which affect the mobility of pesticides in soils. Mobility is closely linked to the physical and hydrological properties of a soil. Different soil types exhibit different characteristics and thus pesticide mobility will vary both in time and in a spatial sense. The potential pathways which water and soluble pesticide may follow in soil can determine the ultimate fate and degree of water source contamination which occurs.

The likelihood of a pesticide leaching out of the soil zone and the extent of any subsequent environmental contamination is determined by a range of environmental factors incorporating usage patterns, climate, soil and hydro-geological conditions. Techniques for quantifying the risk associated with pesticide mobility in soil range from simple assessments based upon physicochemical properties of the pesticide to complex mechanistic models which are being linked to sizeable environmental databases. The use of a given technique will depend upon the requirements of the assessment and upon the quality and quantity of data available both for environmental parameters and from monitoring studies.

Treatment of water supplies to remove pesticides is extremely costly to the customer and protection of water resources from contamination is a logical, long-term option. In order for such protection to be enforced, the risk of contamination by a given pesticide must be assessed, taking into account local conditions. Similarly, regulatory decisions should be based upon the recognition that particular areas or usage patterns may correspond

Application rate = 0.5 kg ha^{-1}; Half life = 60–200 d; K_{oc} = 500–2000 ml g^{-1}

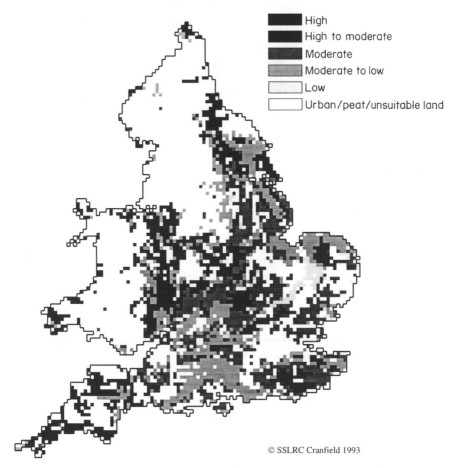

© SSLRC Cranfield 1993

Figure 14   Surface water vulnerability assessment for the use of test pesticide 1 on land well or moderately suitable for winter cereals

to increased risk of contamination of water sources and should thus be treated on an individual basis. This recognition of different levels of risk is reflected in, for example, the EC directive to harmonize registration of pesticides which will see active ingredients accepted by one member country accepted throughout the EC, but requires approval of actual products by individual countries based upon knowledge of local conditions. The current availability of powerful desktop computing facilities makes the use of

Application rate $= 0.5$ kg ha^{-1}; Half-life $= 60-200$ d; $K_{oc} = 500-2000$ ml g^{-1}

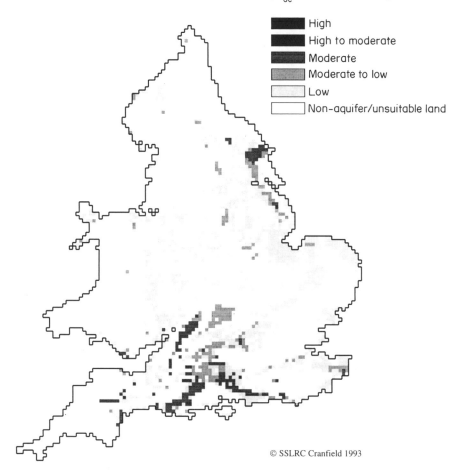

© SSLRC Cranfield 1993

Figure 15 Aquifer vulnerability assessment for the use of test pesticide 1 on land well or moderately suitable for winter cereals

integrated, GIS-based methods of environmental risk assessment a realistic option provided that suitably comprehensive environmental databases are also available. It is in the interests of industry, regulators, water suppliers and the general public alike that tools for accurately assessing the risk to waters, posed by the use of a given pesticide in a given situation, be developed and used. This will allow decisions to be taken to minimize the risk to the environment from pesticide usage whilst maximizing the safe benefit to be gained from pesticides in scenarios where the level of risk is held to be acceptable.

REFERENCES

Addiscott, T. M. (1977). 'A simple computer model for leaching in structured soils', *J. Soil Sci.*, **28**, 554–563.

Addiscott, T. M., and Wagenet, R. J. (1985). 'Concepts of solute leaching in soils: a review of modelling approaches', *J. Soil Sci.*, **36**, 411–424.

Addiscott, T. M., and Whitmore, A. P. (1991). 'Simulation of solute leaching in soils of differing permeabilities', *Soil Use and Management*, **2**, 94–102.

Aller, L., Bennet, T., Lehr, J. H., Petty, R. J., and Hackett, G. (1987). *DRASTIC: A Standardised System for Evaluating Groundwater Contamination Potential using Hydrogeologic Settings*, NWWA/EPA report, US Govt. Print Office, Washington, DC.

Battaglia, G. (1991). 'Experiences and plans for protection zones and buffer zones in Italy', in *Pesticides in Surface Waters. Mechanisms of Transport and Control. Proceedings of a Workshop, Hamburg, August 1991* (eds A. J. Dobbs, W. J. G. M. Peijnenburg, T. R. Roberts and M. Waldman), pp. 95–99, WRc, Marlow, Bucks.

Baver, L. D. (1956). *Soil Physics*, John Wiley and Sons, New York.

BBA (1993). *Criteria for Assessment of Plant Protection Products in the Registration Procedure*, Abteilung für Pflanzenschutzmittel und Anwendungstechnik, Biologische Bundenstalt für Land- und Fortstwirtschaft, Heft 285, Berlin.

Bergström, L. F., and Jarvis, N. J. (1994). 'Evaluation and comparison of pesticide leaching models for registration purposes', *J. Environ. Sci. Health A* **29**, 1061–1072.

Beven, K., and Germann, P. (1982). 'Macropores and water flow in soils', *Water Resource Res.*, **18**, 1311–1325.

Boorman, D. B., and Hollis, J. M. (1990). 'Hydrology of soil types: a hydrologically-based classification of the soils in England and Wales', in *Proceedings of the MAFF Conference of River and Coastal Engineers, Loughborough University, 10–12 July 1990*.

Boorman, D. B., Hollis, J. M., and Lilly, A. (1991). 'The production of the hydrology of soil types (HOST) data set', in *Proceedings of the British Hydrological Society Third National Hydrology Symposium, Southampton University, 16–18 September 1991*.

Bouma, J. (1989a). 'Using soil survey data for quantitative land evaluation', *Adv. Soil Sci.*, **9**, 177–213, Springer Verlag, New York, NY.

Bouma, J. (1989b). 'Land qualities in space and time', in *Land Qualities in Space and Time. Proceedings of a Symposium, Wageningen, August 1988* (eds J. Bouma and A. K. Bregt), pp. 3–13, Pudoc, Wageningen.

Breach, B., and Porter, M. (1993). 'A water supplier's view—an integrated strategy for dealing with pesticide pollution of drinking water catchments', *Pestic. News*, **22**, 6–9.

BRGM (1975–1979). *Cartes de Vulnerabilite à la Contamination des Eaux Souterraines* (FF. 1:250 000)', BRGM, Orleans.

Brown, C. D., Hodgkinson, R. A., Rose, D. A., Syers, J. K., and Wilcockson, S. J. (1995). 'Movement of pesticides to surface waters from a heavy clay soil', *Pestic. Sci.*, **43**, 131–140.

Carsel, R. F., Imhoff, J. C., Kittle, J. L., Jr, and Hummel, P. R. (1991). 'Development of a database and model parameter analysis system for agricultural soils (DBAPE)', *J. Environ. Qual.*, **20**, 642–647.

CEC (1985). *Soil Map of the European Communities, 1:1,000,000*, Directorate-

General for Agriculture Coordination of Agricultural Research, Office for Official Publications of the European Communities, Luxembourg.

Cheng, H. H. (1990). 'Pesticides in the soil environment—an overview', in *Pesticides in the Soil Environment: Processes, Impacts and Modeling* (ed. H. H. Cheng), pp. 1–5, Soil Sci. Soc. Am., Madison, Wisconsin, USA.

Civita, M. (1993). 'Ground water vulnerability maps: a review', in *Mobility and Degradation of Xenobiotics. Proceedings of the IX symposium on pesticide chemistry, Piacenza, October 1993* pp 587–631 (The Proceedings were published by the Universita Cattolica del Sacro Cuore, Piacenza, Italy.

Clark, L., Gomme, J., and Hennings, S. (1991). 'Study of pesticides in waters from a chalk catchment, Cambridgeshire', *Pestic. Sci.*, **31**, 15–33.

Davis, R. P., Garthwaite, D. G., and Thomas, M. R. (1991). *Pesticide Usage Survey Report 85. Arable Farm Crops in England and Wales 1990*, MAFF Publications, London.

Dean, J. D., Huyakorn, P. S., Donigan Jr, A. S., Voos, K. A., Schanz, R. W., Meeks, Y. J., and Carsel, R. F. (1989). *Risk of Unsaturated/Saturated Transport and Transformation of Chemical Concentrations (RUSTIC). Volume I: Theory and Code Verification*, US Environmental Protection Agency, Athens, Georgia.

DoE (1991). *The Use of Herbicides in Non-agricultural Situations in England and Wales*, Department of the Environment Report No. FR/D 0002, Foundation for Water Research, Marlow, Bucks.

DWI (1991). *Drinking Water 1990. A Report by the Chief Inspector Drinking Water Inspectorate*, HMSO, London.

DWI (1992). *Drinking Water 1991. A Report by the Chief Inspector Drinking Water Inspectorate*, HMSO, London.

European Community (1991). 'Council Directive of 15/7/91 concerning the placing of plant protection products on the market', *Official Journal of the European Communities*, **34**, Brussels.

Flury, M., Leuenberger, J., Studer, B., Fluhler, H., Jury, W. A., and Roth, K. (1994). *Pesticide Transport Through Unsaturated Field Soils: Preferential Flow*, ETH Zurich.

Fogg, P., Carter, A. D., and Brown, C. D. (1994). 'A comparison of the use and regulatory interpretation of a predictive model, lysimeter and field studies to determine the leaching potential of a seed dressing', *Brighton Crop Protection Conference—Pests and Diseases—1994*, Volume 3, pp 1283–1288, BCPC, Farnham, Surrey.

Foster, S. S. D. (1987). 'Fundamental concepts in aquifer vulnerability, pollution risk and protection strategy', in *Vulnerability of Soil and Groundwater to Pollutants* (eds. W. van Duijvenbooden and H. G. van Waegeningh), pp. 69–86, RIVM Proceedings and Information **38**, The Hague.

Führ, F., and Hance, R. (eds.) (1992) *Lysimeter Studies of the Fate of Pesticides in the Soil*, Monograph No. 53, BCPC, Farnham.

Greenwood, D. J., Neeteson, J. J., and Draycott, A. (1985). 'Response of potatoes to N fertiliser: dynamic model', *Plant and Soil*, **85**, 185–203.

Gustafson, D. I. (1989). 'Groundwater ubiquity score: a simple method for assessing pesticide leachability', *Environ. Toxic. Chem.*, **8**, 339–357.

Guth, J. A., and Mani, J. T. (1991). 'Experimental methods for measuring movement of pesticides in soil', in *Pesticides in Soil and Water* (ed. A. Walker), p. 141, BCPC Technical Monograph No. 47, Lavenham, Surrey.

Hall, D. G. M. (1994). 'Simulation of dichlorprop leaching in three texturally distinct soils using the PLM model', *J. Environ. Sci. Health A*, **29**, 1211–1230.

Hall, D. G. M., Reeve, M. J., Thomasson, A. J., and Wright, V. (1977). *Water Retention, Porosity and Density of Field Soils*, Soil Survey Technical Monograph No 9, Harpenden.

Hance, R. J. (1987). 'Herbicide behaviour in the soil, with particular reference to the potential for groundwater contamination', in *Progress in Pesticide Biochemistry and Toxicology, Vol. 6, Herbicides* (eds D. H. Hutson and T. R. Roberts), Chapter 7, John Wiley & Sons Ltd, Chichester.

Hance, R. J. (1990). 'Accuracy and prediction in pesticide analysis with reference to the EC "Water Quality" Directive', *Pestic. Outlook*, 1, 23–26.

Harris, G. L., Bailey, S. W., and Mason, D. J. (1991). 'The determination of pesticide losses to water courses in an agricultural clay catchment with variable drainage and land management', *Brighton Crop Protection Conference, Weeds 1991*, pp. 1271–1278, BCPC, Farnham, Surrey.

Hodgson, J. M. (1976). *Soil Survey Field Handbook*, Technical Monograph No. 3, Soil Survey of England and Wales, Harpenden.

Hollis, J. M. (1990). 'Mapping the vulnerability of aquifers and surface waters to pesticide contamination at the national/regional scale', in *Pesticides in Soil and Water: Current Perspectives* (ed. A. Walker), pp. 165–174, BCPC Monograph No. 47, Lavenham Press Limited, Lavenham, Surrey.

Hollis, J. M., and Brown, C. D. (1993). 'An integrated approach to aquifer vulnerability mapping', in *Mobility and Degradation of Xenobiotics. Proceedings of the IX Symposium on Pesticide Chemistry, Piacenza, October 1993*, pp. 633–644.

Hollis, J. M., and Carter, A. D. (1991). 'A comparison of laboratory, lysimeter and field study techniques to determine environmental fate of pesticides in the UK', pp. 1005–1010, *Brighton Crop Protection Conference, Pests and Diseases*, BCPC, Farnham.

Hollis, J. M., Hallett, S. H., and Keay, C. A. (1993). 'The development and application of integrated databases for modelling the environmental fate of herbicides', in *Brighton Crop Protection Conference, Weeds 1993*, Volume 3, pp 1355–1364, BCPC, Farnham, Surrey.

Howard, P. H. (1991). *Handbook of Environmental Fate and Exposure Data for Organic Chemicals. Volume III: Pesticides*, Lewis Publishers Inc., Chelsea, Michigan.

Hutchinson, M. F. (1986). 'Methods of generation of weather sequences', in *Agricultural Environments* (ed A. H. Bunting), pp. 149–157, C.A.B. International, Wallingford, Oxon.

Hutchinson, M. F. (1991). 'Climatic analysis in data sparse regions', in *Climatic Risk in Crop Production* (eds R. C. Muchow and J. A. Bellamy), pp. 55–71, C.A.B. International, Wallingford, Oxon.

Hutson, J. L. (1992). 'The use of models in the regulatory decision making process', in *Brighton Crop Protection Conference—Pests and diseases—1992*, Volume 3, pp. 1253–1260, BCPC, Farnham, Surrey.

Hutson, J. L., and Wagenet, R. J. (1992). 'LEACHM: Leaching Estimation and Chemistry Model, Version 3', *Research Series No. 92-3*, Department of Soil, Crop and Atmospheric Sciences, New York State College of Agriculture and Life Sciences, Cornell University, Ithaca, New York, USA.

IGS (1977). *Hydrogeological Map of England and Wales (scale 1:625 000)*, Institute of Geological Sciences, National Environment Research Council, London.

IUPAC (1987). 'IUPAC report on pesticides (23). Potential contamination of ground water by pesticides', *Pure Appl. Chem.*, 10, 1419–1446.

Jarvis, N. J. (1994). 'The MACRO model (version 3.1). 'Technical description and sample simulations', *Reports and Dissertations* no. 19, Department of Soil Sciences, Swedish University of Agricultural Sciences, Uppsala, Sweden.

Jarvis, N. J., Bergström, L., and Dik, P. E. (1991). 'Modelling water and solute transport in macroporous soil. II. Chloride breakthrough under non-steady flow', *J. Soil Sci.*, **42**, 71–81.

Jarvis, N. J., and Leeds-Harrison, P. B. (1987). 'Modelling water movement in drained clay soil. I. Description of the model, sample output and sensitivity analysis', *J. Soil Sci.*, **38**, 487–498.

Johnson, A. C., Haria, A., Batchelor, C. H., Bell, J. P., and Williams, R. J. (1993). *Fate and Behaviour of Pesticides in Structured Clay Soils*. 1st Interim Report, Institute of Hydrology, Wallingford.

Jones, R. J. A., and Thomasson, A. J. (1985). *An Agroclimatic Databank for England and Wales*, Soil Survey of England and Wales Technical Monograph No. 16, Soil Survey of England and Wales Harpenden.

Jury, W. A. (1982). 'Simulation of solute transport using a transfer function model', *Water Resource Res.*, **18**, 363–368.

Jury, W. A., Focht, D. D., and Farmer, W. J. (1987). 'Evaluation of pesticide groundwater pollution potential from standard indices of soil-chemical adsorption and biodegradation', *J. Environ. Qual.*, **16**, 422–428.

Jury, W. A., and Ghodrati, M. (1989). 'Overview of organic chemical environmental fate and transport modelling approaches', in *Reactions and Movement of Organic Chemicals in Soils* (eds B. L. Sawhney and K. Brown), pp. 271–304, SSSA Special Publication No. 22, Madison, Wisconsin.

Jury, W. A., Spencer, W. F., and Farmer, W. J. (1983). 'Behaviour assessment model for trace organics in soil: I. Model description', *J. Environ. Qual.*, **12**, 558–564.

Khan, M. A., and Liang, T. (1989). 'Mapping pesticide contamination potential', *Environ. Mgt*, **13**, 233–242.

Kitching, R., and Day, J. B. W. (eds) (1979). 'Two-day meeting on lysimeters', *Report 79/6*. Institute of Geological Sciences, HMSO, London.

Knighton, R. E., and Wagenet, R. J. (1987). 'Simulation of solute transport using a continuous time Markov process: II. Application to transient field conditions', *Water Resource Res.*, **23**, 1917–1925.

Knisel, W. G. (1980). *A Field-scale Model for Chemical, Runoff, and Erosion from Agricultural Management Systems*, Conserv. Rep. No. 26, US Dept. Agric., Sci. Educ. Adm, Tifon, GA.

Kreuger, J. (1991). Occurrence of pesticides in Nordic surface waters, in *Nordic Seminar: Pesticides in the Aquatic Environment—Appearance and Effect*. Tune Landboskole, Denmark.

Kung, K-J. S. (1990). 'Preferential flow in a sandy vadose zone: 1. Field observation', *Geoderma*, **46**, 51–58.

Lawes, J. B., Gilbert, J. H., and Warrington, R. (1981). 'On the amount and composition of rain and drainage waters at Rothamsted. Part II', *J. R. Agric. Soc. Eng.*, **17**, 249–279 and 311–350.

Lees, A., and McVeigh, K. (1988). *An Investigation of Pesticide Pollution in Drinking Water in England and Wales*, Friends of the Earth Special Report, London.

Leonard, R. A., and Knisel, W. G. (1988). 'Evaluating groundwater contamination potential from herbicide use', *Weed Technol.* **2**(2), 207–216.

Leonard, R. A., Knisel, W. G., and Still, D. A. (1987). 'GLEAMS: groundwater loading effects of agricultural management systems', *Trans. ASAE*, **30**, 1403–1418.

Lewis, K. J., and Dyson, J. S. (1994). 'Potential for pesticide degradation in the unsaturated zone of a sandy soil. Abstract in environmental behaviour of pesticides and regulatory aspects', p. 73, *COST 5th International Workshop*, Brussels.

Millar, C. E., and Turk, L. M. (1943). *Fundamentals of Soil Science*. John Wiley and Sons, New York.

Morgan, D. R. (1992). 'Pesticides and public health—a case for scientific and medical concern?', *Pestic. Outlook*, **3**, 24–29.

Mueller, T. C., Jones, R. E., Bush, P. B., and Banks, P. A. (1992). 'Comparison of PRZM and GLEAMS computer model predictions with field data for alachlor, metribuzin and norflurazon leaching', *Environ. Toxic. Chem.*, **11**, 427–436.

Mullins, J. A., Carsel, R. F., Scarborough, J. E., and Ivery, A. M. (1993). *PRZM-2, a Model for Predicting Pesticide Fate in the Crop Root and Unsaturated Soil Zones: Users Manual for Release 2.0*, Environmental Research Laboratory, US EPA, Athens, Georgia.

Nash, R. G., and Leslie, A. R. (eds) (1991). *Ground Water Residue Sampling Design*, ACS Symposium Series 465, pp. 165–181, American Chemical Society, Washington, DC.

Nicholls, P. H. (1988). 'Factors influencing the entry of pesticides into soil water', *Pestic. Sci.*, **22**, 123–137.

Nicholls, P. H., Evans, A. A., Bromilow, R. H., Howse, K. R., Harris, G. L., Rose, S. C., Pepper, T. J., and Mason, D. J. (1993). 'Persistence and leaching of isoproturon and mecoprop in the Brimstone Farm plots', pp. 849–854, *Brighton Crop Protection Conference, Weeds*, BCPC, Farnham.

Nicholls, P. H., Evans, A. A., and Walker, A. (1987). 'The behaviour of chlorsulfuron and metsulfuron in soils in relation to incidents of injury to sugar beet', pp. 549–556, *Brighton Crop Protection Conference, Weeds*, BCPC, Farnham.

NRA (1992). *Policy and Practice for the Protection of Groundwater*, National Rivers Authority, Almondsbury, Bristol.

Pennell, K. D., Hornsby, A. G., Jessup, R. E., and Rao, P. S. C. (1990). 'Evaluation of five simulation models for predicting aldicarb and bromide behaviour under field conditions', *Water Resource Res.*, **26**, 2679–2693.

Pestemer, W., Günther, P., Wischnewsky, M.-B., Novopaschenny, I., Wang, K., and Zhao, J. (1993). 'Development of expert systems to aid herbicide use regarding their behaviour in soil', in *Brighton Crop Protection Conference—Weeds—1993*, Volume 3, pp 1365–1372, BCPC, Farnham, Surrey.

Ragg, J. M., Beard, G. R., George, H., Heaven, F. W., Hollis, J. M., Jones, R. J. A., Palmer, R. C. Reeve, M. J., Robson, J. D., and Whitfield, W. A. D. (1984). *Soils and their Use in Midland and Western England*. Soil Survey of England and Wales Bulletin No. 12, Soil Survey of England and Wales, Harpenden.

Rao, P. S. C., Hornsby, G., and Jessup, R. E. (1985). 'Indices for ranking the potential of pesticide contamination of groundwater', *Proc. Soil Crop Sci. Soc. Fla*, **44**, 1–8.

Rao, P. S. C., and Wagenet, R. J. (1985). 'Spatial variability of pesticides in field soils: methods for data analysis and consequences', *Weed Sci.*, **33** (Suppl. 2), 18–24.

Reeve, M. J., and Carter, A. D. (1991). 'Water Release Characteristic', in *Soil Analysis—Physical Methods* (eds K. A. Smith and C. E. Mullins), pp. 111–160, Marcel Dekker, New York.

Richardson, S. J., and Wright, D. A. (1984). *WGEN: a Model for Generating Daily Weather Variables*, US Department of Agriculture, Agricultural Research Service ARS-8, Washington, DC.

Ritter, W. F. (1990). 'Pesticide contamination of groundwater in the United States—a review', *J. Environ. Sci. Health*, **B25**(1), 1–29.

Rose, D. A. (1991). 'Testing goodness of fit', *J. Agric. Sci.*, **117**, 132–133.

Russell (*et al.*) (1995) ???.

Simmons, N. D. (1991). 'Residues in ground water', pp. 1259–1269, *Brighton Crop Protection Conference, Weeds*, BCPC, Farnham.

Soil Conservation Service (1972). 'Hydrology', in *SCS National Engineering Handbook*, Section 4, US Department of Agriculture, Washington, DC.

Soil Survey Staff (1975). *Soil Taxonomy: a Basic System of Soil Classification for Making and Interpreting Soil Surveys*, Soil Conservation Service, US Department of Agriculture, Handbook No. 436, Washington, DC.

Steenhuis, T. S., Pacenka, S., and Porter, K. S. (1987). MOUSE: A management model for evaluating groundwater contamination from diffuse surface sources aided by computer graphics. *Appl. Agr. Res.*, **2**, 277–289.

Taylor, A. W., Freeman, H. P., and Edwards, W. M. (1971). 'Sample variability and the measurement of dieldrin content of a soil in the field', *J. Agric. Food Chem.*, **19**, 832–836.

Thomasson, A. J., and Jones, R. J. A. (1991). 'An empirical approach to crop modelling and the assessment of land suitability', *Agric. Syst.* **37**, 351–367.

Trafford, B. D. (1970). 'Field drainage', *J. R. Agric. Soc.* **131**, 129–152.

US EPA (1992). *Pesticides in Ground Water Database. A Compilation of Monitoring Studies: 1971–1991. National Summary*, United States Environmental Protection Agency, Washington, DC.

US EPA (1994). *Conditional registration of acetochlor*, United states Environmental Protection Agency, Washington, DC.

Van Genuchten, M. Th., Leij, F. J., and Lund, L. J. (1992). *Indirect Methods for Estimating the Hydraulic Properties of Unsaturated Soils*, University of California, Riverside, CA.

Wagenet, R. J., and Rao, P. S. C. (1985). 'Basic concepts of modelling pesticide fate in the crop root zone', *Weed Sci.*, **33** (Suppl. 2), 25–32.

Wagenet, R. J., and Rao, P. S. C. (1991). 'Modeling pesticide fate in soils', in *Pesticides in the Soil Environment: Processes, Impacts and Modeling* (ed. H. H. Cheng), pp. 351–399, SSSA Book Series No. 2, Madison, Wisconsin.

Walker, A. (1987). Evaluation of a simulation model for prediction of herbicide movement and persistence in soil. *Weed Res.*, **27**, 143–152.

Walker, A., and Brown, P. A. (1985). 'The relative persistence in soil of five acetanilide herbicides', *Bull. Environ. Contam. Toxicol.*, **34**, 143–149.

Walker, A., Welch, S. J., and Turner, I. J. (1995). 'Studies of time-dependent sorption processes in soils', in *Pesticide movement to water* (eds. A. Walker *et al.*), pp. 13–18, BCPC Monograph No. 62, BCPC, Farnham, Surrey.

Wauchope, R. D. (1978). 'The pesticide content of surface water draining from agricultural fields—a review', *J. Environ. Qual.*, **7**, 459–472.

Wauchope, R. D., Buttler, T. M., Hornsby, A. G., Augustijn-Beckers, P. W. M., and Burt, J. P. (1992). 'The SCS/ARS/CES pesticide properties database for environmental decision-making', in *Rev. Environ. Contam. Toxicol.*, **123**, 1–164.

White, I., Colombera, P. M., and Philip, J. R. (1977). 'Experimental studies of wetting front instability induced by gradual change of pressure gradient and by heterogeneous porous media', *Soil Sci. Soc. Am. J.*, **41**, 483–489.

Whitmore, A. P. (1991). 'A method for assessing the goodness of computer simulation of soil processes', *J. Soil Sci.*, **42**, 289–299.

Williamson, A. R., and Carter, A. D. (1993). 'Determination of the potential for the pollution of ground and surface water by alachlor when used in oil seed rape and fodder maize; summary and regulatory conclusions', *Brighton Crop Protection Conference—Weeds 1993*, BCPC, Lavenham Press, Lavenham, Surrey.

Worthing, C. R., and Hance, R. J. (1991). *The Pesticide Manual. A World Compendium*, 9th edition, British Crop Protection Council, Unwin Brothers Ltd, Surrey.

Yon, D. A. (1992). 'The use of lysimeters to study the fate of an experimental pesticide following autumn and spring application to winter cereals', In *Lysimeter Studies of the Fate of Pesticides in the Soil*, (eds F. Führ and R. Hance), pp. 145–151, Brighton Crop Protection Conference Monograph No. 53, BCPC, Bracknell.

CHAPTER 4

Pesticide leaching models and their use for management purposes

N. J. Jarvis, L. F. Bergström and C. D. Brown

INTRODUCTION

Leaching of pesticides in agricultural soils to groundwater has long been considered a potentially serious environmental problem. Accordingly,

Environmental Behaviour of Agrochemicals
Edited by T. R. Roberts and P. C. Kearney   © 1995 John Wiley & Sons Ltd

criteria have been introduced to regulate standards for drinking water quality with regard to acceptable pesticide concentrations (e.g. EEC directive 80/778/EEC). Together with this concern regarding the occurrence of pesticides in groundwater, effective means of assessing the risks of pesticide leaching need to be developed. As a first step in screening and registration programmes, it has been common practice to estimate pesticide mobility in short-term laboratory tests. It is even more common to simply rely on easily measurable physicochemical properties of the compound (e.g. sorption constants, water solubilities) and degradation rates in soil incubation studies. If these relatively simple tests indicate that unacceptably large leaching losses of a particular compound are to be expected, then additional experiments conducted under more natural field conditions may be required. However, the time needed to obtain a complete picture of pesticide fate and mobility may be up to 1–2 years or more, while the results of such a field test are also site-specific. The costs of conducting field experiments at a number of different sites are becoming prohibitive, while at the same time there is an urgent need to characterize fate and mobility of pesticides for a wide range of relevant environmental scenarios.

For these reasons, the use of simulation models as a potentially effective and inexpensive screening method in pesticide registration programmes has been discussed. A number of pesticide leaching models applicable to field conditions have been developed in the past, ranging from detailed data-intensive models intended as research tools to help in understanding the complex interplay of processes governing leaching (e.g. Hutson and Wagenet, 1992), to simplified models intended as educational tools and/or as a guide to decision-making (Nofziger and Hornsby, 1986). An excellent review of the fundamental principles and application of pesticide leaching models was presented earlier by Wagenet and Rao (1990). They concluded that model development had proceeded much faster than model testing, commenting that '. . . limited availability of data sets from field studies . . . continues to be a major constraint in evaluating the validity of these models before they can be used with confidence for the purposes for which they were developed, i.e. management and regulatory guidance'. Nevertheless, these models are now being used as tools in the registration process in several countries (Brouwer, 1994; Russell et al., 1994). We therefore consider it timely and desirable to review the extent of progress made in recent years in model testing and evaluation and to assess existing documented models in terms of suitability for management purposes. To this end, we first present some background information on the nature of leaching models. We then describe eight existing documented models, emphasizing the main similarities and differences between the alternative modelling approaches, before discussing some general principles governing the use of pesticide leaching models for registration and other management purposes, including aspects such as parameter estimation, calibration, validation and prediction. We then summarize the results of published model validation studies. Finally, some conclusions and recommendations are made for future model developments and use.

TYPES OF LEACHING MODELS

Leaching models consist of a number of submodels each describing a separate component of the complete soil-plant system. In the specific case of pesticide leaching models, submodels at three different levels can be identified. A top-level submodel calculates water flow, a second level submodel uses the calculated water flows to predict rates of solute transport, while the third level comprises additional submodels to account for transformations of pesticide (e.g. degradation and volatilization). The more detailed simulation models may even include subroutines to describe additional component parts of the soil-plant system (e.g. plant uptake, canopy interception).

Addiscott and Wagenet (1985) presented a hierarchical scheme for classification of leaching models. The primary distinction in this scheme is between *deterministic* and *stochastic* models. A *deterministic* model presumes a single set of driving variables (e.g. climatic variables) and input parameter values (e.g. soil and pesticide properties) and, consequently, a single unique outcome. By contrast, a *stochastic* model considers the outcome to be uncertain and is structured to account for this uncertainty. For example, soil properties are known to be spatially variable, which may, in turn, cause spatial variations in pesticide leaching. Thus, in a stochastic model approach, parameters describing soil properties may be specified as statistical distributions of values, rather than as a single value. The output from such a model would consist, for example, of the mean pesticide leaching load together with statistical limits for the distribution. Of the models dealt with in this paper, all are essentially deterministic. However, deterministic models can be used in a stochastic or probabilistic way by, for example, incorporating the model into a shell program to run Monte Carlo simulations (Carsel *et al.*, 1988a, b; Petach *et al.*, 1991; Mullins *et al.*, 1993; Zhang *et al.*, 1993; Nofziger *et al.*, 1994).

A second distinction may be made between *mechanistic* and *functional* models. A *mechanistic* model utilizes current scientific understanding to incorporate the most fundamental (i.e. mechanistic) descriptions of the important or relevant processes. A *functional* (often called empirical) model uses simplified, less mechanistic, treatments of some or all the relevant processes (e.g. water flow, pesticide transport and/or transformations). The reason for adopting a simplified functional approach is to reduce the number of input parameters required by the model. A clear distinction between mechanistic and functional models may be difficult to make in the case of complex simulation models which usually comprise several submodels. Thus, some component submodels may be mechanistic, whilst others may be simplified functional treatments of the processes. It may be the case that higher-level submodels are treated mechanistically, whilst lower-level submodels are dealt with in a simplified functional way. Furthermore, in some cases, a mechanistic treatment of a particular process may not be possible or even desirable, either because fundamental knowledge of the process is lacking or because models which incorporate descriptions of the current understanding of the process may be too difficult to parameterize.

Finally, simulation models may also be classified according to use, with a distinction often being made between *research* models and *management* models. The main aims of a *research* model are usually considered to be to improve our understanding of the complex interplay of the processes governing leaching, to identify gaps in knowledge and to generate new hypotheses which can be tested experimentally. The main use of a *management* model is as a tool to guide the management of agricultural/natural resources. In the specific case of pesticide leaching models, we are interested in the use of models as management tools in the registration process. The division between research and management models is often considered to parallel the division between mechanistic and functional models (Addiscott and Wagenet, 1985; Wagenet and Rao, 1990), such that mechanistic models are confined to research uses only. However, this is largely a result of the way mechanistic models have been developed in the past, with little thought given to the end-user of the model. This has resulted in difficult-to-use models with poor transferability, such that model testing and validation has rarely been performed by independent scientists. In principle, however, there is no fundamental barrier to prevent detailed mechanistic models being used for management purposes. The first requirement for a management model is a user-friendly interface between the model and the user, if for no other reason than that, by definition, the user of a management model will not be the person who developed it. In the past, mechanistic models have not been user-friendly, although new-generation models are certainly becoming more so. The second requirement of a management-oriented pesticide leaching model is that it be linked to soils/climate/pesticide databases in order to minimize problems of parameter estimation for the non-specialist and to allow straightforward application/extrapolation to a variety of different environmental scenarios. If detailed, data-intensive, mechanistic models are to be used for management purposes, it is necessary that the accumulated research knowledge concerning parameter estimates be easily accessible to the model user. At the simplest level, this can be achieved by user-friendly help menus and by allowing access to default parameter values within the model. Much greater flexibility may be afforded by making extensive use of pedo-transfer functions to derive parameter estimates from more easily obtained soils data. For example, mechanistic models require estimates for parameters describing the soil hydraulic properties, which can often be predicted with reasonable accuracy from soil survey data such as texture, structure, organic matter content, bulk density and soil morphology (Hollis and Woods, 1989; Vereecken et al., 1989; McKenzie et al., 1991; Rawls et al., 1992). Such pedo-transfer functions are included in several pesticide leaching models, including PRZM2, PELMO and LEACHP, either automated within the user interface, or simply contained in the users' manual. Common to all models is the requirement for pesticide properties, such as degradation half-lives and sorption constants. Extensive computerized databases of pesticide properties are now available (e.g. Herner, 1992; Wauchope et al., 1992) and such on-line databases have been coupled to simulation models. Thus, users of the GLEAMS and CMLS models can select from menus containing a wide range of commonly used pesticides (Nofziger and Hornsby, 1986; Knisel et al., 1992). Similarly, climate databases with long time-series weather data can also be linked to models (Hollis et al., 1993). Efforts to link pesticide leaching models to geographical databases containing soils, cropping and climate information are now well underway (e.g. Carsel et al., 1988a, 1991; Petach et al., 1991; Wagenet et al., 1991; Hollis et al., 1993) and should lead to regionally or nationally based model systems which are both flexible and easy to use.

One other potential barrier to the use of the more detailed mechanistic models for management purposes can be mentioned. This is the often short time steps and long execution times required by such models in order to ensure stability of the numerical solution. This may limit the use of mechanistic models for long-term or multiple (stochastic) simulations (Petach et al., 1991). However, such limitations should become less important in the future, given the increasingly rapid development of still more powerful, but affordable, personal computers and work stations.

COMPARISON OF EIGHT PESTICIDE LEACHING MODELS

For the purposes of this review, eight pesticide leaching models have been selected for comparison: CALF (Nicholls et al., 1982) PRZM/PRZM2 (Carsel et al., 1985; Mullins et al., 1993), LEACHM (Hutson and Wagenet, 1992, the pesticide version called LEACHP), CMLS (Nofziger and Hornsby, 1986), GLEAMS (Leonard et al., 1987; Knisel et al., 1992), PESTLA (Boesten and van der Linden, 1991), MACRO (Jarvis, 1991, 1994) and PELMO (Klein, 1991). Methods of calculating soil water flow, pesticide transport and transformations are summarized for each model in Tables 1–3. It should be remembered that simulation models are usually under more or less continuous development with new versions introduced at frequent intervals. The summary descriptions of the models given in Tables 1–3 refer to model versions available in 1994. It is of course possible, and indeed likely, that some of these model descriptions may quickly become out of date.

The selected models can be classified into mechanistic and functional groups on the basis of these model process descriptions, particularly the higher-level model components of water flow and solute transport (Tables 1 and 2). The models may be broadly divided into three groups. The first

Table 1 Model treatments of soil water flow

Model	Soil water flow		Preferential flow
	Richards' equation	Capacitance	
PRZM		✔	
PELMO		✔	
CALF		✔	(✔)
GLEAMS		✔	
CMLS		✔	
MACRO	✔[a]		✔
PESTLA	✔		
LEACHP	✔		(✔)

[a]Richards' equation is used in micropores only. A capacitance approach is used in the macropores.

Table 2 Model treatments of transport processes

Model	Pesticide transport		Sorption isotherm		Canopy interception and wash-off
	Convection–dispersion equation	Mass flow	Freundlich	Linear	
PRZM	✔			✔	✔
PELMO	✔		✔		
CALF		✔		✔	
GLEAMS		✔		✔	✔
CMLS		✔		✔	
MACRO	✔[a]			✔	✔
PESTLA	✔		✔		
LEACHP	✔[b]	✔[b]	✔[c]		

[a]The CDE is used in micropores only. Pesticide is moved by mass flow in the macropores.
[b]Both options are available.
[c]Two-site sorption included as an option in version 3.1.

Table 3 Model treatments of transformation processes

Model	Degradation model		Soil water and temperature effects on degradation	Degradation of sorbed and liquid phases treated separately	Formation and transport of metabolites	Volatilization
	First-order	Others				
PRZM	✓	✓[b,c]		✓[b]	✓[b]	✓[b]
PELMO	✓	✓[a]	✓			
CALF	✓		✓			
GLEAMS	✓				✓	
CMLS	✓					
MACRO	✓		✓	✓		
PESTLA	✓		✓			
LEACHP	✓		✓	✓	✓	✓

[a]The order of kinetics can be specified.
[b]PRZM2 only.
[c]Option to use microbial growth model (Soulas, 1982).

group, comprising LEACHP, PESTLA and MACRO, use mechanistic treatments for both water flow and solute transport. The second group, the functional models CALF, GLEAMS and CMLS, employ simplified treatments to describe both water flow and solute transport. PRZM and PELMO adopt a mixed mechanistic/functional approach, utilizing a simple capacitance-type water flow model, but a physically based approach to predict solute transport. PELMO was developed from the original PRZM model and differs from its parent model largely in the more detailed treatment of sorption and transformation processes and in the nature of the user interface. In the following sections, alternative ways of modelling soil-related processes affecting pesticide leaching are discussed and the various approaches used in the different models are compared.

Soil water flow

The models LEACHP, PESTLA and MACRO use Richards' equation based on Darcy's law to calculate soil water flow (see Table 1):

$$\frac{\partial \theta}{\partial t} = -\frac{\partial}{\partial z}\left(K\left(\frac{\partial \psi}{\partial z} + 1 \right) \right) - S_w \qquad (1)$$

where θ is the soil water content, t is time, z is vertical distance, K is the unsaturated hydraulic conductivity, ψ is the soil water potential and S_w is a sink term accounting for root water uptake. This physically based approach requires information on the two fundamental soil hydraulic functions, the water retention curve $\theta(\psi)$ and the unsaturated hydraulic conductivity function $K(\theta)$. Empirical equations are used to describe $\theta(\psi)$ (e.g. Brooks and Corey, 1964; van Genuchten, 1980), the parameters of which may be obtained by least-squares fitting to measured water retention data. A minimum of three (but usually four) parameters are needed to characterize properly the shape of the water retention curve in a mathematical form. In the mechanistic approach, the soil water retention function is then combined with a $K(\theta)$ function derived from a theoretical treatment of the relationship between soil pore size distribution and conductivity (e.g. Mualem, 1976). The result is a 'closed-form' model of soil hydraulic properties which, in addition to the retention curve, requires two more parameters, the first being the saturated hydraulic conductivity and the second a parameter related to pore tortuosity which is often treated as a constant.

The remaining models (PRZM, PELMO, CALF, GLEAMS and CMLS) treat water flow in a simplified way, thereby reducing input parameter requirements. The term 'capacitance', used in Table 1 to describe the water flow component in such models, is a generic term suggesting that water flow is predicted as a function of water content, in contrast to the physically based method in which water potential drives flow. Specifically, in the functional model approach, water flow is calculated, usually on a daily basis, using the tipping-bucket analogy, whereby flow is assumed zero until a 'field capacity' water content is reached. At larger water contents, the excess water is routed downwards to the next layer in the soil profile and so on. The advantages of using such a simplified approach are to reduce computer processing time and to considerably reduce data requirements, with only easily obtained soil water constants, such as field capacity and wilting point, required to model water flow. There are two main disadvantages with the capacitance approach. First, the rate of water redistribution depends on the time step in the model. Thus, the capacitance approach is best suited to sandy soils which may drain to field capacity during one day (i.e. the time step employed in these models). To allow for slower water redistribution, the GLEAMS model employs a storage-routing method (this is also available as an option in PRZM), where the flow rate is given as a function of the water content in excess of field capacity. However, the parameters in such an approach are not physically based so that differences between soil types are not easy to represent in any fundamental way. A second major disadvantage is that upward water flow (and pesticide transport) to the soil surface during dry periods is not easy to deal with in a physically realistic way. For example, CMLS, PRZM and PELMO do not attempt to model upward water flow. The GLEAMS model redistributes pesticide (but apparently not water) upwards in the soil profile in proportion to the amount of water extracted from each layer. The CALF model redistributes 'immobile' water, but not pesticide, in response to evaporation, depending on water content differences between adjacent layers. In reality, it is the hydraulic head gradient and not water content differences which drive water flow, so that this approximate method is, in principle, incorrect and will probably tend to overestimate upward water flow due to capillarity.

Solute transport

Two physical processes, mass flow (convection) and diffusion, control the transport of solutes in soil. Mass flow refers to the convective transport of solutes dissolved in flowing water, generally a much faster process than diffusion. Water in different soil pores or flow pathways moves at different rates depending on pore size and continuity. This spatial variation in water flow rates results in spreading, or dispersion, of the solute. The convection–dispersion equation (CDE), which accounts for these processes, is used in the more mechanistic models (PESTLA, LEACHP, MACRO, PRZM and PELMO) to model solute transport (Table 2):

$$\frac{\partial(\theta c + \gamma s)}{\partial t} = \frac{\partial}{\partial z}\left(D\theta\frac{\partial c}{\partial z} - qc\right) - S_t \tag{2}$$

where c and s are the pesticide concentrations in liquid and sorbed phases, S_t is a sink term accounting for pesticide transformation (e.g. degradation), γ is the soil bulk density, q is the water flow rate and D is the dispersion coefficient usually given by:

$$D = D_v v + D_0 f \tag{3}$$

where D_0 is the diffusion coefficient in free water, f is the pore continuity (or tortuosity) factor, D_v is the soil dispersivity and v is the pore water velocity which is simply given by q/θ. In PRZM and PELMO, equation (3) is not used. Instead, the dispersion coefficient is set to a fixed value independent of the water flow rate. It can be seen from equations (2) and (3) that if diffusion is neglected (it is usually overshadowed in importance by mass flow), only one parameter, the dispersivity (or dispersion coefficient), is needed as input to the CDE.

Only mass flow is accounted for in CALF and GLEAMS (Table 2). Nevertheless, dispersion is inevitably included in these models in an approximate way, in the form of inherent numerical dispersion due to discretization of the soil profile. The degree of numerical dispersion depends on the number and thickness of layers in the soil profile. In the CALF model, the layer thickness is fixed at 1 cm, ensuring that dispersion is constant for all simulations. In GLEAMS, where layer thicknesses are chosen by the user, an undesirable discretization-dependent variation in simulation results may occur, which would outweigh the minor advantages gained by model simplification. CMLS uses an alternative simplified method to calculate pesticide transport which, at the same time, avoids problems of numerical dispersion. Instead of calculating the transport of a distributed mass of pesticide in the soil, CMLS assumes that all the pesticide is located at one depth in the soil, keeping track of the location of this 'solute front' by assuming simple piston displacement. Finally, it should be noted that modelling physical dispersion with the CDE does not in itself eliminate numerical dispersion. The dispersion coefficient in the CDE can be corrected to account for numerical dispersion (van Genuchten and Wierenga, 1974), but this is not always done. For example, numerical dispersion was present in the original version of the PRZM model (Carsel *et al.*, 1984; Loague *et al.*, 1989b). An alternative numerical solution which eliminates this problem has now been included as an option in the new version of the model, PRZM2 (Mullins *et al.*, 1993).

Preferential flow

One important assumption which underpins the mechanistic approach to modelling water flow and solute transport represented by Richards' equation and the CDE is that rates of lateral 'mixing' are large in relation to vertical transport velocities, such that the soil at a given depth can be characterized by single values of water potential (water content) and solute concentration. However, there is now a considerable body of experimental evidence which contradicts this basic assumption of physical equilibrium. In fine-textured soils, structural features termed macropores (e.g. shrinkage cracks, earthworm channels, root holes) may dominate the soil hydrology, operating as high-conductivity flow pathways bypassing or short-circuiting the denser soil matrix (Bouma, 1981; Beven and Germann, 1982). This bypassing flow leads to a more rapid than expected leaching of surface-applied chemicals, including pesticides, since solutes contained in the fast-flowing water in macropores will not have sufficient time to equilibrate with slowly moving or stagnant water in the matrix. Preferential flow is used as a generic term to describe physical non-equilibrium flow processes. Preferential flow has also been observed in sandy soils (Hillel, 1987), although the causes are different, including small-scale variations in bulk density, larger-scale profile heterogeneities such as horizon interfaces, textural boundaries or soil lenses (Kung, 1991), trapped air, or water repellency (Hendrickx *et al.*, 1993).

Preferential flow may be especially critical for pesticide leaching, since the biologically and chemically reactive topsoil is bypassed. The opportunity to limit leaching losses is reduced if pesticide quickly reaches subsoil layers where degradation and sorption processes are generally less effective. This is particularly true for otherwise 'non-leachable' compounds characterized by strong sorption and/or rapid degradation (Jarvis, 1995a). If preferential flow occurs, one fraction of the applied compound may move rapidly to depth in the soil, whilst the bulk of the pesticide may be transported more slowly than expected, being protected in smaller intra-aggregate pores and by-passed by the fast-flowing water in macropores.

Preferential transport of pesticides has been demonstrated in field soils by, among others, Bottcher et al. (1981), Jury et al. (1986), Kladivko et al. (1991), Steenhuis and Parlange (1991) and Ghodrati and Jury (1992). Indeed, it is now increasingly apparent that, under field conditions, preferential flow may be the rule rather than the exception. Thus, from dye tracing experiments conducted at 14 representative sites, Flury et al. (1994) concluded that preferential flow behaviour was to be expected in most Swiss agricultural soils. For these reasons, some of the models listed in Table 1 now attempt to account for preferential flow. The functional description of water flow recently incorporated as an option in LEACHP (Hutson and Wagenet, 1993) builds on the simple model presented by Addiscott (1977). This is also the same simple capacitance-type model which forms the basis of the CALF model, with water considered 'mobile' when the field capacity water content is exceeded. It is uncertain to what extent this modelling approach can account for preferential flow, since the immobile water content must be filled before mobile water moves downwards to the next layer in the profile. The version of Addiscott's model which has been incorporated into CALF and LEACHP also assumes complete equilibration of solute concentrations between mobile and immobile water in each soil layer at the end of each daily time step (later versions of Addiscott's model allow for incomplete equilibration). Nevertheless, simplified mass flow/capacitance models may indeed predict deeper leaching of pesticides than models based on the CDE, since transport effectively occurs through a smaller displaced water content (the excess water above field capacity, rather than field capacity itself), while water redistribution takes place on a daily basis before the calculations of sorption equilibration are performed (Hutson and Wagenet, 1993). Another modelling approach which accounts for non-equilibrium flow behaviour was recently presented by Hutson and Wagenet (1995). In this model, called TRANSMIT, the LEACHM model is run for a series of interacting regions which are conceptualized as either parallel soil columns representing large-scale field heterogeneity or as pore domains to account for preferential flow processes operating at the small scale (i.e. macropores). In such a multi-region approach, estimation of parameter values controlling exchange of water and solute between the regions is necessarily subjective. This means, in turn, that the use of TRANSMIT will probably be restricted to research and education applications (Hutson and Wagenet, 1995).

The dual-porosity MACRO model (Jarvis, 1991, 1994) divides the soil porosity into two flow domains (macropores and micropores), each characterized by a flow rate and solute concentration. Richards' equation and the

CDE are used to model soil water flow and solute transport in the soil micropores, while a simplified capacitance-type approach is used to calculate fluxes in the macropores. An approximate, but physically based, approach is used to calculate mass exchange between the flow domains (see Gerke and van Genuchten, 1993). Compared to the classical one-domain approach based on Richards' equation and the CDE, four extra parameters are required: the fraction of sorption sites in each domain (van Genuchten and Wierenga, 1976), a diffusion path length regulating mass exchange between the domains, a macropore size distribution index, and the saturated conductivity of the soil micropores. The fraction of sorption sites in the macropore region can be fixed as a constant proportion of the ratio between macroporosity and total porosity, while the diffusion path length can be represented by an effective aggregate radius based on soil morphology descriptions. Leaching does not appear to be particularly sensitive to the size distribution index, which may also vary within narrow bounds (Jarvis, 1991), while the conductivity of the micropores can be estimated from the water retention curve parameters (Jarvis, 1995a). Thus, in the future, MACRO may have potential utility as a management model, since the problem of parameter estimation does not appear insurmountable.

Sorption

The retardation of pesticide transport due to sorption is an important component of any pesticide leaching model. Five of the models (PRZM, CALF, GLEAMS, CMLS and MACRO) adopt the simplest possible approach, assuming instantaneous equilibration of the pesticide between solid and liquid phases and a linear sorption isotherm:

$$s = kc \qquad (4)$$

where k, the sorption coefficient, is the slope of the linear isotherm. It is often assumed that k may be derived from the soil organic carbon content f_{oc} and the 'soil organic carbon sorption coefficient' k_{oc}:

$$k = f_{oc}k_{oc} \qquad (5)$$

The assumption underlying this parameter-estimation method is that sorption is dominated by the soil organic matter fraction. This approach is only valid for non-ionic, hydrophobic pesticides and soil horizons with organic carbon contents larger than c. 1% (Green and Karickhoff, 1990) and is certainly not appropriate for those pesticides which exist as weak acids or bases in soil solution.

Measured pesticide sorption isotherms often approximate to the simple linear form, although significant deviations are not uncommon, particularly at larger concentrations. PELMO, PESTLA and LEACHP therefore offer greater flexibility to the user, modelling sorption with the Freundlich isotherm:

$$s = kc_{\text{ref}}\left(\frac{c}{c_{\text{ref}}}\right)^N \tag{6}$$

where N is the Freundlich exponent and c_{ref} is a reference concentration introduced to make the units of k independent of N (Boesten, 1994). This approach introduces one additional parameter, the Freundlich exponent which, when set to unity, reduces the equation to a linear isotherm.

Sorption is relatively rapid in relation to the typical time scales of soil biological and physical processes such as degradation and water flow, so that at the pore scale, an assumption of instantaneous equilibrium may be reasonable. However, at any given time, not all the pesticide will be exposed to all the sorption sites in the soil, so that sorption may be transport-limited by slow rates of diffusion, for example, from large inter-aggregate pores to smaller intra-aggregate pores. Sorption hysteresis and apparent long-term increases in k (Walker, 1987; Calvet, 1993) are commonly observed phenomena caused by diffusion-limited sorption. This suggests the need for non-equilibrium approaches to modelling sorption. An empirical treatment, in which the sorption coefficient increases with the square root of time (Walker, 1987; Nicholls, 1994), has been introduced into CALF and VARLEACH (a later variation of the CALF model). Other, more mechanistic approaches, have been developed. 'Two-site' models (van Genuchten and Wagenet, 1989) divide the sorption sites into two fractions, one equilibrating instantaneously with the solution phase, while the remaining fraction is subject to kinetic sorption, where the rate of sorption depends on the difference in pesticide concentration between solution and sorbed phases. The latest version of LEACHP includes the two-site sorption model as an option (Table 2). It may be less critical to account for kinetic sorption in two- or multi-domain models such as MACRO and TRANSMIT, because the assumption of instantaneous 'local' equilibrium is likely to be reasonable for each pore class or flow domain (Rao et al., 1993). Indeed, Nkeddi-Kizza et al. (1984) demonstrated the exact mathematical equivalence of two-site sorption and two-domain transport models under saturated flow conditions.

Pesticide transformations

Pesticide transformations usually constitute the largest component of the mass balance and are therefore critical in modelling pesticide leaching. These transformations include processes such as microbial degradation, photolysis, chemical hydrolysis and volatilization. For non-volatile pesticides, microbial degradation is normally the most important transformation process, so that most models do not explicitly account for multiple dissipation mechanisms in soil, assuming that the effects of all dissipation processes can be included in a single 'lumped' degradation term. Indeed, of the models listed in Table 3, only LEACHP allows the possibility of treating chemical degradation separately from biodegradation.

Rates of microbial degradation in soil depend on a wide range of interacting factors, including the structure of the compound itself, bioavailability, the amounts, distribution and activity of microorganisms in the soil and soil environmental conditions (water, temperature). Despite, or perhaps because of this complexity, the models generally employ empirical treatments of degradation (Table 3), assuming that simple first-order kinetics can be used to describe the degradation rate S_t:

$$S_t = \mu(\theta c + \gamma s) \tag{7}$$

where μ is a rate coefficient which is inversely proportional to the pesticide half-life. In field soils, soil temperature and water conditions may affect pesticide degradation rates. The limiting effects of low soil temperatures and dry soil are included in five of the eight models (Table 3), the exceptions being PRZM, GLEAMS and CMLS. The effects of soil temperature on degradation are predicted from physical principles with the Arrhenius equation or with simple Q_{10}-type relationships. In LEACHP, PELMO and MACRO, soil temperatures are calculated from air temperatures from physical principles using the heat conduction equation. In the CALF model, soil temperatures are calculated from air temperatures using empirical relationships. The influence of soil water content on degradation is usually accounted for by simple empirical equations or indices (e.g. Walker, 1974; Boesten and van der Linden, 1991).

In many cases, first-order kinetics has been shown to adequately describe measurements of pesticide degradation (Rao and Davidson, 1980). However, it is not too surprising that such a simple empirical approach sometimes fails to describe the complex nature of microbial degradation. There may be many reasons for this. For example, degradation may occur metabolically with exponentially growing microorganism populations utilizing the pesticide compound as an energy source (Soulas, 1982; Stenström, 1992). Indeed, enhanced degradation resulting from long-term population increases induced by repeated pesticide applications appears to be a widespread phenomenon (Walker and Roberts, 1993). Also, the assumption of 'lumped' first-order kinetics often breaks down at small concentrations because mass transfer to active sites by diffusion becomes the rate-limiting step (Rao et al., 1993). This may result in significant underestimates of persistence in soil, particularly when making long-term extrapolations from measurements made during only one or two 'apparent' half-lives (Scow and Hutson, 1992). The bioavailability of pesticides is known to depend strongly on the concentration in soil solution in the vicinity of degrading microorganisms. In this respect, soil microbial populations may be sparse and unevenly distributed, both because of micro-scale variations in environmental conditions (e.g. rhizosphere vs bulk soil, inter- vs intra-aggregate pores) and because small intra-aggregate pores may be physically inaccessible. Thus, research modelling efforts are now increasingly being focused on the complex interactions between sorption, diffusion and degradation processes in heterogeneous soil (Ogram *et al.*, 1985; Scow and Hutson, 1992; Rao *et al.*, 1993; Scow, 1993).

Recognizing the complexity of microbial degradation, PELMO offers an alternative empirical approach, with the order of kinetics specified by the user. An option is available in PRZM2 to calculate pesticide degradation using the mechanistic model of Soulas (1982), which accounts for the growth and activity of four microbial subpopulations. However, it is highly unlikely that such an option will ever be routinely used for management applications, due to the difficulty in determining the large number of required input parameters. In particular, model parameters derived from 'ideal' batch systems or laboratory cultures are not easily transferable to field conditions (Rao *et al.*, 1993). Three of the models (LEACHP, MACRO and PRZM2) account for limited bioavailability of the sorbed pesticide to the microbial biomass, allowing the user to specify different degradation rate coefficients in dissolved and sorbed phases (Table 3). In MACRO, pesticide is stored in four different compartments (macropores/micropores, dissolved/sorbed), so that the user may specify up to four different degradation rate coefficients (van Genuchten and Wagenet, 1989). Thus, in principle, MACRO can be used to investigate the effects of interactions between sorption, diffusion and degradation processes on pesticide leaching under field conditions. Nevertheless, it is likely that for the forseeable future, 'lumped' degradation rate coefficients will continue to be assumed in most model applications for management purposes. This, again, is due to the difficulties in parameterizing such a detailed approach (Gamerdinger *et al.*, 1990).

Three models (GLEAMS, LEACHP and PRZM2) allow the user the possibility of predicting not only the fate and mobility of the parent compound, but also the formation and transport of metabolites. This is clearly a useful feature in cases where the environmental fate of metabolites is of concern. LEACHP and PRZM2 can also be used to describe the behaviour of volatile organic compounds, since these are the only models which account for evaporation and gas phase transport. Both models adopt physically based approaches, utilizing Henry's law to partition pesticide between the liquid and gas phases and Fick's law to describe gaseous diffusion in soil.

PRINCIPLES OF MODEL APPLICATION

In this section, we discuss some of the general concepts and principles concerning the use of models, including parameter estimation, calibration and validation.

Parameter estimation and calibration

For most model applications, the availability of directly measured parameters is limited, usually for reasons of experimental difficulty, lack of money, resources or time. Model calibration is often performed to derive the remaining, unknown or uncertain, parameter values. This is a process of adjusting or fine-tuning model parameter values to improve agreement between model predictions and measurements. Calibration is an iterative process (Hutson and Wagenet, 1991), with literature sources, general knowledge and simple estimation methods being used to arrive at initial guesses for the parameter values followed by calibration against a data set to improve model performance. Calibration can be performed subjectively, relying on the experience or expertise of the user. This may be adequate where the model is used for research purposes, although even here it is debatable whether many researchers have the multi-disciplinary competence (e.g. soil science, hydrology, agronomy, pesticide science, etc.) required to understand and effectively parameterize the more comprehensive simulation models. This is even more likely to be the case where the model is used for management purposes, since the user may not necessarily have the same degree of technical expertise as a specialist researcher. Management models must also be easy to use and efficient in terms of demands on human resources (Russell *et al.*, 1994). Given these considerations, it is clear that parameter estimation procedures, making extensive use of 'default' values and pedo-transfer functions (Petach *et al.*, 1991; Wagenet *et al.*, 1991; Hollis *et al.*, 1993), must be incorporated into data-demanding models if they are to be widely used for management purposes.

Validation

Model validation is the independent testing of a model against a set of measurements not included in any prior model calibration. This constitutes a rigorous test of whether the calibrated parameter set is generally valid. However, such a rigorous model validation is rarely performed, perhaps because model performance will almost inevitably be poorer than during the calibration period. One reason for this lies in the nature of the model itself. By definition, a model is a simplified representation of reality and cannot therefore account for all possible process interactions and feedbacks which exist in nature. This need for simplification means that poorly understood and complex processes are represented by model parameters which are considered as constants. In reality, this is not necessarily the case, so that calibrated parameter values cannot be expected to be generally valid, even for a well defined modelling scenario. A few examples of such model parameters are the degradation rate constant (variations in microbial populations and/or activity), hydraulic properties (effects of tillage, swell/ shrink, freeze/thaw) and root depth and distribution (varying with soil conditions and plant growth).

THE USE OF PESTICIDE LEACHING MODELS FOR REGISTRATION PURPOSES

In this section, we discuss the use of models within the registration process, concentrating on issues of model accuracy, ways of accounting for uncertainty, model flexibility and ease of use.

Predictive accuracy and accounting for uncertainty

There are two main sources of error in simulation model outputs: model error and parameter error (Loague and Green, 1991). Model errors result from incorrect or undue simplification of process descriptions in the model and neglect of significant processes (Russell *et al.*, 1994). Clearly, some degree of model error is inevitable, since by definition models are simplifications of reality. However, in principle, model errors should be minimized when mechanistic process descriptions are used (Wauchope, 1992) and in detailed models which include as many of the relevant processes as possible.

Parameter error is the use of inappropriate parameter values. This may be potentially serious for comprehensive data-demanding simulation models and for those parameters for which the model outcome (e.g. leaching) is especially sensitive. Predictions of leached amounts are highly sensitive to model parameters related to pesticide sorption and degradation (e.g. k or k_{oc} and organic carbon content, N and μ) which are usually both uncertain and variable (Boesten, 1991, 1994; Boesten and van der Linden, 1991; Jarvis, 1991), particularly for subsoil horizons where measured data are scarce. Sensitivity to variability or uncertainty in these parameters is especially large for small leached fractions (Boesten, 1991). It is fortunate that leaching appears to be less sensitive to some of the most spatially variable soil parameters such as saturated hydraulic conductivity (Boesten, 1991; Hutson and Wagenet, 1991; Jarvis, 1991). Parameter error is particularly critical where the models are used predictively, since methods for parameter estimation (e.g. pedo-transfer functions) may introduce additional sources of error. In a comparative test of pedo-transfer functions to predict $\theta(\psi)$, Tietje and Tapkenhinrichs (1993) reported root mean square differences between measured and predicted water contents of the order of $0.05-0.07\,\mathrm{m}^3\,\mathrm{m}^{-3}$. This suggests that soil hydraulic properties derived from pedo-transfer functions may be sufficiently accurate for most management purposes and are likely to be more reliable than the 'guessti-mates' of inexperienced users.

As noted earlier, model accuracy depends on how a model is used, whether it is calibrated or only used predictively. Calibration is the process of utilizing measurements to minimize parameter error and improve accuracy. Thus, the predictive use of models usually leads to reduced accuracy. Recognizing this, limited numbers of fixed soil/climate scenarios, with attendant default parameter settings, have been developed for use by registration authorities together with a selected pesticide leaching model (e.g. PELMO in Germany, Klein, 1991; PESTLA in the Netherlands, Brouwer, 1994). The idea is to calibrate the model for these scenarios for a limited number of compounds. The risk of pesticide leaching can then be assessed in a comparative framework utilizing these fixed scenarios. To account for parameter uncertainty and variability, 'worst-case', 'average-case' and 'best-case' scenarios can be modelled (Russell and Layton, 1992; Wauchope, 1992; Hollis and Brown, 1993; Brouwer, 1994), utilizing any available information on the expected ranges of critical or sensitive model parameters.

The use of fixed scenarios together with calibrated leaching models may produce the most reliable results, but it is also somewhat restrictive. Thus, model users may wish to run a model predictively, for example, to extrapolate the model to combinations of soils, climates and crops for which no prior calibration has been performed and for which no site-specific measurements of model parameter values are available. Examples of model validation tests performed without the benefit of prior calibration and site-specific data are rare. Thus, the predictive accuracy of models used in this way is largely unknown, although it is likely to be poor (Loague and Green, 1991). A stochastic treatment of parameter uncertainty and variability would then be appropriate, providing information to model users concerning the expected probability of pesticide leaching losses exceeding given threshold values. Thus, some deterministic models (e.g. PRZM2, CMLS) are now coupled to shell programs which allow the user to run Monte Carlo simulations (Mullins et al., 1993; Nofziger et al., 1994). Stochastic approaches can also account for significant effects of parameter interdependence (e.g. organic carbon content and degradation rate coefficient) on pesticide leaching predictions, providing sufficient measured data is available concerning the correlation of properties of interest (e.g. Carsel *et al.*, 1988b; Zhang *et al.*, 1993).

Model flexibility

Current developments in model design are towards allowing flexibility in model applications within a single modelling framework (Donigian and Carsel, 1992). Thus, the user should be able to make 'on-line' choices, in user-friendly menus, between alternative process descriptions, making the model as complicated or as simple as the particular application requires. As many relevant processes as possible should be available, and the user should be able to select those deemed necessary. Recent developments in several models illustrate this trend for increased model flexibility. Thus, LEACHP allows the user to choose between a mechanistic description of water flow and solute transport (Richards' equation and the CDE) and the mobile–immobile model as formulated in CALF (Hutson and Wagenet, 1993). Similarly, a number of user options in the MACRO model allow the user to customize the simulation to the required degree of complexity. PELMO and PRZM2 allow the user to specify alternative degradation models.

COMPARISONS OF MODEL PREDICTIONS AND FIELD MEASUREMENTS

Tables 4 and 5 summarize published studies where model simulations have been compared with measurements. Only those studies performed in the field, or under field-like conditions (e.g. lysimeters), are considered. For example, some model tests have been performed in laboratory lysimeter or column experiments with unnaturally large throughputs of water, but these are not listed in Tables 4 and 5. It should be noted that Table 4 cannot be considered a comprehensive summary of all model tests, but rather a summary of those studies revealed by a computer literature search in 1993. It is likely that many more model applications are unpublished or only published as internal, confidential, reports. Nevertheless, in the following sections, we attempt to draw some general conclusions from the published model tests listed in Tables 4 and 5. We first describe the range of soil/pesticide/climate scenarios that the models have been tested for, then discuss the types of measurements used in model comparisons, before summarizing the results of both single and comparative model tests.

Table 4 Summary of studies comparing model predictions with measurements

Model	Pesticides	Type of measurements[a]	Soil types[b]	Ref.
PRZM	Aldicarb, aldoxycarb	Coring, 800 cm	Various, mainly sandy	5
	Metalaxyl	Coring, 90 cm	Sand, sandy loam	1
	EDB	Drilling, 2000 cm	'Fine-textured'	8
	DBCP, EDB, TCP	Drilling, 2000 cm	'Fine-textured'	9
	Atrazine, metolachlor	Coring, 100 cm	Sand	17
	Carbofuran, chlorpyrifos	Lysimeter, 90 cm	Sand	17
	Atrazine	Coring, 26 cm	Sandy clay loam, sandy loam, loam	10
	Aldicarb, metolachlor	Coring, 120 cm	Sandy loam–sandy clay loam	16
	Aldicarb	Coring, 300 cm	Sandy loam, loamy sand	11
CALF	Atrazine, metribuzin	Coring, 15 cm	Sandy loam	12
	Monuron	Coring, 15 cm	Silty clay loam	13
	Simazine	Coring, 15 cm	Silty clay loam	14
	Chlorsulfuron	Coring, 15 cm	Sandy loam, silty clay loam	15
	Propyzamide, isoxaben, linuron, R-40244	Coring, 15 cm	Sandy loam	21
GLEAMS	Cyanazine, alachlor, atrazine	Coring, 10 cm, 40 cm	Silt loam, sand, sandy loam	6
	Fenamiphos	Coring, 30 cm	Sandy clay loam	7
CMLS	Simazine, Carbofuran, Metalaxyl	Coring, 180 cm	Sandy loam–silt loam	18
MACRO	Dichlorprop	Lysimeter, 105 cm	Silty clay	2
	Simazine, metamitron, methabenzthiazuron	Coring, 10 cm	Silt loam	3
	Alachlor	Suction cups, coring, 150 cm	Sandy loam–loamy sand	4
LEACHP	Aldicarb	Coring, 150 cm	Sandy loam	19
	DBCP	Suction cups, 150 cm	Clay loam	20

[a]Numbers refer to maximum depth of comparison between data and model.
[b]Dashes indicate topsoil–subsoil textures.

References
1 Carsel et al., 1986, Environ. Toxicol. Chem., **5**, 345–353.
2 Jarvis et al., 1991, Proc. ASAE Symp., 308–317.
3 Jarvis, 1995b, Ecol. Model., in press.
4 Jarvis, et al., 1995 Proc. BCPC Symposium, 161–170.
5 Jones, et al., 1986, Environ. Toxicol. Chem., **5**, 1027–1037.
6 Leonard et al., 1987, Trans. ASAE, **30**, 1403–1418.
7 Leonard et al., 1990, ASCE J. Irrig. Drain. Eng., **116**, 24–35.
8 Loague et al., 1989b Pacific Sci., **43**, 67–95.
9 Loague et al., 1989a Pacific Sci., **43**, 362–383.
10 Loague & Green, 1991, J. Contam. Hydrol., **7**, 51–73.
11 Lorber & Offutt, 1986, ACS Symp. Ser. **315**, 342–365.
12 Nicholls et al., 1982, Pestic. Sci., **12**, 484–494.
13 Nicholls et al., 1983, Aspects Appl. Biol., **4**, 485–494.
14 Nicholls et al., 1984, Weed Res., **24**, 37–49.
15 Nicholls et al., 1987, Proc. 1987 Brit. Crop Prot. Conf., Weeds, 549–556.
16 Parrish et al., 1992, J. Environ. Qual., **21**, 685–697.
17 Sauer et al., 1990, J. Environ. Qual., **19**, 727–734.
18 Sukop & Cogger, 1992, J. Environ. Sci. Health, **B27(5)**, 565–590.
19 Wagenet & Hutson, 1986, J. Environ. Qual., **15**, 315–322.
20 Wagenet et al., 1989, J. Environ. Qual., **18**, 78–84.
21 Walker, 1987, Weed Res., **27**, 143–152.

Table 5 Summary of comparative model tests

Study[d]	Models compared[a]	Pesticides	Type of measurements[b]	Soil types[c]
1	PRZM, GLEAMS, CMLS, LEACHP	Aldicarb	Coring, 800 cm	Sand
2	PRZM, GLEAMS	Alachlor, metribuzin, norflurazon	Coring, 60 cm	Sandy loam–sandy clay Loamy sand–sandy clay loam
3	LEACHP, LEACHA	Atrazine	Coring, 250 cm	Sandy loam
4	PRZM, CALF, GLEAMS, CMLS, MACRO, PESTLA, PELMO	Dichlorprop, bentazone	Lysimeters, 105 cm	Loamy sand–sand Sandy loam–sand Sandy clay loam–clay Silty clay loam–silt loam Silty clay–clay

[a]Including only those models discussed in this paper.
[b]Numbers refer to maximum depth of comparison between data and model.
[c]Dashes indicate topsoil–subsoil textures.
[d]References

1 Pennell et al., 1990, Water Resources Res., 11, 2679–2693.
2 Mueller et al., 1992, Environ. Toxicol. Chem. 11, 427–436.
3 Hutson & Wagenet, 1993, J. Environ. Qual., 22, 494–499.
4 Bergström & Jarvis (eds), 1994, special issue of J. Environ. Sci. Health, A29, 1061–1072.

Types of scenarios

In total, the models have been compared to measurements of the mobility and persistence of some 26 compounds (Tables 4 and 5). Clearly, this is only a small fraction of the total number of pesticides in use, although many of the model applications have been performed for some of the most widely used pesticides. Not surprisingly, the majority of the model applications relate to pesticides which are known or thought to be 'leachable'. Thus, the screening model described by Jury *et al.* (1987) places 15 of the 26 compounds in the 'high risk' category. PRZM, CMLS and LEACHP, in particular, have been applied to highly 'leachable' pesticides (e.g. aldicarb, DBCP, EDB, carbofuran). In contrast, the CALF and GLEAMS models have been applied to a larger proportion of 'non-leachable' pesticides. This may be due partly to differing purposes for which these models were developed. For example, the early impetus for developing the CALF model was the study of efficacy, persistence, and carry-over effects (Cook *et al.*, 1985; Nicholls *et al.*, 1987). An important component of the GLEAMS model is the estimation of pesticide lost in surface run-off and eroded material, a critical loss pathway for less mobile compounds.

Tables 4 and 5 also show that most model applications have been made in sandy-textured soils. Again, this is not surprising. Among the many reasons for this bias is the fact that sandy soils are often thought to represent a worst-case scenario for leaching, exhibit relatively small variability in leaching, are easy to sample and monitor, and best satisfy various model assumptions and limitations (Smith *et al.*, 1990). For example, structured soils are likely to exhibit preferential flow behaviour and would therefore invalidate many models currently in use. It can be noted here that although sandy soils may represent a worst-case scenario for leaching to groundwater aquifers, surface water vulnerability (via subsurface flow) may be greater in drained, fine-textured, structured soils which exhibit preferential flow (Jarvis, 1995a).

Types of measurements

In most applications, the models have been compared to concentration–depth profiles of the pesticide of interest, usually obtained by soil coring (Tables 4 and 5). This may represent an adequate test of model predictions of leaching for the highly mobile compounds, where a large proportion of the applied amount moves deep into the soil. However, soil coring measurements cannot be considered a generally valid test of the ability of a model to predict pesticide leaching, since it is often the movement of the 'leading edge' and not the main mass of pesticide which is of greatest interest. Environmentally significant amounts of a particular compound may be leached at concentrations smaller than the detection limit which can be obtained from soil core samples. For example, detection limits for core sampling are often in the range in the low parts per billion range (e.g. 1–5 ppb for aldicarb, Jones et al., 1986; 8–30 ppb for metribuzin, norflurazon and alachlor, Mueller et al., 1992). The allowable concentration set by the EC for a single pesticide in drinking water supplies (= 0.1 ppb) is one to two orders of magnitude smaller. Such small concentrations are easier to detect in water samples, where the sample volume is larger. This sampling problem with soil coring may be especially critical for the less mobile compounds, where the overwhelming proportion of the pesticide is broken down and/or remains near the soil surface and it is only the possible occurrence of preferential flow which constitutes a leaching risk. Preferential movement of small amounts of pesticide may not be detected by soil coring techniques.

To properly validate leaching models, flux measurements are generally required, such as water samples drained from either lysimeters or drained field plots. Unfortunately, there are very few published studies of this type (Table 4). Sauer et al. (1990) compared PRZM predictions with measurements of the mass distribution and leaching of carbofuran in undisturbed lysimeters. Except for the near-surface zone where the model underestimated residue concentrations, PRZM produced acceptable simulations of the concentration–depth distribution of carbofuran within the lysimeters. However, measured leaching of the compound from the lysimeters was underestimated by up to one order of magnitude. This emphasizes that leaching models should be validated against leaching measurements.

One further advantage of flux measurements is that the hydrological components of the model can be tested independently of the transport and transformation submodels. Thus, in the comprehensive model test summarized by Bergström and Jarvis (1994), soil water discharges from field lysimeters were compared to predictions of eight different models. All the models were able to reproduce the observed discharges, but this was achieved in some cases by extensive model calibration. Unfortunately, no measurements of soil water content were made in the lysimeters that could have constrained model users during the calibration process. There are few other examples of tests of the water balance and water flow components of pesticide leaching models. Wagenet and Hutson (1986) demonstrated reasonable agreement between limited measurements of soil water content and LEACHP predictions. With calibration of model parameters related to root water uptake, Jarvis (1995b) demonstrated on excellent agreement between MACRO simulations and detailed measurements of soil water content and pressure head made during a 3-year period in a silty soil.

Tests of the PRZM model

Table 4 shows that of the existing models, PRZM has been by far the most widely tested. In a comprehensive study, Jones *et al.* (1986) summarized comparisons of PRZM model predictions of aldicarb and aldoxycarb leaching with measurements made on 34 plots in a range of soils and climates in the USA. A good agreement between observed and predicted leaching depths was found (varying from 0.5 to 8 m), where 'leaching depth' was defined as the deepest core sample containing the compound above the detection limit (= 5 ppb). Several additional aspects of this study are worth noting: first, while many model parameters were independently estimated, degradation rate constants were derived by calibrating against the field measurements of residue concentrations. Second, Jones *et al.* (1986) showed that while leaching depths were reasonably accurately predicted for the data set as a whole, concentration–depth profiles at individual sites were not, at least not without calibration of model transport parameters.

Carsel *et al.* (1986) compared PRZM predictions of metalaxyl distribution in soil with data obtained at two sites in Florida and Maryland. Again, degradation rate coefficients were obtained by calibrating the model against the data. For the Maryland site, degradation was best described by a phased first-order model, with the rate coefficient in the model reduced by an order of magnitude 30 days after application. This probably reflects the lumping of different dissipation pathways into a single first-order degradation rate coefficient in the model. With the model calibrated in this way, PRZM accurately simulated metalaxyl concentration–depth distributions in the soil.

Lorber and Offutt (1986) compared concentration–depth profiles of aldicarb measured at three sites in North Carolina and Wisconsin with PRZM model simulations. Accurate simulations were achieved, but only after calibration of both the pesticide half-life and sorption constant. They noted that the calibrated value of the sorption constant was larger than the calculated value based on the known k_{oc} for aldicarb and the soil organic carbon contents. Several explanations for this discrepancy were offered, including an overestimation of downward water flow in the summer (no upward fluxes of water and pesticide are calculated in PRZM) and kinetic sorption and/or preferential water flow.

Loague *et al.* (1989a, b) compared PRZM model predictions of leaching for the highly mobile, volatile, compounds EDB, DBCP and TCP to concentration–depth profiles measured to 20 m depth in structured fine-textured soils in Hawaii. The model failed to match the observed data, with concentrations generally overpredicted by several orders of magnitude. This generally poorer model performance can probably be attributed to a lack of model calibration. Although a separate preprocessing model was used to calculate initial volatilization losses at the soil surface, significant errors may have arisen because the version of PRZM used could not account for volatilization.

Sauer *et al.* (1990) compared PRZM simulations with measurements of atrazine and metolachlor leaching in a sandy soil in Wisconsin. Once again, estimates of the half-lives were obtained by calibration against the field data. PRZM overestimated leaching of both compounds and failed to predict the observed accumulation of residues near the soil surface 168 days after application. This was attributed to pesticide uptake by the maize plants and subsequent release following leaf fall. Another possible explanation is that PRZM failed to account properly for upward fluxes of water and pesticide to the soil surface during the drier than normal summer.

Loague and Green (1991) tested the ability of the PRZM model to predict atrazine leaching at a field site in Georgia. Objective statistical tests indicated a poor model performance in predicting the shape of the observed concentration–depth profiles. This was mainly attributed to inadequate parameter estimates due to lack of both site-specific data and model calibration, but also to the inability of the model to account for preferential flow processes.

Parrish *et al.* (1992) tested the PRZM model against field measured concentration–depth profiles of aldicarb, metolachlor and bromide. The model performed satisfactorily for the two pesticides, particularly for metolachlor where leaching was restricted to the surface 30 cm. However, model predictions of the movement of the non-reactive bromide tracer were poorer. The measurements indicated a significant loss of bromide below the measurement depth, while the distinct unimodal concentration–depth profile predicted by the model was not matched in the measurements. These observations were tentatively attributed to the effects of preferential flow.

Tests of other models

Table 4 shows that both CALF and GLEAMS have largely been compared to measurements of pesticide residues in the surface soil to depths of c. 30 cm. This reflects the original purposes for which the models were developed, including degradation processes, pesticide efficacy, carry-over effects, and surface run-off/soil erosion. Both models have been shown to simulate successfully the amounts and distribution of pesticides in the topsoil at a number of field sites (Nicholls *et al.*, 1982; Leonard *et al.*, 1987, 1990). However, neither model has been extensively tested for leaching predictions.

Wagenet and Hutson (1986) demonstrated good agreement between field measurements and LEACHP predictions of aldicarb residues in a sandy loam soil. The compound was largely confined to the surface 30 cm, with only small concentrations detected to 60 cm depth. The sampling depth intervals were also rather coarse (surface–30 cm and 30–60 cm). Thus, the model was validated for degradation processes, but not really for leaching predictions. Wagenet *et al.* (1989) presented simulations of the fate of the volatile compound DBCP using the LEACHP model and comparisons with limited data obtained with solution samplers. However, it was not possible to evaluate the performance of LEACHP due to uncertainties in the experimental data.

Sukop and Cogger (1992) compared field measurements of leaching depths of carbofuran, metalaxyl and simazine with CMLS model predictions. An excellent agreement was found for all three compounds when experimentally determined sorption constants were used as model input. Using published k_{oc} values and the known soil organic carbon content also proved adequate to predict carbofuran and simazin transport, but considerably overestimated leaching of metalaxyl since sorption of this compound was not controlled by organic matter.

Jarvis *et al.* (1995) attempted to quantify the level of accuracy to be expected in purely predictive applications of MACRO without site-specific data and model calibration. Alachlor concentrations measured by core and suction cup samplers in a sandy loam soil were compared with MACRO model predictions based on parameter estimates derived from a combination of default values supplied with the model, mean properties for the soil mapping unit (Hall series) obtained from the SEISMIC database system (Hollis *et al.*, 1993), pedo-transfer functions, and compound properties (k_{oc} and degradation half-life) obtained from experiments carried out at the sites. The model performed surprisingly well. Residue concentrations measured at 0–10 cm depth were closely matched, while predicted and measured concentrations in suction cups at 25 and 40 cm depth were mostly within a factor of two. However, the model failed to match the rapid breakthrough of the compound to suction cups at 150 cm, although it was not clear whether this reflected shortcomings in the model itself, in the parameterization, or even errors in the measurements caused by disturbance of the soil.

Comparative model tests

Only PRZM has been extensively compared to field measurements. Nevertheless, most of the models listed in Tables 1–3 have been included in a limited number of recent comparative model tests. These studies are summarized in Table 5. Pennell *et al.* (1990) compared LEACHP, GLEAMS, CMLS and PRZM predictions with concentrations of bromide, aldicarb and its major metabolites measured to 8 m depth in a sandy soil in Florida. Objective statistical criteria were used to judge model performance.

Only a limited test could be performed for GLEAMS since it only predicts leaching at the base of the root zone (in this case at 1.8 m depth). Nevertheless, GLEAMS underpredicted both bromide and aldicarb leaching from the root zone, with too large concentrations predicted close to the soil surface 30 days after application. Compared to GLEAMS, the remaining models gave better predictions of solute leaching from the root zone. LEACHP and PRZM gave similar predictions of solute distributions. These predictions were better than GLEAMS at early times, but still did not accurately reproduce the measurements, especially later in the experiment. In the case of aldicarb, this error seemed largely due to an underestimate of degradation. Although underestimated late in the experiment, the movement of the centres of mass of bromide, aldicarb and metabolites were reasonably well predicted by CMLS, LEACHP and PRZM.

Mueller et al. (1992) compared PRZM and GLEAMS model predictions with soil coring measurements made to 60 cm depth for three pesticides (norflurazon, metribuzin and alachlor) in two soils (Table 5). The models gave similar predictions of concentration–depth profiles, slightly overestimating downward movement and underestimating concentrations near the soil surface. This was tentatively ascribed to hysteresis in pesticide sorption. However, as discussed above, model errors in the calculations of soil water flow may be another explanation. Nevertheless, both models gave statistically adequate predictions of both the total pesticide residues and also the depth to the limit of detection. With respect to the latter, the PRZM model performed marginally better than GLEAMS.

Bergström and Jarvis (1994) summarized the results of a comprehensive study to compare and evaluate the performance of eight models against a common data set consisting of lysimeter leaching experiments with two herbicides (bentazone, dichlorprop) applied to five soils of contrasting texture (Table 5). It was concluded that, in general, most of the models produced strikingly similar results and also experienced similar problems. In particular, preferential flow apparently occurred in all five soils, not only in the structured soils but also in two sandy soils. Not surprisingly, those models which do not account for preferential flow (e.g. PRZM, CALF, PELMO) could not reproduce the early breakthrough of small quantities of herbicide observed in all the lysimeters. This was most apparent for the structured soils and for the herbicide dichlorprop. However, all the models produced reasonable predictions of the total mass of bentazone leached from one sandy soil, despite the occurrence of preferential flow. This observation supports one conclusion drawn from a desk study with the MACRO model (Jarvis, 1995a), that preferential flow may have only limited effects on the mass balance of highly mobile compounds. The second major problem experienced by the model users was that the degradation rates measured in laboratory incubation experiments did not seem to reflect conditions in the lysimeters. For dichlorprop, it was concluded that the rapid metabolic degradation observed in the laboratory could not have occurred in the field. For bentazone, the dissipation rates in the lysimeters appeared larger than those measured in the laboratory, perhaps due to additional loss pathways (photolysis).

CONCLUDING REMARKS

Research into the interplay of factors affecting pesticide fate and mobility in soils has developed at a rapid pace in the last 15 years. The development of simulation models has proved useful, both in guiding experimentation and in integrating, testing and improving our understanding of the dynamic and complex nature of the soil–plant–pesticide system. Pesticide leaching models are also now being used for management purposes, by both industry and public authorities, within the registration process. The use of models for management purposes implies a different set of needs and priorities. Most important of these are ease of parameterization and predictive reliability. To some extent, these requirements are conflicting, since reliability implies the need for mechanistic models which minimize 'model error', but at the cost of increasing data demands. This conflict can partly be resolved by incorporating 'on-line' help menus and built-in parameter estimation procedures in mechanistic models. Nevertheless, in general, a management model will lie some way behind the 'research front', attempting to find a reasonable balance or compromise between mechanistic treatments of processes and ease of use of the model in terms of parameterization. What is clear, however, is that where mechanistic treatments of relevant processes are possible for little extra difficulty in parameter estimation, then these should be included in management models in preference to simple empirical approaches. If not, models may suffer from limited applicability and poor reliability. One example of this is the effect of temperature on degradation. Soil temperature is relatively simple to calculate mechanistically, requiring no additional parameters (both thermal conductivity and heat capacity can be predicted from widely available soil physical properties). Despite this, soil temperature is not calculated at all in PRZM, GLEAMS and CMLS and is only treated empirically in CALF. This may make these models unreliable, or at best difficult to use, under widely varying climatic conditions.

Current research in the field of non-equilibrium processes, such as preferential flow and kinetic sorption, and in the micro-scale interactions between diffusive transport, sorption and degradation, points to the need for two- and multi-region models which recognize the complex, heterogeneous, nature of the soil system. However, it remains to be seen to what extent such models can be used for management purposes. There are certainly considerable difficulties in deriving efficient and reliable parameter estimation procedures, but these may not be insurmountable in the long term. For example, carefully designed batch and column experiments conducted under controlled laboratory conditions (Ogram *et al.*, 1985; Gamerdinger *et al.*, 1993) may offer one solution to the problem of determining parameter values required by non-equilibrium models.

To date, most model applications have involved some calibration, typically either of the hydrology submodel or of pesticide sorption and degradation parameters. Therefore, the predictive accuracy of existing pesticide leaching models is still not well documented. However, it is likely to be inherently poor, because we are usually trying to predict the fate of a small fraction of the applied substance when estimates of many sensitive model parameters are uncertain and spatially variable. This is particularly true for the parameters controlling sorption and degradation processes. Thus, environmental concentrations predicted by deterministic models must be qualified by large safety factors. Alternatively, probabilistic approaches which explicitly account for uncertainty in the input parameters can provide estimates of the likelihood of exceeding given critical concentrations.

A pesticide leaching model to be used for management purposes should include the following desirable attributes and features:

(1) mechanistic descriptions of all relevant processes affecting leaching, provided that parameterization of such descriptions is not too difficult (i.e. no unnecessary empiricism).
(2) flexibility in customizing simulations to the particular requirements of each application.
(3) automatic procedures for estimating model parameters from widely available soils information (pedo-transfer functions).
(4) a user-friendly interface, with 'on-line' help, and access to databases on soils, weather, cropping patterns and pesticide properties.
(5) procedures to account for the uncertainty of predictions, making estimates of the probability of defined critical environmental concentrations being exceeded.

Unfortunately, modelling systems which satisfy all these criteria do not exist at present. However, in the near future, it is likely that pesticide leaching models, incorporated into computer-based decision-support systems will include many, if not all, of these features.

REFERENCES

Addiscott, T. M. (1977). 'A simple computer model for leaching in structured soils', *J. Soil Sci.*, **28**, 554–563.

Addiscott, T. M., and Wagenet, R. J. (1985). 'Concepts of solute leaching in soils: a review of modelling approaches', *J. Soil Sci.*, **36**, 411–424.

Bergström, L., and Jarvis, N. J. (1994). 'Evaluation and comparison of pesticide leaching models for registration purposes: overview'. *J Environ. Sci. Health*, **A29**, 1061–1072.

Beven, K. and Germann, P. (1982). 'Macropores and water flow in soils', *Water Resources Res.*, **18**, 1311–1325.

Boesten, J. J. T. I. (1991). 'Sensitivity analysis of a mathematical model for pesticide leaching to groundwater', *Pestic. Sci.*, **31**, 375–388.

Boesten, J. J. T. I. (1994). 'Simulation of bentazon leaching in sandy loam soil from Mellby (Sweden) with the PESTLA model', *J. Environ. Sci. Health*, **A29**, 1231–1253.

Boesten, J. J. T. I., and van der Linden, A. M. A. (1991). 'Modeling the influence of sorption and transformation on pesticide leaching and persistence', *J. Environ. Qual.*, **20**, 425–435.

Bottcher, A. B., Monke, E. J., and Huggins, L. F. (1981). 'Nutrient and sediment loadings from a subsurface drainage system', *Trans. ASAE*, **24**, 1221–1226.

Bouma, J. (1981). 'Soil morphology and preferential flow along macropores', *Agric. Water Manage.*, **3**, 235–250.

Brooks, R. H., and Corey, A. T. (1964). 'Hydraulic properties of porous media', *Colorado State Univ. Hydrol. Paper 3*, Ft. Collins, CO.

Brouwer, W. W. M. (1994). 'Use of simulation models for registration purposes: evaluation of pesticide leaching to groundwater in the Netherlands', *J. Environ. Sci. Health*, **A29**, 1117–1132.

Calvet, R. (1993). 'Comments on the characterization of pesticide sorption in soils', *Proceedings of the IX Symposium Pesticide Chemistry, Mobility and Degradation of Xenobiotics*, Piacenza, Oct. 1993, 277–288.

Carsel, R. F., Imhoff, J. C., Kittle, J. L., and Hummel, P. R. (1991). 'Development of a database and model parameter analysis system for agricultural soils', *J. Environ. Qual.*, **20**, 642–647.

Carsel, R. F., Jones, R. L., Hansen, J. L., Lamb, R. L., and Anderson, M. P. (1988a). 'A simulation procedure for groundwater quality assessments of pesticides', *J. Contam. Hydrol.*, **2**, 125–138.

Carsel, R. F., Mulkey, L. A., Lorber, M. N., and Baskin, L. B. (1985). 'The pesticide root zone model (PRZM): a procedure for evaluating pesticide leaching threats to groundwater', *Ecol. Model.*, **30**, 46–69.

Carsel, R. F., Nixon, W. B., and Ballantine, L. G. (1986). 'Comparison of pesticide root zone model predictions with observed concentrations for the tobacco pesticide metalaxyl in unsaturated zone soils', *Environ. Toxicol. Chem.*, **5**, 345–353.

Carsel, R. F., Parrish, R. S., Jones, R. L., Hansen, J. L., and Lamb, R. L. (1988b). 'Characterizing the uncertainty of pesticide leaching in agricultural soils', *J. Contam. Hydrol.*, **2**, 111–124.

Carsel, R. F., Smith, C. N., Mulkey, L. A., Dean, J. D., and Jowise, P. (1984). *Users Manual for the Pesticide Root Zone Model (PRZM) Release 1*, EPA/600/3-84/109, US Environmental Protection Agency, Athens, GA, USA.

Cook, D. A., Bromilow, R. H., and Nicholls, P. H. (1985). 'The extent and efficacy of granular pesticide usage to control ectoparasitic nematodes on sugar beet', *Crop Protect.*, **4**, 444–457.

Donigian, A. S., and Carsel, R. F. (1992). 'Developing computer simulation models for estimating risks of pesticide use: research vs. user needs', *Weed Technol.*, **6**, 677–682.

Flury, M., Flühler, H., Jury, W. A., and Leuenberger, J. (1994). 'Susceptibility of soils to preferential flow of water', *Water Resources Res.*, **30**, 1945–1954.

Gamerdinger, A. P., Dowling, K. C., and Lemley, A. T. (1993). 'Miscible displacement and theoretical techniques for simultaneous study of pesticide sorption and degradation during transport', in *Sorption and Degradation of Pesticides and Organic Chemicals in Soil*, SSSA Special Publ. no. 32, Soil Science Society of America, Madison, WI, USA, pp. 115–123.

Gamerdinger, A. P., Wagenet, R. J., and van Genuchten, M. T. (1990). 'Application of two-site/two-region models for studying simultaneous nonequilibrium transport and degradation of pesticides', *Soil Sci. Soc. Am. J.*, **54**, 957–963.

Gerke, H. H., and van Genuchten, M. T. (1993). 'Evaluation of a first-order water transfer term for variably-saturated dual-porosity flow models', *Water Resources Res.*, **29**, 1225–1238.

Ghodrati, M., and Jury, W. A. (1992). A field study of the effects of soil structure and irrigation method on preferential flow of pesticides in unsaturated soil', *J. Contam. Hydrol.*, **11**, 101–125.

Green, R. E., and Karickhoff, S. W. (1990). 'Sorption estimates for modeling', in *Pesticides in the Soil Environment: Processes, Impacts and Modeling*' (ed. H. H. Cheng), SSSA Book Series, no. 2, Soil Science Society of America, Madison, WI, USA, pp. 79–101.

Hendrickx, J. M., Dekker, L. W., and Boersma, O. H. (1993). 'Unstable wetting fronts in water-repellent field soils', *J. Environ. Qual.*, **22**, 109–118.

Herner, A. (1992). 'The USDA-ARS pesticide properties database: a consensus data set for modelers', *Weed Technol.* **6**, 749–752.

Hillel, D. (1987). 'Unstable flow in layered soils: a review', *Hydrol. Proc.*, **1**, 143–147.

Hollis, J. M., and Brown, C. D. (1993). 'An integrated approach to aquifer vulnerability mapping', in *Proceedings of the IX Symposium on Pesticide Chemistry, Mobility and Degradation of Xenobiotics*, Piacenza, Oct. 1993, pp. 633–644.

Hollis, J. M., Hallett, S. H., and Keay, C. A. (1993). 'The development and application of an integrated database for modelling the environmental fate of herbicides', in *Proceedings of the 1993 BCPC Conference—Weeds, Brighton*, BCPC, Lavenham Press, Lavenham, Surrey, pp. 1355–1364.

Hollis, J. M., and Woods, S. M. (1989). 'The measurement and estimation of saturated hydraulic conductivity', *SSLRC Research Report*, Soil Survey and Land Research Centre, Silsoe, UK, 19 pp.

Hutson, J. L., and Wagenet, R. J. (1991). 'Simulating nitrogen dynamics in soils using a deterministic model', *Soil Use Manag.*, **7**, 74–78.

Hutson, J. L., and Wagenet, R. J. (1992). LEACHM: Leaching estimation and chemistry model. A process-based model of water and solute movement, transformations, plant uptake and chemical reactions in the unsaturated zone. Version 3', Department of Soil, Crop and Atmospheric Sciences Research Series no. 92-3, New York State College of Agriculture and Life Sciences, Cornell University, Ithaca, NY, USA.

Hutson, J. L., and Wagenet, R. J. (1993). 'A pragmatic field-scale approach for modeling pesticides', *J. Environ. Qual.*, **22**, 494–499.

Hutson, J. L., and Wagenet, R. J. (1995). 'A multi-region model describing water flow and solute transport in heterogeneous soils', *Soil Sci. Soc. Am. J.*, in press.

Jarvis, N. J. (1991). 'MACRO—a model of water movement and solute transport in macroporous soils', *Reports and Dissertations no. 9*, Department of Soil Sciences, Swedish University of Agricultural Sciences, Uppsala, Sweden, 58 pp.

Jarvis, N. J. (1994). 'The MACRO model (Version 3.1). Technical description and sample simulations', *Reports and Dissertations no. 19*, Department of Soil Sciences, Swedish University of Agricultural Sciences, Uppsala, Sweden, 51 pp.

Jarvis, N. J. (1995a). 'The implications of preferential flow for the use of simulation models in the registration process', in *Proceedings of the 5th International Workshop, Environmental Behaviour of Pesticides and Regulatory Aspects*, (eds. A. Copin, G. Houins, L. Pussemier, J. F. Salombier) Brussels, April 1994, 464–469.

Jarvis, N. J. (1995b). 'Simulation of soil water dynamics and herbicide persistence in a silt loam soil using the MACRO model', *Ecol. Model.*, in press.

Jarvis, N. J., Bergström, L., and Stenström, J. (1991). 'A model to predict pesticide transport in macroporous field soils', in *Proceedings of the National Symposium on Preferential Flow* (eds. T. Gish and A. Shirmohammadi), ASAE, St. Joseph, MI, USA, pp. 308–317.

Jarvis, N. J., Larson, M., Fogg, P., and Carter, A. D. (1995). 'Validation of the dual-porosity model MACRO for assessing pesticide fate and mobility in soils', in *Proceedings of the BCPC Symposium, Pesticide Movement to Water*, (eds. A. Walker, R. Allen, S. W. Bailey, A. M. Blair, C. D. Brown, P. Günther, C. R. Leake and P. H. Nicholls) Warwick, April 1995, 161–170.

Jones, R. L., Black, G. W., and Estes, T. L. (1986). 'Comparison of computer model predictions with unsaturated zone field data for aldicarb and aldoxycarb', *Environ. Toxicol. Chem.*, **5**, 1027–1037.

Jury, W. A., Elabd, H., and Resketo, M. (1986). 'Field study of napropamide movement through unsaturated soil. *Water Resources Res.*, **22**, 749–755.

Jury, W. A., Focht, D. D., and Farmer, W. J. (1987). 'Evaluation of pesticide groundwater pollution potential from standard indices of soil-chemical adsorption and biodegradation', *J. Environ. Qual.*, **16**, 422–428.

Kladivko, E. J., van Scoyoc, G. E., Monke, E. J., Oates, K. M., and Pask, W. (1991). 'Pesticide and nutrient movement into subsurface tile drains on a silt loam soil in Indiana', *J. Environ. Qual.*, **20**, 264–270.

Klein, M. (1991). *PELMO: Pesticide Leaching Model*. Fraunhofer Institute, Schmallenberg, Germany.

Knisel, W. G., Davis, F. M., and Leonard, R. A. (1992). *GLEAMS Version 2.0, Part III: User Manual*, USDA-ARS, Southeast Watershed Research Lab., Tifton, GA, USA.

Kung, K-J. S. (1991). 'Preferential flow in a sandy vadose zone. I. Field observation', *Geoderma*, **46**, 51–58.

Leonard, R. A., Knisel, W. G., Davis, F. M., and Johnson, A. W. (1990). 'Validating GLEAMS with field data for fenamiphos and its metabolites', *ASCE J. Irrig. Drain. Eng.*, **116**, 24–35.

Leonard, R. A., Knisel, W. G., and Still, D. A. (1987). 'GLEAMS: groundwater loading effects of agricultural management systems', *Trans. ASAE*, **30**, 1403–1418.

Loague, K. M., Giambelluca, T. W., Green, R. E., Liu, C. C. K., Liang, T. C., and Oki, D. S. (1989a). 'Simulation of organic chemical movement in Hawaii soils with PRZM: 1. Predicting deep penetration of DBCP, EDB, and TCP', *Pacific Sci.*, **43**, 362–383.

Loague, K. M., and Green, R. E. (1991). 'Statistical and graphical methods for evaluating solute transport models: overview and application', *J. Contam. Hydrol.*, **7**, 51–73.

Loague, K. M., Green, R. E., Liu, C. C. K., and Liang, T. C. (1989b). 'Simulation of organic chemical movement in Hawaii soils with PRZM: 1. Preliminary results for ethylene dibromide', *Pacific Sci.*, **43**, 67–95.

Lorber, M. N., and Offutt, C. K. (1986). 'A method for the assessment of ground water contamination potential. Using a pesticide root zone model (PRZM) for the unsaturated zone', in *Evaluation of Pesticides in Groundwater* (eds. W. Y. Gardner, R. C. Honeycutt and H. N. Nigg), ACS Symposium Series 315, American Chemical Society, Washington, DC, pp. 342–365.

McKenzie, N. J., Smettem, K. R. J., and Ringrose-Voise, A. J. (1991). 'Evaluation of methods for inferring air and water properties of soils from field morphology', *Aust. J. Soil Res.*, **29**, 587–602.

Mualem, Y. (1976). 'A new model for predicting the hydraulic conductivity of unsaturated porous media'. *Water Resources Res.*, **12**, 513–522.

Mueller, T. C., Jones, R. E., Bush, P. B., and Banks, P. A. (1992). 'Comparison of PRZM and GLEAMS computer model predictions with field data for alachlor, metribuzin and norflurazon leaching', *Environ. Toxicol. Chem.*, **11**, 427–436.

Mullins, J. A., Carsel, R. F., Scarbrough, J. E., and Ivery, A. M. (1993). *PRZM-2, a Model for Predicting Pesticide Fate in the Crop Root and Unsaturated Soil Zones: Users Manual for Release 2.0*, EPA/600/R-93/046, US Environmental Protection Agency, Athens, GA, USA.

Nicholls, P. H. (1994). 'Simulation of the movement of bentazon in soils using the CALF and PRZM models', *J. Environ. Sci. Health*, **A29**, 1157–1166.

Nicholls, P. H., Briggs, G. G., and Evans, A. A. (1983). 'Simulation of herbicide movement in soils in winter', *Aspects Appl. Biol.*, **4**, 485–494.

Nicholls, P. H., Briggs, G. G., and Evans, A. A. (1984). 'The influence of water solubility on the movement and degradation of simazine in a fallow soil', *Weed Res.*, **24**, 37–49.

Nicholls, P. H., Walker, A., and Baker, R. J. (1982). 'Measurement and simulation of the degradation of atrazine and metribuzin in a fallow soil', *Pestic. Sci.*, **12**, 484–494.

Nicholls, P. H., Walker, A., and Evans, A. A. (1987). 'The behaviour of chlorsulfuron and metsulfuron in soils in relation to incidents of injury to sugar beet', in *Proceedings of the 1987 BCPC Conference—Weeds*, BCPC, Farnham, Surrey, pp. 549–556.

Nkeddi-Kizza, P., Biggar, J. W., Selim, H. M., van Genuchten, M. T., Wierenga, P. J., Davidson, J. M., and Nielsen, D. R. (1984). 'On the equivalence of two conceptual models for describing ion exchange during transport through an aggregated oxisol', *Water Resources Res.*, **20**, 1123–1130.

Nofziger, D. L., and Hornsby, A. G. (1986). 'A micro-computer based management tool for chemical movement in soil', *Appl. Agric. Res.*, **1**, 50–56.

Nofziger, D. L., Chen J.-S., and Haan, C. T. (1994).'Evaluating the chemical movement in layered soil model as a tool for assessing risk of pesticide leaching to groundwater', *J. Environ. Sci. Health*, **A29**, 1133–1155.

Ogram, A. V., Jessup, R. E., Ou, L. T., and Rao, P. S. C. (1985). 'Effects of sorption on biological degradation rates of (2,4-dichlorophenoxy)acetic acid in soils', *Appl. Environ. Microbiol.*, **49**, 582–587.

Parrish, R. S., Smith, C. N., and Fong, F. K. (1992). 'Tests of the pesticide root zone model and the aggregate model for transport and transformation of aldicarb, metolachlor, and bromide', *J. Environ. Qual.*, **21**, 685–697.

Pennell, K. D., Hornsby, A. G., Jessup, R. E., and Rao, P. S. C. (1990). 'Evalution of five simulation models for predicting aldicarb and bromide behaviour under field conditions', *Water Resources Res.*, **26**, 2679–2693.

Petach, M. C., Wagenet, R. J., and DeGloria, S. D. (1991). 'Regional water flow and pesticide leaching using simulations with spatially distributed data', *Geoderma*, **48**, 245–269.

Rao, P. S. C., Bellin, C. A., and Brusseau, M. L. (1993). 'Coupling biodegradation of organic chemicals to sorption and transport in soils and aquifers: paradigms and paradoxes', in *Sorption and Degradation of Pesticides and Organic Chemicals in Soil*, SSSA Special Publ. no. 32, Soil Science Society of America, Madison, WI, USA, pp. 1–26.

Rao, P. S. C., and Davidson, J. M. (1980). 'Estimation of pesticide retention and transformation parameters required in nonpoint source pollution models', in *Environmental Impact on Nonpoint Source Pollution* (eds. M. R. Overcash and J. M. Davidson), Ann Arbor Sci. Publ., Ann Arbor MI, pp. 23–68.

Rawls, W. J., Ahuja, L. R., and Brakensiek, D. L. (1992). 'Estimating soil hydraulic properties from soils data', in *Indirect Methods for Estimating the Hydraulic Properties of Unsaturated Soils* (eds. M. T. van Genuchten, F. J. Leij and L. J. Lund), Univ. California, Riverside, CA, USA, pp. 329–340.

Russell, M. H., and Layton, R. J. (1992). 'Models and modeling in a regulatory setting: considerations, applications and problems', *Weed Technol.*, **6**, 673–676.

Russell, M. H., Layton, R. J., and Tillotson, P. M. (1994). 'The use of pesticide leaching models in a regulatory setting: an industrial perspective', *J. Environ. Sci. Health*, **A29**, 1105–1116.

Sauer, T. J., Fermanich, K. J., and Daniel, T. C. (1990). 'Comparison of the pesticide root zone model simulated and measured pesticide mobility under two tillage systems', *J. Environ. Qual.*, **19**, 727–734.

Scow, K. M. (1993). 'Effects of sorption–desorption and diffusion processes on the kinetics of biodegradation of organic chemicals in soil', in *Sorption and Degradation of Pesticides and Organic Chemicals in Soil*, SSSA Special Publ. no. 32, Soil Science Society of America, Madison, WI, USA, pp. 73–114.

Scow, K. M., and Hutson, J. L. (1992). 'Effect of diffusion and sorption on the kinetics of biodegradation: theoretical considerations', *Soil Sci. Soc. Am. J.*, **56**, 119–127.

Smith, C. N., Parrish, R. S., and Brown, D. S. (1990). 'Conducting field studies for testing pesticide leaching models', *Int. J. Environ. Anal. Chem.*, **39**, 3–21.

Soulas, G. (1982). 'Mathematical model for microbial degradation of pesticides in the soil', *Soil Biol. Biochem.*, **14**, 107–115.

Steenhuis, T. S., and Parlange, J.-Y. (1991). 'Preferential flow in structured and sandy soils', in *Proceedings of the National Symposium on Preferential Flow* (eds. T. Gish and A. Shirmohammadi), ASAE, St. Joseph, MI, USA, pp. 12–21.

Stenström, J. (1992). 'Rate determining factors for the decomposition of pesticides in soil', in *Proceedings of International Symposium, Environmental Aspects of Pesticide Microbiology*', Sigtuna, Sweden, Department of Microbiology, Swedish University of Agricultural Sciences, Uppsala, Sweden, pp. 141–146.

Sukop, M., and Cogger, C. G. (1992). 'Adsorption of carbofuran, metalaxyl, and simazine: k_{oc} evaluation and relation to soil transport', *J. Environ. Sci. Health*, **B27**, 565–590.

Tietje, O., and Tapkenhinrichs, M. (1993). 'Evaluation of pedo-transfer functions', *Soil Sci. Soc. Am. J.*, **57**, 1088–1095.

van Genuchten, M. T. (1980). 'A closed-form equation for predicting the hydraulic conductivity of unsaturated soils', *Soil Sci. Soc. Am. J.*, **44**, 892–898.

van Genuchten, M. T., and Wagenet, R. J. (1989). 'Two-site/two-region models for pesticide transport and degradation: theoretical development and analytical solutions', *Soil Sci. Soc. Am. J.*, **53**, 1303–1310.

van Genuchten, M. T., and Wierenga, P. J. (1974). 'Simulation of one-dimensional solute transfer in porous media', *Agricultural Experimental Station Bulletin 628*, New Mexico State Univ., Las Cruces, New Mexico, USA, 40 pp.

van Genuchten, M. T., and Wierenga, P. J. (1976). 'Mass transfer in sorbing porous media. I. Analytical solutions', *Soil Sci. Soc. Am. J.*, **40**, 473–480.

Vereecken, H., Maes, J., Feyen, J., and Darius, P. (1989). 'Estimating the soil moisture retention characteristic from texture, bulk density, and carbon content', *Soil Sci.*, **148**, 389–403.

Wagenet, R. J., Bouma, J., and Grossman, R. B. (1991). 'Minimum data sets for use of soil survey information in soil interpretive models', in *Spatial Variabilities of Soils and Landforms*, SSSA Special Publication no. 28, Soil Science Society of America, Madison, WI, USA, pp. 161–182.

Wagenet, R. J., and Hutson, J. L. (1986). 'Predicting the fate of nonvolatile pesticides in the unsaturated zone', *J. Environ. Qual.*, **15**, 315–322.

Wagenet, R. J., Hutson, J. L., and Biggar, J. W. (1989). 'Simulating the fate of a volatile pesticide in unsaturated soil: a case study with DBCP', *J. Environ. Qual.*, **18**, 78–84.

Wagenet, R. J., and Rao, P. S. C. (1990). 'Modeling pesticide fate in soils', in *Pesticides in the Soil Environment: Processes, Impacts and Modeling*' (ed. H. H. Cheng), SSSA Book Series, no. 2, Soil Science Society of America, Madison, WI, USA, pp. 351–399.

Walker, A. (1974). 'A simulation model for prediction of herbicide persistence', *J. Environ. Qual.*, **3**, 396–401.

Walker, A. (1987). 'Evaluation of a simulation model for prediction of herbicide movement and persistence in soil', *Weed Res.*, **27**, 143–152.

Walker, A., and Roberts, S. J. (1993). 'Degradation, biodegradation and enhanced biodegradation', in *Proceedings of the IX Symposium on Pesticide Chemistry, Mobility and Degradation of Xenobiotics*, Piacenza, Oct. 1993, pp. 357–370.

Wauchope, R. D. (1992). 'Environmental risk assessment of pesticides: improving simulation model credibility. *Weed Technol.*, **6**, 753–759.

Wauchope, R. D., Buttler, T. M., Hornsby, A. G., Augustijn-Beckers, P. W. M., and Burt, J. P. (1992). 'The SCS/ARS/CES pesticide properties database for environmental decision-making', *Rev. Environ. Contam. Toxicol.*, **123**, 1–164.

Zhang, H., Haan, C. T., and Nofziger, D. L. (1993). 'An approach to estimating uncertainties in modeling transport of solutes through soils', *J. Contam. Hydrol.*, **12**, 35–50.

CHAPTER 5

Pesticides in run-off and surface waters

B. Burgoa and R. D. Wauchope

INTRODUCTION

Although the leaching of pesticides into ground water resources has been emphasized in pesticide water quality research for the last 10 years, several recent studies (Parsons and Witt, 1989; US EPA, 1990) suggest that ground water contamination is not as extensive as has been feared. In contrast, the run-off of pesticides—the transport of pesticides with water and sediment

Environmental Behaviour of Agrochemicals
Edited by T. R. Roberts and P. C. Kearney © 1995 John Wiley & Sons Ltd

into surface waters by rainfal draining from the surface of agricultural fields—has, after a lapse in attention, begun to attract concern again (Fawcett, 1991; World Wildlife Fund, 1991).

In general, the concentrations of pesticides observed in run-off at the edge of agricultural fields are much higher than those found in ground water. Fortunately, the health risk of pesticides in run-off is not as direct to humans as it is in ground water. Although many surface water bodies are sources of drinking water, they are typically large-volume systems in which rapid dilution in streams and lakes (Mayeux *et al.*, 1984; Leitch and Fagg, 1985; Willis *et al.*, 1987), and degradation, in many cases, results in relatively low levels of pesticides in tap water (Baker and Richards, 1991). It is aquatic ecosystems in small streams, ponds, and estuaries exposed in some cases to run-off with little dilution, that are considered to be at greatest risk.

Reviews of pesticide run-off more than a decade ago (Baker, 1980; Caro, 1976; Wauchope, 1978) made four generalizations:

(1) Run-off losses are usually less than 5% of applied amounts, even under worst-case conditions. Concentrations can vary by orders of magnitude within and between run-off events. Total losses, however, for a given event, are generally proportional to amounts applied (e.g. Hall, 1974; Wauchope *et al.*, 1989). For this reason losses are best expressed as a percentage of the amounts applied and are typically 0.5% or less, although losses of up to 5% or even higher are possible under worst-case conditions.

(2) Storm timing is a critical determinant of pesticide run-off losses. That component of pesticide residues in the field which is most susceptible to run-off, namely on the surfaces of foliage and soil, is quickly dissipated after application by photodegradation, volatilization and other weathering processes.

(3) Application target and formulation are important factors. Application target tends to effect losses in the order: pesticide incorporated into soil < applied to soil surface < applied to foliage. Factors can also combine to produce large differences, for example, field studies suggest that a wettable powder applied to the soil surface would result in an average of 30 times higher losses than an incorporated EC (Wauchope and Leonard, 1980).

(4) The majority of pesticide lost is in the water phase of run-off, except for extremely soil-bound pesticides. Pesticide concentrations in run-off water are usually lower than in the sediment phase of run-off because of the preference of the chemical for the sediment surfaces. However, sediment is such a small fraction of run-off even in highly erosive storms, that the bulk of pesticides loss will be in the water phase. So, soil conservation practices *per se* will not greatly affect pesticide amounts in run-off.

In Table 1 the characteristics of the sites of pesticide run-off experiments conducted since 1978, are listed in order of plot size. The studies can be divided into small-scale ($0.5-70 \text{ m}^2$, 1–20 in the table) which are referred to as 'plot' studies, and larger-scale ($0.05-100 \text{ ha}$) studies referred to as 'field' or sometimes 'watershed' studies. The primary reason for this division is that plot studies typically utilize simulated rainfall, and thus rainfall timing and intensity can be controlled. Because of their size, field and watershed studies measure run-off from natural rainfall only.

There is considerable controversy as to whether the results of plot/simulated rainfall studies can be extrapolated to the field. This is due in part to the large size difference between the largest plots (70 m^2) and the smallest fields (500 m^2). Coody *et al.* (1990) have recently developed techniques for applying realistic simulated rainfall to plots up to 1000 m^2 in area. Coody's technique has considerable potential as an intermediate-scale approach, which allows controlled precipitation on plots large enough to examine tillage practices, crop cover, etc., and as a calibration tool for models (Wauchope, 1992). Only preliminary results with pesticides have been published, however (Wauchope *et al.*, 1993).

Table 2 gives the results of the run-off event in each of the experiments in Table 1 where pesticide losses were greatest.

In a review of the literature from 1978 to 1985, Leonard (1990) indicated that these generalizations have been confirmed in more recent research. He concluded that annual run-off losses of surface soil-applied pesticides averaged 2% and soil-incorporated or foliar-applied pesticides averaged < 1%. Leonard also reviewed the use of computer simulation models to describe and analyse non-point pollution and the effects of management practices on pollution potential. This is a technology which has developed in the 1980s. It has become a major tool for dealing with the complexities of the environmental fate of agricultural chemicals (Cheng, 1990; DeCoursey, 1990; Ertel, 1990).

In this chapter run-off studies and surface water monitoring results from 1978 through 1991 are reviewed. We will (1) examine run-off studies for insights into the pesticide run-off process; (2) analyse run-off results in terms of two concepts, namely a 'pesticide deposit release curve' and a 'characteristic chemograph'; (3) relate pesticide concentration in the run-off to levels found in surface waters.

PESTICIDES IN EDGE-OF-FIELD RUN-OFF

The data

Almost without exception, these are the first run-off event from each site. Most of the data show that run-off concentrations will decrease

Table 1 Experimental sites and characteristics

No.	Plot size (ha)	Location	Slope (%)	Soil classification	Texture	O.M. (%)	pH	Crop	Reference
1	0.000 05	Tifton, GA	2	Plinthic Kandiudults	s	1.39	6.4	Bare	Burgoa & Wauchope, 1991
2	0.000 05	Tifton, GA	2	Plinthic Paleudults	s	–	–	Bare	Hubbard et al., 1989
3	0.000 05	Tifton, GA	2	Rhodic Kandiudults	ls	–	–	Bare	Hubbard et al., 1989
4	0.000 05	Tifton, GA	2	Rhodic Kandiudults	scl	–	–	Bare	Hubbard et al., 1989
5	0.000 14	Arlington, WI	6	Typic Argiudolls	sil	3	–	Corn	Sauer & Daniel, 1987
6	0.000 27	Tifton, GA	1.3	Mollic Hapludalfs	sil	0.3	–	Bare	Wauchope, 1987
7	0.000 27	Tifton, GA	1.3	Mollic Hapludalfs	sil	0.3	–	Bermuda	Wauchope, 1987
8	0.0011	LSU, LA	0.2	Aeric Fluvaquents	sil	–	–	Sugarcane	Smith et al., 1983
9	0.00136	Ames, IA	5	Typic Hapludolls	sil	2.2	–	Bare	Baker et al., 1982
10	0.00238	Peterborough, Ontario	9	Boralfs	c–l	2.7–4.7–	–	Corn	Buttle, 1990
11	0.0028	Tifton, GA	3	Typic Hapludults	ls	0.5	–	Soybean	Rohde et al., 1980
12	0.0030	Monmouth, IL	7–11	Typic Arguidolls	sil	2.2	–	Corn Soybean	Felsot et al., 1990
13	0.00396	Pennsylvania	14	Typic Hapludalfs	sic	2.3	6.6	Corn	Hall et al., 1984
14	0.004	North Carolina	3.0	Aquic Paleudults	sil	1.0	6.0	Bare	Wiese et al., 1980
15	0.004	Mississippi	0.2	Mollic Hapludalfs	sl	0.7	7.3	Bare	Wiese et al., 1980
16	0.004	Tenneesse	2.0	Humic Hapludults	sl	1.6	5.6	Bare	Wiese et al., 1980
17	0.004	Texas	0.1	Torrertic Paleustolls	cl	1.7	7.3	Bare	Wiese et al., 1980
18	0.004	South Carolina	1.0	Typic Hapludults	sl	1.9	6.1	Bare	Wiese et al., 1980
19	0.004	Puerto Rico	4.0	Aquic Chromuderts	c	3.6	5.6	Bare	Wiese et al., 1980
20	0.00696	Walnut Gulch, AZ	7.7–8.0	Aridic Calciustolls	ls	1.5	–	Rangeland	Morton et al., 1989
21	0.0549	Baton Rouge, LA	0.15	Typic Fluvaquents	sicl	–	–	Cotton	Carroll et al., 1981
22	0.26	Coshocton, OH	11.3	Aquultic Hapludalfs	sil	–	–	Corn	Edwards et al., 1980
23	0.26 & 0.36	Howard Co, MD	6 & 7	Typic Dystrochrepts	l	2.7	5.9	Corn	Glenn & Angle, 1987

No.		Location		Soil classification				Crop	Reference
24	0.34	Tifton, GA	3	Typic Hapludults	ls	0.5	—	Soybean	Rohde et al., 1980
25	0.55	Coshocton, OH	6.0	Aquic Hapludalfs	sil	—	—	Corn	Edwards et al., 1980
26	0.58	Coshocton, OH	15.8	Typic Dystrochrepts	sil	—	—	Fescue	Edwards et al., 1980
27	0.63	Coshocton, OH	14.8	Typic Dystrochrepts	sil	—	—	Fescue	Edwards et al., 1980
28	1	Chattahoochee, GA	—	Typic Hapludults	sl	—	—	Harwood/Pine	Mayack et al., 1982
29	2–4	Baton Rouge, LA	0.1	—	—	—	—	Corn	Southwick et al., 1990
30	2.6	Bear Creek, MS	0.2	—	—	—	—	Cotton	Cooper et al., 1987
31	3.07	Coshocton, OH	15.8	Typic Hapludults	sil	—	—	Fescue	Edwards et al., 1980
32	3.07	Coshocton, OH	15.8	Typic Hapludults	sil	—	—	Fescue	Edwards et al., 1980
33	3.1	Oxford, MS	—	—	l	—	—	Soybean	Willis et al., 1987
34	4–5	Saskatchewan, Can.	1–4	Boralfs	—	—	—	Fallow	Nicholaichuk & Grover, 1983
35	7.3	Bear Creek, MS	0.2	—	—	—	—	Wheat	Cooper et al., 1987
36	8	Riesel, TX	2.1	Udic Pellusterts	c	—	—	Cotton	Mayeux et al., 1984
37	10	Arkansas	40	Typic Dystrochrepts	sil	—	—	Bermuda	Lavy et al., 1989
38	11.5	Fleming Creek, AK	30	Lithic Hapludults	sl	—	—	Mixed Pine	Bouchard et al., 1985
39	15.6	Clarksdale, MS	0.2	Vertic Haplaquepts	sic	2.5	—	Oaks	McDowell et al., 1981
40	18	Shelby County, TN	—	Aeric Fluviquents	sil	—	—	Cotton	Klaine et al., 1988
41	18.7	Clarksdale, MS	0.2	Fluvaquentics Udults	sicl	—	—	Corn	Willis et al., 1983
42	20	Wye River, MD	0–2	—	si	—	—	Grass	Glotfelty et al., 1984
43	39.6	Pennsylvania	14	Typic Hapludalfs	—	—	—	Cotton	Hall et al., 1983
44	56	Archerton, Victoria	<20	—	c–l	—	—	Pine	Leitch & Fagg, 1985
45	70	Henderson Co, NC	—	—	—	—	—	Apple	Penrose & Lenat, 1982
46	113	Flagstaff, AZ	—	Typic Chromusterts	c	—	—	Juniper/Pinyon	Johnsen, 1980

Table 2 Single largest-loss events of pesticides from microplots

Site no.	Pesticide	Application rate ($kg\,ha^{-1}$)	Simulated rainfall		Run-off total (cm)	Sediment yield ($mg\,ha^{-1}$)	Average pesticide conc.		Percent loss
			Days after application	Total (cm)			Water ($\mu g\,l^{-1}$)	Sediment ($\mu g\,kg^{-1}$)	
1	Alachlor	2.80	1	7.3	4.8	1.71	7.8	120	1.0
5	Alachlor	2.80	21	7.3	2.0	0.52	2.3	70	0.15
5	Alachlor	2.80	60	13.6	4.8	1.15	19.7	300	0.02
5	Alachlor	2.80	<7	13.6	7.3	2.03	236	1060	0.36
5	Alachlor	2.80							6.0
12	Alachlor (L-4E)	3.36	<2	6.4	2.77	32.4	651	1199	6.52
12	Alachlor (L-4E)	3.36	<2	5.6	3.24	24	152	1189	2.3
9	Alachlor (EC)	2.1	<3	127	62.8	11.4	220	2240	8.57
1	Atrazine		1						3.0
5	Atrazine	2.80	<7	13.6	7.3	2.03	351	1580	8.7
5	Atrazine	2.80	21	7.3	3.2	0.90	30.8	540	0.38
5	Atrazine	2.80	21	13.6	4.8	1.15	16.5	1680	0.34
5	Atrazine	2.80	60	7.3	2.0	0.52	12.0	1430	0.11
6	Atrazine (Aa-4L)	3.24	1	0.3	0.66	0.51	4.0	–	8.4
6	Atrazine (Aa-80W)	3.43	1	0.3	0.77	0.69	4.1	–	8.7
6	Atrazine (Aa-9-O)	3.19	1	0.3	0.81	0.70	5.0	–	11.6
6	Atrazine (EC)	2.69	1	0.3	0.84	1.1	1.8	–	5.1
9	Atrazine (WP)	2.1	<3	127	62.8	11.4	141	1101	5.71
12	Carbofuran (F-15G)	1.12	<2	5.6	3.24	24	374	609	12.0
5	Chlorpyrifos	1.34	60	7.3	2.0	0.52	0.90	520	0.04
5	Chlorpyrifos	2.80	21	7.3	3.2	0.90	1.79	680	0.10
5	Chlorpyrifos	1.34	21	13.6	6.0	2.94	0.38	300	0.08
2	Cyanazine	4.48	0.04	8.6	1.34	–	<100–5000	–	0.4
2	Cyanazine	4.48	0.04	25.0	17.45	–	<100–14000	–	15.8
2	Cyanazine	4.48	0.04	15.0	4.43	–	<100–5000	–	2.6

3	Cyanazine	4.48	0.04	25.0	12.05	—	—	—	1.8
3	Cyanazine	4.48	0.04	15.0	2.60	—	—	—	0.1
3	Cyanazine	4.48	0.04	8.6	0.00	—	—	—	0.0
4	Cyanazine	4.48	0.04	15.0	11.54	—	<100–150	—	11.4
4	Cyanazine	4.48	0.04	25.0	19.65	—	<100–12 000	—	22.5
4	Cyanazine	4.48	0.04	8.6	5.77	—	<100–150	—	2.6
7	Cyanazine (Bx-4L)	5.1	1	1.6	0.19	0.028	7	—	2.8
7	Cyanazine (Bx-90DF)	4.7	1	1.6	0.19	0.028	3	—	1.5
14	Fluometuron	4.4	<2	—	1.0	—	660	—	1.5
15	Fluometuron	4.4	<2	—	1.0	—	280	—	0.6
16	Fluometuron	4.4	<2	—	1.0	—	280	—	0.6
17	Fluometuron	4.4	<2	—	1.0	—	90	—	0.2
18	Fluometuron	4.4	<2	—	1.0	—	0.0	—	0.0
19	Fluometuron	4.4	<2	—	1.0	—	380	—	0.9
10	Metolachlor (broadcast)	2.64	22	1.5	0.07	1.8	270	634	0.12
9	Propachlor	2.1	<3	127	62.8	11.4	173	840	6.08
7	Sulfometuron (75DG)	0.41	1	2.4	0.19	0.008	0.43	—	1.3
7	Sulfometuron (EC)	0.21	1	1.6	0.19	0.028	0.09	—	0.8
2	Sulfometuron-methyl	0.60	0.04	25.0	17.45	—	—	—	24.3
2	Sulfometuron-methyl	0.60	0.04	8.6	1.34	—	—	—	0.2
2	Sulfometuron-methyl	0.60	0.04	15.0	4.43	—	—	—	2.5
3	Sulfometuron-methyl	0.60	0.04	25.0	12.05	—	—	—	15.9
3	Sulfometuron-methyl	0.60	0.04	8.6	0.00	—	—	—	0.0
3	Sulfometuron-methyl	0.60	0.04	15.0	2.60	—	—	—	1.0
4	Sulfometuron-methyl	0.60	0.04	8.6	5.77	—	—	—	4.2
4	Sulfometuron-methyl	0.60	0.04	25.0	19.65	—	—	—	34.7
4	Sulfometuron-methyl	0.60	0.04	15.0	11.54	—	—	—	24.5
20	Tebuthiuron	1.0	<0.04	4.0	2.0	0.421	320	160	48
12	Terbufos (C-15G)	1.12	<2	6.4	2.77	32.4	39	2237	7.43
11	Trifluralin	1.12	10	11.01	4.92	—	12	—	0.53
11	Trifluralin	1.12	1	10.18	5.57	—	25.45	—	1.26

exponentially as the time between application and a run-off-generating storm occurs, with half-lives of between several days to 2 weeks. For this reason the majority of pesticides losses during a season are usually due to a single storm (Wauchope and Leonard, 1980).

Effect of application rate and storm timing

Examination of the data in Tables 2–4 confirms that concentrations of pesticide in run-off increase with application rate, and if event or seasonal losses are compared to application rates, the fractional (percentage) losses are consistent. Edwards *et al.* (1980) and Glotfelty *et al.* (1984) indicate that concentrations in run-off of glyphosate and atrazine, respectively, from fields were proportional to application rates. In two microplot studies (Hubbard *et al.*, 1989; Wauchope *et al.*, 1989), dispersable-granule formulations of cyanazine and sulfometuron-methyl were applied at 4.48 kg/ha and 0.6 kg/ha rates, respectively, to microplots prior to simulated rainfall. In most cases the percentage losses of the two pesticides in the same event were quite similar, in spite of the seven-fold difference in application rate. This suggests that losses of some of the newer pesticides that are applied in gram-per-hectare quantities will be proportional. Leonard *et al.* (1979) presents a plot showing a correlation of 0.93 between pesticide concentrations in the surface 1 cm of soil and the concentration in run-off.

Although factors such as pesticide properties, soil properties and application deposit placement are important, the overwhelming determinants of run-off losses are storm timing and intensity. Donigian *et al.* (1977), in comparing ARMS model predictions with field data, stated that if the model successfully predicted the hydrology and erosion, then pesticide loss predictions, though based on a simple extraction theory, were good.

Run-off events which occur within a few days of pesticide application account for a large portion of the total seasonal pesticide losses in run-off water (Carroll *et al.*, 1981; Smith *et al.*, 1983; Spencer *et al.*, 1985; Glenn and Angle, 1987; Willis *et al.*, 1987). The concentration of pesticides in run-off generally decreases with increasing time after application, which parallels pesticide dissipation in soil (Buttle, 1990). However, losses also decrease rapidly with successive run-off events (Mayeux *et al.*, 1984; Leitch and Fagg, 1985; Klaine *et al.*, 1988). In Figure 1, losses of first events are plotted vs the period of time between application and the event, from all the studies listed in Table 1. Note that there is no apparent difference between 'plot' and 'field' results. In general, losses decrease with a half-life of about 1 week. This is similar to the decrease rate for succeeding storms on the same field (Wauchope and Leonard, 1980). This is surprising because on small plots where controlled events have been applied it is clear that during each event the chemical that is available for run-off is rapidly depleted, giving a quasi-exponential decrease in concentration during the event (Baker *et al.*, 1978, 1982; Baker and Laflen, 1979; Wauchope and Street, 1987 a, b; Hubbard *et al.*, 1989; Wauchope *et al.*, 1989; Burgoa and Wauchope, 1991). Baker *et al.* (1978) give excellent data from a sequence of run-off events from the same plots. Each succeeding run-off event exhibits an initial concentration which is equal to the final concentration of the previous event.

Effects of site hydrology

'Infiltration' is the term applied to the process of water entry into the soil, generally by downward flow through all or part of the soil surface (Hillel, 1980). The bottom diagram in Figure 2 schematically illustrates the hydrology at a point in a field which is initially at low soil moisture, then subjected to an idealized rain of constant intensity (Freeze and Cherry, 1979; Viessman *et al.*, 1989). Initially, the dry soil absorbs all moisture and the infiltration rate and precipitation rate are equal. As the soil near the surface approaches saturation, however, infiltration slows and run-off begins. A steady state may eventually be reached as shown in Figure 2. Hillel (1980) lists the following factors controlling infiltration rate: (1) time after onset of rain; (2) initial soil water content; (3) hydraulic conductivity; (4) soil surface conditions, particularly crust formation; (5) impeding layers within the profile.

The relevance of this concept is that the amount of pesticide which will be lost in run-off depends on the amount which is available, at or near the soil surface, to be mixed into the flowing water during the time run-off is occurring, but the rainfall generating the run-off can also drastically affect pesticide availability.

In an agricultural field or crop there are three main compartments where agrochemicals may be applied and which are the initial sources of agrochemicals for run-off. These are (1) crop foliage, (2) the soil surface, and (3) the soil below the surface (Wauchope, 1978; Baker, 1980; Leonard, 1990). Before and during run-off, rainfall can transfer agrochemicals between these compartments—from foliage to soil surface (wash-off) or from the soil surface to the soil profile and then downward (leaching).

This infiltration–leaching effect has been observed in field studies. Southwick *et al.* (1990) reduced atrazine and metolachlor run-off by more than half by increasing infiltration with subsurface drains. They observed both the leaching and decreased-run-off volume effects: the decrease in pesticide loss with drains was greater in early events after application. With fenvalerate, which does not leach, Smith *et al.* (1983) still showed increased run-off losses in a year with greater antecedent moisture. Morton *et al.* (1989) obtained more than twofold greater losses in run-off water and sediment from wet plots than from dry plots. The former also exhibited a

Table 3 Seasonal losses of pesticides from field/watershed scales with natural rainfall

Site no.	Pesticide	Application Rate (kg ha⁻¹)	Application Date	Rainfall Date	Rainfall Total (cm)	Run-off total (cm)	Sediment yield (mg ha⁻¹)	Average pesticide conc. Water (µg l⁻¹)	Average pesticide conc. Sediment (µg kg⁻¹)	Percent loss
34	2,4-D	0.42	1976–81	—	—	3.1	—	0–45	—	1.8
29	Atrazine	1.63	4/22/87	—	—	41.2	—	—	—	3.17
29	Atrazine	1.63	4/22/87	—	—	25.4	—	—	—	1.4
40	Atrazine	0.4	5/17/85	—	—	—	—	<0.1–250	—	1.5
43	Atrazine (80WP)	2.2	5/22/72	5/1–7/31	54.8	3.0	9.1	—	—	1.3
43	Atrazine (80WP)	4.5	5/22/72	—	—	3.5	4.3	—	—	0.6
23	Atrazine (Consv. Till.)	2.20	6/8/79	6/08–7/2	6.1	0.07	—	14–47	—	<0.02
23	Atrazine (Consv. Till.)	2.20	5/14/82	5/20–6/17	9.84	0.36	—	3–60	—	0.03
23	Atrazine (Consv. Till.)	2.20	6/8/79	6/18–10/9	52.35	1.37	—	5–1332	—	1.58
23	Atrazine (No Till.)	2.20	5/14/82	5/20–6/17	9.84	0.34	—	0	—	0
23	Atrazine (No Till.)	2.20	5/11/81	6/08–7/2	6.1	0.00	—	0	—	0
23	Atrazine (No Till.)	2.20	6/8/79	6/18–10/9	52.35	1.16	—	0–975	—	1.12
42	Atrazine (WP)	1.68	5/25/81	—	—	—	—	0–300	—	2–3
44	Clopyralid (L-L)	2.5	4/21/83	4/23–4/24	7.2	—	—	2–17	—	0.01
41	DDT	7.84	1972	1973–78	801	326	129.4	3.1–5.4	—	1.7
30	DDT, DDE, DDD	Residual	—	7/77–4/78	137.9	—	—	0.14–1.39	—	—
35	DDT, DDE, DDD	Residual	—	7/77–4/78	137.9	—	—	0.2–3.1	—	—
33	Endosulfan	0.56	1984	10/6–10/20	13.54	13.19	1.39	0.7–18.7	8700	3.2
45	Endrin	—	5/77	12/5–3/17	—	—	—	—	2.9–22.7	
22	Glyphosate	1.12	4/18/73	4/27–6/17	—	2.0	—	5–90	—	0.45
25	Glyphosate	1.12	4/30/73	5/10–6/6	—	0.78	—	5–94	—	0.58

No.										
26	Glyphosate	1.12	4/26/74	5/12	—	0.05	—	3	—	<0.01
27	Glyphosate	3.36	4/26/74	5/12–6/22	5.0	0.15	—	2–55	—	0.013
31	Glyphosate	1.12	4/26/74	5/12–5/22	—	0.35	—	6–7	—	0.02
32	Glyphosate	8.96	4/29/75	4/30–8/19	—	0.98	—	3–5153	—	1.85
28	Hexazinone	16.8	4/23/79	—	—	—	—	1–44	—	—
37	Hexazinone	1.36	5/02/84	5/84–5/86	291.4	—	—	0.03–3.85	0.00	4.7
38	Hexazinone	2.0	4/82	—	—	—	—	<3–14	—	2–3
33	Methoxychlor	1.12	1984	10/6–10/20	13.54	13.19	1.39	0.6–5.9	207370	8.2
29	Metolachlor	2.16	4/22/87	—	—	25.4	—	—	—	1.07
29	Metolachlor	2.16	4/22/87	—	—	41.2	—	—	—	2.44
36	Picloram	1.12	4/21/78	6/6	12.3	2.0	—	0–38	—	1.2
36	Picloram	1.00	4/11/79	5/1–5/29	29.5	15.2	—	16–250	—	3.68
46	Picloram	2.8	8/21/68	2/69–2/71	92	3.95	—	2–200	—	1.1
23	Simazine	2.20	6/8/79	6/18–10/9	52.35	1.37	—	4–456	—	0.53
23	Simazine (Consv. Till.)	2.20	5/14/82	5/20–6/17	9.84	0.36	—	0–1	—	<0.01
23	Simazine (Consv. Till.)	2.20	5/11/81	6/08–7/2	6.1	0.07	—	1–2	—	<0.01
23	Simazine (Consv. Till.)	2.20	5/11/81	6/08–7/2	6.1	0.00	—	0	—	0
23	Simazine (No Till.)	2.20	6/8/79	6/10–10/9	52.35	1.16	—	0–210	—	0.36
23	Simazine (No Till.)	2.20	5/14/82	5/20–6/17	9.84	0.34	—	0	—	0
42	Simazine (WP)	1.68	5/25/81	—	—	—	—	0–300	—	2–3
30	Toxaphene	Residual	—	7/77–4/78	137.9	—	—	0.04–2.02	—	—
35	Toxaphene	Residual	—	7/77–4/78	137.9	—	—	0.9–4.2	—	—
39	Toxaphene	20.16	1972	1973–75	485	220	68.6	9–16	—	2.1
41	Toxaphene	60.43	1972–77	1973–78	801	326	129.4	7.5–20.9	—	3.7
24	Trifluralin	1.12	7/13/74	1974	124.3	5.2	—	1.0	—	0.15
24	Trifluralin	1.12	5/15/75	1975	131.2	6.2	—	0.62	—	0.03
41	Trifluralin	7.84	1972–78	1973–78	801	326	129.4	0.1–0.8	—	0.9

Table 4 Single worst event loss of pesticides from field/watershed scale

Site no.	Pesticide	Application		Rainfall		Run-off total (cm)	Sediment yield (mg ha⁻¹)	Average pesticide conc.		Percent loss
		Rate ($kg\,ha^{-1}$)	Date	Days after application	Total (cm)			Water ($\mu g\,l^{-1}$)	Sediment ($\mu g\,kg^{-1}$)	
29	Atrazine	1.63	4/22/87	31	–	1.8	–	270	–	1.6
23	Atrazine	2.20	6/8/79	10	3.81	0.2	–	1332	–	1.2
33	Endosulfan	0.56	9/5–10/4	2	11.9	3.6	0.44	5.9	98 900	1.64
27	Glyphosate	3.36	4/26/74	36	–	0.05	–	55	–	0.01
22	Glyphosate	1.12	4/18/73	9	–	0.35	–	90	–	0.3
31	Glyphosate	1.12	4/26/74	16	–	0.18	–	6	–	0.01
32	Glyphosate	8.96	4/29/75	1	–	0.32	–	5153	–	1.8
26	Glyphosate	1.12	4/26/74	16	–	0.05	–	3	–	0.002
25	Glyphosate	1.12	4/30/73	10	–	0.68	–	94	–	0.6
37	Hexazinone	1.36	5/2/84	68	6.25	–	–	16	–	0.10
33	Methoxychlor	1.12	9/5–10/4	2	11.9	3.6	0.44	18.7	8570	0.19
29	Metolachlor	2.16	4/22/87	31	–	1.8	–	420	–	1.5
36	Picloran	1.00	4/11/79	30	12.6	8.7	–	64	–	0.55
23	Simazine	2.20	6/8/79	10	3.81	0.2	–	456	–	0.4
24	Trifluralin	1.12	7/74	1	15.4	2.0	–	38	–	0.2
24	Trifluralin	1.12	5/75	10	5.4	0.2	–	23	–	0.02

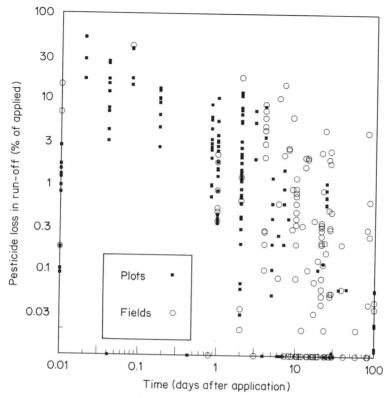

Figure 1 Pesticide losses (as % of applied amounts) shown against the time elapsed between pesticide application and run-off event, for plots (≤ 0.007 ha) and fields (> 0.007 ha)

sealing (crusting) effect. Wauchope *et al.* (1989) obtained a strong correlation between total cyanazine and sulfometuron-methyl losses from small plots and the length of time between the beginnings of rainfall and run-off, indicating that leaching rapidly decreased the amount of the two herbicides available for run-off.

Clearly, more intense storms will quickly overwhelm the infiltration rate and increase the relative amounts of water in run-off. Rainfall intensities from 4 to 125 cm h^{-1} were used by Hubbard *et al.* (1989) in a 0.5 m^2 plot rainfall simulator experiment to simultaneously determine run-off and leaching losses of bromide, cyanazine, and sulfometuron-methyl. An increase in rainfall intensity increased the fraction of both water and pesticide loss in the run-off. However, the movement of bromide, which is extremely mobile, was not affected by rainfall intensity. It appears that bromide was essentially completely leached away from the run-off mixing zone in all

On soil surface, highly (and rapidly) soluble chemicals are immediately available but leach out of interaction zone before run-off occurs

On foliage surface, soluble or emulsifiable chemicals flush off; run-off availability depends on timing

On soil surface, slowly soluble or controlled-release chemicals maintain an intermediate availability somewhat independent of mobility

Idealized constant-intensity rain: as soil saturates, infiltration decreases to less than rainfall rate and run-off occurs

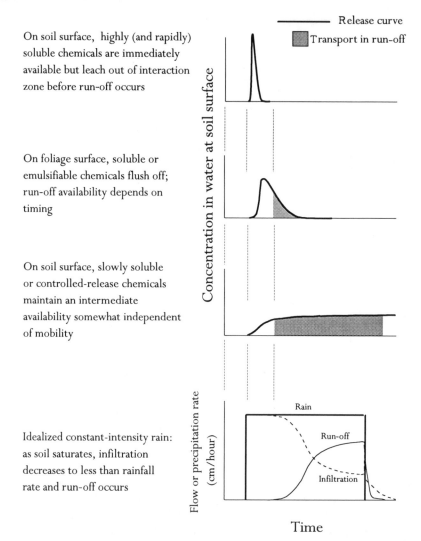

Figure 2 Hypothetical 'release curves' (plots of pesticide concentration in the soil water generated by rain at the soil surface) under various conditions, shown on the same time axis as rainfall, infiltration, and run-off

cases. A similar result was obtained by Burgoa and Wauchope (1991) using the same apparatus.

The storm-intensity effect is most often observed on small plots on bare soil. In large heterogeneous field studies this effect is generally overwhelmed by other factors (Mayack *et al.*, 1982; Smith *et al.*, 1983; Lavy *et al.*, 1989).

Water which is unavailable for run-off as the result of infiltration can return by horizontal subsurface flow to the surface and contribute to run-off further downslope, a process referred to as interflow. This process is very important in hillslope sites. If pesticides are removed from interflow water by adsorption by soil, the overall result will be a partial removal of pesticides from the run-off water. In an important study, Edwards et al. (1980) demonstrated this process with glyphosate. A low intensity storm which exhibited quite slow run-off response, characteristic of interflow, resulted in lower glyphosate concentrations than a higher intensity storm, even though the latter occurred later in the season. The more intense storm simply resulted in a higher proportion of run-off as direct or overland flow.

Effects of formulation

All other factors being equal, formulation can effect the movement of pesticides from the soil surface. Larger losses of tebuthiuron were observed with pellets than with wettable powder (Morton et al., 1989). Additives that reduced the leaching of atrazine applied to the soil surface increased the amount of run-off losses (Burgoa and Wauchope, 1991). In simulated storms, total atrazine losses decreased from emulsion and dispersable-liquid to dispersible-granule and wettable-powder formulations (Wauchope, 1987). However, losses of cyanazine and sulfometuron-methyl applied as suspension concentrates, dispersible granules, and emulsifiable concentrates, under conditions of similar run-off volumes, were fairly constant (Wauchope et al., 1989).

It appears that formulation effects are not as large as the effects of storm timing and application rate. Given that storms occurring immediately after application are most important, however, it is on these storms that formulation will have the most effect.

Effects of active ingredient chemical and physical properties: run-off indices

A recent compilation of pesticide physical properties (Wauchope et al., 1992; Augustijn–Beckers et al., 1994) indicates that soil organic carbon sorption coefficients (K_{oc}) for pesticide active ingredients in current use range over six orders of magnitude, and half-lives range from 1 to 1000 days (with a few exceptions). These two properties—persistence and soil organic carbon sorption coefficient or 'K_{oc}'—have been used to screen pesticides for their leaching tendency (and therefore potential as ground water contaminants) with some success (Jury et al., 1987; Becker et al., 1989; Gustafson, 1989; Hollis, 1990; Goss, 1992; Hornsby, 1992; Wauchope et al., 1992). The main reason this approach has been accepted is because Gustafson (1989) and earlier Cohen *et al.* (1984) demonstrated it would separate 'leachers' from 'nonleachers' found in several state and regional monitoring programs. The concept of giving pesticide active ingredients a relative 'environmental impact' score based on summation of their physical/chemical/toxicological properties was probably first suggested by Weber (1977).

Several of the same workers have also used combinations of persistence and K_{oc} to attempt to develop a run-off index (Hollis, 1990; Goss, 1992; Hornsby, 1992), but with the exception of Hollis (1990) the approach has not been compared with monitoring results. One possible shortcoming of the approach is that the persistence data used, which are estimates of lifetime in the soil below the surface, appear to bear little relevance to the lifetimes of those residues which are most available for run-off (Wauchope, 1978; Wauchope and Leonard, 1980; Leonard, 1990; Wauchope *et al.*, 1992). Leonard (1990) in a statistical modeling study using 50 years of rainfall data on a clay and sandy soil showed that, on average, run-off losses were less sensitive to persistence than leaching losses.

Effects of erosion control

Pesticides which are very strongly adsorbed by soil surfaces (K_{oc} on the order of 2000 or more) exhibit run-off concentrations proportional to sediment concentration (McDowell *et al.*, 1981; Willis *et al.*, 1983, 1987). Thus erosion control practices will prevent losses of those pesticides. Tillage effects on pesticide losses have been studied by a large number of authors. It is well established that crop residue considerably reduces soil loss; hence, crop residues can considerably reduce the loss of herbicide in sediment (Baker *et al.*, 1982). Pesticide run-off concentrations are highest during periods of minimum vegetative cover on the soil (McDowell *et al.*, 1981; Cooper *et al.*). Buffer strips or vegetated waterways can be used to reduce the concentration of some pesticides in run-off water. Small grain strips and buffer zones reduced water and soil losses by 70% compared with nonstripped areas during 11 erosion events (Hall *et al.*, 1983), and herbicide losses were reduced by 65%. More than 86% of trifluralin lost in run-off was retained by grassed waterways although half of the latter was attributed to infiltration (Rohde *et al.*, 1980).

A significant number of pesticides currently in use, however, are not strongly adsorbed to soil, and run-off losses of these chemicals will not necessarily be controlled by erosion control. For example, only 29% of the active ingredients in the SCS/SRS/CES database have K_{oc} equal to or greater than 2000 (Wauchope *et al.*, 1992). For weakly adsorbed compounds, no conservation tillage system will completely eliminate pesticide run-off, but losses can be minimized by controlling *both* water and sediment movement (Hall *et. al.*, 1984; Glen and Angle, 1987; Sauer and Daniel,

1987; Buttle, 1990; Felsot *et. al.*, 1990). The use of grass cover and plant residues increased time to run-off and decreased water run-off by increasing the infiltration rate of the surface horizon. When the run-off volumes are the same, however, losses of non-strongly-adsorbed pesticides are not significantly different between grass cover and bare soil (Wauchope *et. al.*, 1989).

Pesticide concentrations in run-off water and sediment from conservation tillage systems (chisel, ridge-plant, and no-till) can actually be higher than for conventional, moldboard plow tillage (Wauchope, 1987). Higher concentrations in some corn conservation tillage systems were thought to be due to wash-off from corn residues or soil surface and the concentration of erosion forces down the smooth, residue-free row of the ridge-plant systems (Ahuja, 1986; Sauer and Daniels, 1987). Large losses of alachlor were attributed to high concentrations in run-off associated with extensive wash-off of the herbicide from soybean crop residues (Felsot *et. al.*, 1990).

SOME DETAILS OF THE PHYSICAL PROCESSES OF PESTICIDE RUN-OFF

Application deposit release to water

During run-off, pesticide deposits on foliage, soil surfaces, and incorporated will exhibit a characteristic rate and timing of release to run-off which we will call a 'deposit release curve'. This release will depend on the intensity and timing of water flows, the chemistry and formulation of the chemical, and the affinities of the surfaces and phases within the compartment for the chemical. Some cases are illustrated in Figure 2.

Pesticide at or near the soil surface

Consider the top two diagrams in Figure 2—situations in which pesticides are at or near the soil surface as for example a preplant herbicide to a bare soil. This case is assumed by the CREAMS (Knisel, 1988; Leonard and Wauchope, 1980) and HSPF (Donigan and Crawford, 1976; Johanson *et. al.*, 1984) models (although corrections are possible for other situations). Three mechanisms of transfer between soil and run-off water are considered (Ahuja, 1986; Emmerich *et. al.*, 1989).

(1) mixing between run-off water and soil solution containing pesticide;
(2) dissolution of the pesticide crystals or granules to the run-off water;
(3) desorption of pesticides from the organic matter, soil particles (including sediments) and plant residues.

If a pesticide deposit at or near the surface is rapidly and highly soluble and is also very mobile (usually, but not always, quite soluble pesticides are also quite mobile), much of the deposit will be leached away from near the soil surface by infiltration, thus making it less available to be entrained into run-off. For this reason, although it is often assumed that the more soluble pesticides are more likely to be lost in run-off, *quite soluble pesticides may under certain conditions exhibit very small run-off losses.* Conversely, Figure 2 shows the case of a slowly soluble or insoluble pesticide at the soil surface. This case may exhibit large run-off losses because such pesticides are typically strongly bound to sediment, are not leached into the soil by infiltration, and will run-off with sediment. Thus, *under erosive conditions, insoluble or slowly soluble pesticides and/or soil-sorbed pesticides may exhibit larger run-off losses than soluble pesticides.*

Controlled-release formulations at the soil surface

The effect of controlled release pesticide applied at the soil surface is illustrated in Figure 2. Slow-release formulations, which are being considered as a possible way to decrease leaching may, by slowing leaching, increase run-off potential (Burgoa and Wauchope, 1991).

Incorporated pesticides

It is well known that incorporation drastically reduces run-off losses (Wauchope, 1978; Leonard, 1990), and several recent studies have confirmed this (Buttle, 1990). A maximum reduction of water, soil, and atrazine losses from a hillside was achieved with a conventional-tillage management system that combined preplant incorporation of atrazine residues with strip cropping on plot tiers. Incorporation of atrazine into the top 5 cm of the soil by lightly rototilling reduced losses by 90% (Hall *et. al.*, 1983).

Such incorporation studies dramatically demonstrate the thinness of the 'interaction zone' at the soil surface which is extracted by run-off. Typical incorporation depths are 7–10 cm, and usually much of the pesticide remains in the top 3 cm (Barrentine, 1988), yet 90% reductions in run-off losses are typical for pesticides after incorporation.

Pesticides on crop foliage

Foliar residues, which give a sharp spike of pesticide in foliar wash-off water (McDowell *et. al.*, 1984, 1987; Willis *et. al.*, 1986) may show a very sensitive dependence on the timing of wash-off relative to the timing of run-off. If, as shown in Figure 2, the breakthrough of wash-off (which occurs at the time the capacity of a plant to hold water is exceeded) happens before run-off, subsequent leaching of a mobile chemical before run-off could lead to low losses. The other extreme is possible, in which the wash-off occurs after run-off has begun, leading to essentially all wash-off pesticide becoming part of run-off.

Foliar wash-off is only crudely handled in current mainstream models (Leonard and Wauchope, 1980; Smith and Carsel, 1984; Leonard *et. al.*, 1987; Leavesley *et. al.*, 1990). Functions are used that relate wash-off to rainfall volume or intensity and empirical 'wash-off' or 'dislodgable' fractions. These functions are based mainly on the pioneering work of McDowell and Willis (e.g. McDowell *et. al.*, 1984, 1987; Willis *et. al.*, 1986) who obtained the important result that, with highly insoluble pesticides, the wash-off fraction was relatively independent of rainfall intensity, but dependent on wash-off volume.

Wash-off experiments with more soluble pesticides have focused on rainfall effects on post-emergent herbicide activity (Caseley and Coupland, 1980; Bryson, 1987; Wauchope and Street, 1987a, b). Results indicated that a large fraction of soluble agrochemicals can quickly become non-washable. This fraction may, however, still be larger than for very low solubility pesticides.

Field data suggest that water-soluble *foliar-applied* pesticides result in some of the highest run-off losses (Wauchope, 1978; Wauchope and Leonard, 1980). This may not be true for turf: rainfall simulation experiments comparing turf to bare soil indicate that applications to turf may have losses because less run-off occurs for a given amount of rainfall (Wauchope *et. al.*, 1989).

Pesticide entrainment into run-off water

Once a pesticide is released from deposits on foliage and on or in the soil, transport into overflowing run-off water must occur. A model for the mechanism and rates of transport from plant foliage to the soil surface has not been dealt with in detail. For pesticides at or just below the soil surface two approaches have been taken.

The first approach consists of an 'extraction' of a thin layer of soil at the soil surface, brought about by the pumping action of raindrop impact (Donigian *et. al.*, 1977; Leonard and Wauchope, 1980). The mixing depth was assumed constant with time and ranged between 0.2 and 1.0 cm. In this model the soil/water distribution coefficient (as obtained by a laboratory slurry mixing experiment) determines the concentration of chemicals in the mixing water, which re-emerges to join overflowing water as it is forced to the surface by other impacting raindrops nearby. The extraction model is at the centre of the major non-point pesticide run-off models such as CREAMS (Knisel, 1980) and HSPF (Johanson *et. al.*, 1984). Leonard (1990) showed a positive relation between herbicide concentration in the run-off and herbicide concentration in the top 1 cm of soil. Ahuja (1986) presented a review dealing with mixing zones, measurement of the 'effective depth of interaction' (EDI), and transfer of chemicals to run-off.

Ahuja *et. al.* (1981) conducted a series of elegant studies in which an empirical soil phosphate desorption kinetic model was determined for a soil slurry, then combined in a model with rainfall data and infiltration/run-off rates to predict run-off concentrations. The model had the EDI as a parameter. Using small tilted beds packed with soil for which the phosphate desorption model calibrated, the EDI was determined and was found to agree with an EDI value determined experimentally by conducting run-off experiments in which the soil was treated with a tracer at various depths. However, Ahuja later (1986) concluded that a permanent zone of uniform mixing and constant thickness does not exist. EDI was used as an approximation of the mixing zone.

The second approach began with the assumption that run-off chemical concentration is controlled by solute molecular diffusion upward from the soil solution to the soil surface and thence into run-off water (Wallach and van Genuchten, 1990). Wallach *et. al.* (1988) used a solute diffusion model in the soil body with a rate-limiting chemical transfer process across a laminar boundary layer at the soil surface/run-off water interface. The mass transfer coefficient was found to be proportional to the diffusion coefficient, soil surface roughness, run-off hydraulic gradients, and hydraulic radius, and inversely proportional to the thickness of the laminar boundary layer at the soil surface (Wallach *et. al.*, 1988, 1989). This model showed good agreement in comparison with experimental data from Ahuja and Lehman (1983) under a no-infiltration condition.

Ahuja and Lehman (1983) suggested that the overall mechanism of transfer may be described mathematically as an accelerated-diffusion process, in which diffusion is accelerated by raindrop pumping. Ahuja (1990) used a general diffusion-type equation to describe Ahuja and Lehman's data (1983). Others have shown that mixing can play an important role in the release of chemicals from the soil surface into the flow (Peyton and Sanders, 1990).

The Ahuja/Sharpley kinetic model for run-off is being incorporated into the ARS Root Zone Water Quality Model (DeCoursey, 1990).

PESTICIDES IN SURFACE WATER BODIES

Monitoring studies

In the past decade, four large monitoring projects have been conducted in the US and Canada. The main objective was to quantify agricultural impact on regional surface water quality. Pereira *et. al.* (1990), and Pereira and

Rostad (1990) studied the occurrence and distribution of pesticides along the lower Mississippi River and its major tributaries. Seventeen locations were selected for sampling during five different time periods between 1987 and 1989. Representative depth-integrated water samples were collected and analyzed for different triazine and chloroacetanilide herbicides and their degradation products. Table 5 presents some of the data collected in May–June 1988. The estimated annual transport of atrazine was approximately 0.4% and 1.7% of the total amount of atrazine applied annually in the midwest in 1987 and 1989, respectively (Pereira and Rostad, 1990).

Another monitoring project was conducted in the Lake Erie Basin by Baker (1988). The watershed sizes used in this study ranged from 1130 to 16 395 ha. The main agricultural activity in the watershed was intensive corn and soybean production. Collected water samples were analyzed for herbicides and insecticides used in those two crops. In chemograph patterns for pesticides in watershed tributaries Baker observed:

(1) pesticide concentration increased in association with run-off events;
(2) hydrological factors had a greater influence on pesticide concentration than pesticide breakdown in the soil;
(3) multiple storms causing high pesticide concentrations can occur in the same watershed in the same year; and
(4) the shapes of the pesticide chemographs were rather broad, corresponding more closely to chemographs for nitrates than for sediments.

Using these chemograph patterns, Baker (1988) deduced that pesticides were exported from fields, behind the sediments, throughout the period of surface water run-off whereas nitrate entered streams via tile drainage and interflow.

Glotfelty et. al. (1984) investigated the water quality in the Chesapeake Bay by sampling the Wye River which flows westward into Eastern Bay. The main agricultural activity of the watershed was corn and soybean production. The results of samples from 15 sites on the river showed a decline in herbicide concentrations as one moved downstream. The difference in herbicide concentrations and losses in the sample years reflected the relation between rainfall timing and application dates. Pesticides losses were 3% in years when rainfall occurred less than 2 weeks after application as compared to 0.2% when rainfall occurred later.

The Southern Ontario/Great Lakes area was studied by the group of Frank and Braun (Braun and Frank, 1980; Frank et. al., 1982). Braun and Frank (1980) sampled 11 watersheds for organophosphorus and organo-chlorine insecticides between 1975 and 1977. The watershed sizes ranged from 1860 to 7913 ha where 94% of the area was in agricultural production (corn, fruits, tobacco, and vegetables). More than 75% of the insecticide losses were associated with run-off caused by storms and spring thaws.

Table 5 Pesticide concentration in surface water bodies

Pesticides	Concentration (μg l^{-1})	Location	Date	Reference
2,4,5-T	ND-0.41	SFWMD, FL	1986-88	Pfeuffer, 1991
2,4-D	ND-1.1	Padilla Bay, WA	1987	Mayer & Elkins, 1990
2,4-D	ND-0.32	S. Ontario, Can.	1975-77	Frank et al., 1982
Alachlor	0.006-0.90	Mississippi R.	1988	Pereira et al., 1990
	2700¶	North Carolina		Skaggs et al., 1980
	ND-0.5	W. Virginia, rainfall	1985	Richards et al., 1987
	ND-1.0¶	New York, rainfall	1985	Richards et al., 1987
	0.472	Maumee R., OH	1985	Baker, 1988
	ND-6.0¶	Ohio, rainfall	1985	Richards et al., 1987
	ND-0.009	S. Ontario, Can.	1975-77	Frank et al., 1982
	ND-1.0¶	Indiana, rainfall	1985	Richards et al., 1987
Aldrin	ND	Ontario, Can.	1980-81	Johnson et al., 1988
Atrazine	ND-1.5	Ohio, rainfall	1985	Richards et al., 1987
	<0.0001-0.0328	S. Ontario, Can.	1975-77	Frank et al., 1982
	ND->1.0	W. Virginia, rainfall	1985	Richards et al., 1987
	1-15¶	Wye R., MD	1981	Glotfelty et al., 1984
	0.067-0.694	Mississippi R.	1988	Pereira et al., 1990
	1.902	Maumee R., OH	1985	Baker, 1988
	ND-1.0	Indiana, rainfall	1985	Richards et al., 1987
	ND-13.2¶	SFWMD, FL	1986-88	Pfeuffer, 1991
	<0.5	W. Virginia, rainfall	1985	Richards et al., 1987
Azinphos-methyl	ND	S. Ontario, Can.	1975-77	Braun & Frank, 1980
Azinphos-methyl	ND-0.025	Greece	1984-85	Albanis et al., 1986
BHC	<0.0011¶¶	Ontario, Can.	1980-81	Johnson et al., 1988
Butachlor	<0.10	Koise R., Japan,	1985-86	Shiraishi et al., 1988
Carbaryl	ND-38	Greece	1984-85	Albanis et al., 1986
Carbofuran	ND-0.042	Greece	1984-85	Albanis et al., 1986
	ND-0.001 87	S. Ontario, Can.	1975-77	Frank et al., 1982
	0.046	Maumee R., OH	1985	Baker, 1988
	<0.18¶¶	Koise R., Japan	1985-86	Shiraishi et al., 1988

Pesticide	Concentration	Location	Date	Reference
Chlordane	ND–0.000007	Ontario, Can.	1980–81	Johnson et al., 1988
	<0.0004–0.047	S. Ontario, Can.	1975–77	Braun & Frank, 1980
Chlomitrofen	<0.42¶¶	Ishikari R., Japan	1984	Ohyama et al., 1986
Chlorfenvinphos	ND¶¶	S. Ontario, Can.	1975–77	Braun & Frank, 1980
Chlorpyrifos	ND–0.0016	S. Ontario, Can.	1975–77	Frank et al., 1982
	<0.01–1.6E-6	S. Ontario, Can.	1975–77	Braun & Frank, 1980
CNP	<0.42¶¶	Ishikari R., Japan	1984	Ohyama et al., 1986
Cyanazine	ND–0.647	Mississippi R.	1988	Pereira et al., 1990
	<1.0	Indiana, rainfall	1985	Richards et al., 1987
	<0.5	W. Virginia, rainfall	1985	Richards et al., 1987
	<0.5	W. Virginia, rainfall	1985	Richards et al., 1987
	0.032	Maumee R., OH	1985	Baker, 1988
	<0.4	Ohio, rainfall	1985	Richards et al., 1987
Cyprazine	ND–0.0003¶¶	S. Ontario, Can.	1975–77	Frank et al., 1982
DDT	<0.01–0.70	Bear Creek, MS	1977–79	Cooper et al., 1987
	0.0031–0.347	S. Ontario, Can.	1975–77	Frank et al., 1982
Demeton	ND–0.00008	Ontario, Can.	1980–81	Johnson et al., 1988
Diazinon	ND	S. Ontario, Can.	1975–77	Braun & Frank, 1980
	<0.18	Koise R., Japan	1985–86	Shiraishi et al., 1988
	ND–0.00015	S. Ontario, Can.	1975–77	Frank et al., 1982
	ND–0.057	Greece	1984–85	Albanis et al., 1986
	<0.01–140E-6	S. Ontario, Can.	1975–77	Braun & Frank, 1980
Dicamba	ND–0.0007	S. Ontario, Can.	1975–77	Frank et al., 1982
	ND–160	Padilla Bay, WA	1987	Mayer & Elkins, 1990
Dicofol	ND	S. Ontario, Can.	1975–77	Braun & Frank, 1980
Dieldrin	ND–0.00008	Ontario, Can.	1980–81	Johnson et al., 1988
	ND–0.12	S. Ontario, Can.	1975–77	Frank et al., 1982
Dimethoate	ND	S. Ontario, Can.	1975–77	Braun & Frank, 1980
Dinoseb	0.3–18.6¶	British Columbia	1985	Wan, 1989
	<10¶¶	Israel	1980–84	Wynne, 1986
Dorsan	ND–0.00008	Ontario, Can.	1980–81	Johnson et al., 1988
Endosulfan	ND–0.173	S. Ontario, Can.	1975–77	Frank et al., 1982
	<0.0004–0.049	S. Ontario, Can.	1975–77	Braun & Frank, 1980
	<10¶	Israel	1980–84	Wynne, 1986

continued overleaf

Table 5 (*continued*)

Endrin	0.7 & 2.1¶	Pond, W. Virginia	1985	Weaver et al., 1990
	0.2	Ravine, W. Virginia	1985	Weaver et al., 1990
	ND	Ontario, Can.	1980–81	Johnson et al., 1988
	0.2–0.4	N. Carolina	1977	Penrose & Lenat, 1982
	0.5	Stream, W. Virginia	1985	Weaver et al., 1990
EPTC	ND–0.0002	S. Ontario, Can.	1975–77	Frank et al., 1982
Ethion	<0.01–0.04E-6	S. Ontario, Can.	1975–77	Braun & Frank, 1980
Fonofos	<0.001	Maumee R., OH	1985	Baker, 1988
Heptachlor	ND	Ontario, Can.	1980–81	Johnson et al., 1988
Heptachlor epoxide	ND	Ontario, Can.	1980–81	Johnson et al., 1988
	<0.0004–0.370	S. Ontario, Can.	1975–77	Braun & Frank, 1980
	ND–0.370	S. Ontario, Can.	1975–77	Frank et al., 1982
Iprobenfos	<4.63¶¶	Koise R., Japan	1985–86	Shiraishi et al., 1988
Isoprothiolane	<0.28¶¶	Koise R., Japan	1985–86	Shiraishi et al., 1988
Leptophos	ND¶¶	S. Ontario, Can.	1975–77	Braun & Frank, 1980
Linuron	0.013	Maumee R., OH	1985	Baker, 1988
	<0.2	Chesapeake B.	1977–78	Zahnow & Riggleman, 1980
Malathion	<0.011–1.8E-6	S. Ontario, Can.	1975–77	Braun & Frank, 1980
MCPA	ND–0.0003	S. Ontario, Can.	1975–77	Frank et al., 1982
Methoxychlor	ND¶¶	Ontario, Can.	1980–81	Johnson et al., 1988
	ND¶¶	S. Ontario, Can.	1975–77	Braun & Frank, 1980
Metolachlor	<2.5	Rainfall, OH	1985	Richards et al., 1987
	ND	W. Virginia, rainfall	1985	Richards et al., 1987
	1.316	Maumee R., OH	1985	Baker, 1988
	<0.5	W. Virginia, rainfall	1985	Richards et al., 1987
	0.046–0.273	Mississippi R.	1988	Pereira et al., 1990
	ND–1.0	Indiana, rainfall	1985	Richards et al., 1987
Metribuzin	0.254	Maumee R., OH	1985	Baker, 1988
	ND–0.00014	S. Ontario, Can.	1975–77	Frank et al., 1982

Pesticide	Concentration	Location	Year	Reference
Mirex	ND¶¶	Ontario, Can.	1980–81	Johnson et al., 1988
	<0.000013¶¶	Lawrence R., Can.	1985–87	Kaiser et al., 1990
Molinate	<0.36	Koise R., Japan	1985–86	Shiraishi et al., 1988
Oxadiazon	<0.33	Koise R., Japan	1985–86	Shiraishi et al., 1988
Parathion	ND	S. Ontario, Can.	1975–77	Braun & Frank, 1980
	<10¶	Israel	1980–84	Wynne, 1986
Parathion-methyl	ND–32	Greece	1984–85	Albanis et al., 1986
Phosalone	ND	S. Ontario, Can.	1975–77	Braun & Frank, 1980
Phosmet	ND	S. Ontario, Can.	1975–77	Braun & Frank, 1980
Picloran	ND–25.5	Coweeta, NC	1978	Neary et al., 1985
Prometron	ND–0.0001	S. Ontario, Can.	1975–77	Frank et al., 1982
Propaphos	<0.08¶¶	Koise R., Japan	1985–86	Shiraishi et al., 1988
Simazine	0.165	Maumee R., OH	1985	Baker, 1988
	ND–0.13	Mississippi R.	1988	Pereira et al., 1990
Simazine	ND–0.0011	S. Ontario, Can.	1975–77	Frank et al., 1982
Simetryn	<0.96¶¶	Koise R., Japan	1985–86	Shiraishi et al., 1988
Terbufos	<0.001	Maumee R., OH	1985	Baker, 1988
Thiobencarb	<0.41¶¶	Koise R., Japan	1985–86	Shiraishi et al., 1988
Toxaphene	<0.01–1.07	Bear Creek, MS	1977–79	Cooper et al., 1987
Trifluralin	<10¶	Israel	1980–84	Wynne, 1986

¶Pesticide concentration above Health Advisory Level (HAL) (Hornsby, 1992).
¶¶Not listed or discontinued use.
ND, not detectable.

Monitoring studies have been carried out in watersheds of different sizes. Scale could have had an effect on pesticide concentration and water quality of streams. Baker (1988) observed that peak run-off chemical concentrations were higher and the duration of exposure was shorter in small (compared with large) watersheds. The annual variability in material exported from the watershed was more accentuated in smaller watersheds. However, Baker (1988) concluded that the seasonal export of soluble chemicals was not affected by watershed size.

Routes of entry

The routes of entry of pesticides from the agricultural fields or watersheds into surface water were related to point and non-point sources. Point sources confirmed application of pesticides to the stream water by spills, losses from spray equipment during filling or cleaning, drift, and direct application (Frank et. al., 1982). Under such conditions, chemographs showed a significant increase in pesticide concentration with no change in the stream flow hydrograph (Braun and Frank, 1980). Point sources accounted for 82% of the pesticide losses into surface water during pesticide application season, but only 22% on an annual basis (Frank et. al., 1982).

Non-point routes of pesticides to the streams in a watershed were associated with run-off, groundwater or base flow, and rainfall. Run-off sources were storm or snowmelt run-off (overland flow) and interflow (Braun and Frank, 1980; Baker, 1988). The characteristic chemograph of run-off losses showed a significant increase in pesticide concentration concurrent with an increase in stream flow (Baker, 1988). Pesticides transported by infiltrating water could reach the stream water through base flow. Pesticides lost in run-off represented 18% of that applied during spray season, but 60% annually (Frank et. al., 1982). Base flow loss represented only 18% of the pesticides applied annually (Frank et. al., 1982).

Rainfall was shown to be an important route of surface water contamination (Richards et. al., 1987; Johnson et. al., 1988). Richards et. al. (1987) collected rainwater from Indiana, West Virginia, New York, and Ohio throughout 1985. The samples were analyzed for 19 herbicides and insecticides. Rainwater samples from New York and West Virginia stations showed measurable concentrations of all compounds less frequently, and in lower concentration, than did the samples from Indiana and Ohio. Seasonal variation in the pesticide concentration of the rainwater was related to pesticide application in the fields (Richards et. al., 1987). These authors explained that pesticide concentrations were below detection limits in early spring before application, reached maxima in the rainfall events following application in May, and declined to non-detectable levels by the end of July.

Parallel to their work, Johnson *et. al.* (1988) studied precipitation and run-off loading to two Ontario lake systems. They found that direct precipitation contributed more than 90% of the organochlorine pesticides loadings to the lakes.

Monitoring results and patterns

A list of pesticides detected in surface water bodies is presented in Table 5. It is important to mention that a large number of pesticides were analyzed for but not detected in these or other sites. This information was not included in this table. This indicates that their presence in surface water depended not only on their use.

The concentrations of pesticides found in surface water bodies were smaller than those in the edge-of-field run-off (see Tables 2–4). Such differences in concentration were related to sampling distance and seasonal variation. Several authors have shown a decrease in pesticide concentrations with sampling distance downstream which they have attributed to dilution effect (Johnson, 1980; Mayeux *et. al.*, 1984; Willis *et. al.*, 1987). Dilution factor, the ratio between run-off from treated and untreated areas, was correlated with a decrease in pesticide concentration (Willis *et. al.*, 1987). Other factors, like degradation and adsorption to sediment or organic matter particles in suspension, could decrease pesticide concentration in the column of water (Mayeux *et. al.*, 1984; Willis *et. al.*, 1987). Most of the data presented were also found to follow seasonal fluctuations. Highest pesticide concentrations were associated with field application of the pesticides (Albanis *et. al.*, 1986; Ohyama *et. al.*, 1986; Richards *et. al.*, 1987; Shiraishi *et. al.*, 1988; Wan, 1989).

A comparison has been made of the solubility, half-life, and K_{oc} values of pesticides found in surface water bodies (Table 5) (Wauchope *et. al.*, 1992). The minimum and maximum values of solubility and K_{oc} for the pesticides presented in Table 5 were from 0.1 to $500\,000\,\text{mg}\,\text{l}^{-1}$ and from 2 to $180\,000\,\text{mg}\,\text{l}^{-1}$, respectively. The shortest half-lives were for malathion (1 day), terbufos (5 days), and EPTC (6 days). Malathion and EPTC were considered point source pollutants from spills (Frank *et. al.*, 1982). With the exception of these three pesticides, half-lives ranged from 9 to 180 000 days.

SUMMARY AND CONCLUSION

Pollution of surface water by agrochemicals has not only a detrimental effect on a source of drinking water, but on the survival of important flora and fauna populations in streams/rivers, ponds/lakes, wetlands, and estuaries.

The non-point sources of agrochemicals are related to essential agricultural activities. This review describes several large monitoring studies the objective of which was to determine the effect of agricultural activity on water quality of rivers and lakes.

The monitoring results showed that pesticides were found in surface water bodies. However, their concentrations were 1000-fold smaller than those in run-off from the edge of agricultural fields. This decrease in pesticide concentration was related mainly to a downstream dilution process. In general, highest pesticide concentrations were found in surface water when rainfall occurred during or after pesticide application (spring). The largest pesticide losses occurred mainly during the first storm causing run-off. However, details of several storms causing high pesticide concentrations occurring in the same year were also found in literature. The amounts of losses were independent of the application rate for the pesticide.

Rainfall was found to be an important non-point source of pesticides. High pesticide concentration in the rainfall was detected during pesticide application. Spills from spray equipment handling, drift during application, and direct application were the major point sources leading to pesticide pollution of surface water.

Pesticide chemographs were used to describe the changes in concentration of pesticides with time at any one point downstream from pesticide entry. In conjunction with hydrographs and other chemical data, pesticide chemographs were used to determine their routes of entry to surface water and their sources. Pesticides arrived from a non-point source to a downstream point with overland flow, behind sediment (i.e. early portion of overland flow), and in front of an almost non-reactive nitrate (i.e. interflow or tile drainage). Factors that influenced pesticide concentrations in run-off (e.g. storm timing, site hydrology, formulation, pesticide physical and chemical properties) also had an effect on pesticide chemographs.

Increasing the understanding of the processes that affect pesticide movement in the soil will improve our capability to accurately simulate their concentration at the edge of the field and in surface water bodies. Although the sorption/desorption kinetics of pesticides on soil particles have been studied extensively (e.g. Brusseau and Rao, 1989), the kinetic release of pesticides from plant tissues and pesticide formulation has not been described. Release curves are used as a mechanistic approach to describe the relationship between rainfall, run-off, infiltration, and wash-off processes. Release time of the crop-target pesticides from the plants could have a considerable effect on the run-off concentration of pesticides. A pesticide released slowly by the plant (i.e. reactive with plant tissues) could increase pest protection but also increase pesticides losses with run-off. The same release kinetics would also be applied to slow release soil-applied pesticides and fertilizers are being developed to decrease leaching losses and ground water contamination. However, such pesticide formulations remain longer on the soil surface and under the influence of run-off water. The interaction between pesticides and run-off water on the soil surface has been changed from a simple zone of interaction to a diffusion control process.

New interactions between pesticides and soil are being included in a new generation of models. Such models require a larger number of parameters and a new experimental expertise. More than one component (i.e. soil, water, crop, weather, etc.) will be studied simultaneously. The multidisciplinary research group will be an important part of the success of future studies. Study sites will have to be more closely monitored and better characterized. Extrapolation to different scales (i.e. plots, fields, and watersheds) is an important component in modeling the new process.

The data presented in this and other reviews show that pesticide losses under 'normal' rainfall conditions account for 0.5% of the amount applied. Although the concentration of the pesticides in the run-off water was different, the same amount of pesticide loss was observed from plots, fields, and different sizes of watersheds. This demonstrates that plots are realistic simulators of fields or watersheds. Research plots could be used as part of pesticide registration, to study run-off–pesticide interaction, and patterns of surface water contamination. A better understanding of pesticide movement and factors that control it could prevent surface and ground water contamination.

REFERENCES

Ahuja, L. R. (1986). 'Characterization and modeling of chemical transfer to runoff', *Adv. Soil Sci.*, **4**, 149–188.

Ahuja, L. R. (1990). 'Modeling soluble chemical transfer to runoff with rainfall impact as a diffusion process', *Soil Sci. Soc. Am. J.*, **54**, 312–321.

Ahuja, L. R., and Lehman, O. R. (1983). 'The extent and nature of rainfall–soil interaction in the release of soluble chemicals to runoff', *J. Environ. Qual.*, **12**, 34–40.

Ahuja, L. R., Sharpley, N., Yamamoto, M., and Menzel, R. G. (1981). 'The depth of rainfall–runoff soil interaction as determined by 32p', *Water Resource Res.*, **17**, 969–974.

Albanis, T. A., Pomonis, P. J., and Sdoukos, A. Th. (1986). 'Organophosphorus and carbamate pesticides residues in the aquatic system of Ioannia basin and Kalamas river (Greece)', *Chemosphere*, **15**, 1023–1034.

Augustijn-Beckers, P. W. M., Hornsby, A. G., and Wauchope, R. D. (1994). 'The SCS/ARS/CES pesticide properties database for environmental decision-making. II. Additional compounds', *Rev. Environ. Contam. Toxicol.*, **137**, 1–82.

Baker, D. B. (1988). *Sediment, Nutrients and Pesticide Transport in Selected Lower Great Lakes Tributaries.* USEPA-905/4-88-001. Great Lakes National Program Office, Chicago, IL.

Baker, D. B., and Richards, R. P. (1991). 'Herbicides concentrations in Ohio's drinking water supplies: a quantitative exposure assessment', in *Pesticides in the Next Decade: the Challenges Ahead* (ed Weigmann D. L.), pp 9–30. *Proc. 3rd Nat. Conf. on Pesticides, Richmond, VA Nov., 1990*.

Baker, J. L. (1980). 'Agricultural areas as nonpoint sources of pollution' in *Environmental Impact of Noupoint Source Pollution*, (eds M. R. Overcash and J. M. Davidson), pp 275–310. Ann Arbor Science Publ., Ann Arbor, Michigan.

Baker, J. L., and Laflen, J. M. (1979). 'Runoff losses of surface-applied herbicides as affected by wheel track and incorporation', *J. Environ. Qual.*, **8**, 602–607.

Baker, J. L., Laflen, J. N., and Hartwing, R. O. (1982). 'Effects of corn residue and herbicides placement on herbicide runoff losses', *Trans. ASAE*, **25**, 340–343.

Baker, J. L., Laflen, J. N., and Johnson, H. P. (1978). 'Effect of tillage systems on runoff losses of pesticides, a rainfall simulation study', *Trans. ASAE*, **21**, 886–892.

Barrentine, W. L. (1988). 'Incorporation and injection of herbicides into the soil', in *Methods of Applying Herbicides*, (eds C. G. McWhorter and M. R. Gebhardt) Weed Sci. Soc. Am. Monogr. No. 4, pp. 231–254, WSSA, Champaign, IL.

Becker, R. L., Herzfeld, D., Ostlie, R. R., and Stamm-Katovich, E. J. (1989). 'Pesticides: surface runoff, leaching, and exposure concerns', *U. Minn. Extension Serv Bull AG-BU-3911*, St. Paul, MN, 32 pp.

Bouchard, D. C., Lavy, T. L., and Lawson, E. R. (1985). 'Mobility and persistence of hexazinone in a forest watershed', *J. Environ. Qual.*, **14**, 229–233.

Braun, H. E., and Frank, R. (1980). 'Organochloride and organophosphorus insecticides. Their use in eleven agricultural watersheds and their loss to stream waters in southern Ontario, Canada 1975–1977', *Sci. Total Environ.*, **15**, 169–192.

Brusseau, M. L., and Rao, P. S. C. (1989). Sorption nonideality during organic contaminants transport in porous media', *CRC Crit. Rev. Environ. Control*, **9**, 33–99.

Bryson, C. T. (1987). 'Effects of rainfall on foliar herbicides applied to rhizome johnsongrass', *Weed Sci.*, **35**, 115–119.

Burgoa, B. and Wauchope, R. D. (1991). 'The effect of formulations on runoff and leaching losses of atrazine and alachlor from tilted beds', *Abstr. Weed Sci. Soc. Am.*, **31**, 63.

Buttle, J. N. (1990). 'Metolachlor transport in surface runoff', *J. Environ. Qual.*, **19**, 531–538.

Caro, J. (1976). 'Pesticides in agricultural runoff', in *Control of Water Pollution from Cropland. Vol. II—An overview* (ed. B. A. Stewart), pp. 91–119. US EPA Rep. EPA-600/2-75-026b or USDA Rep. no. ARS-H5-2, 187 pp.

Carrol, B. R., Willis, G. H., and Graves, J. B. (1981). 'Permethrin concentration on cotton plants, persistence in soil, and loss in runoff', *J. Environ. Qual.*, **10**, 497–500.

Caseley, J. C., and Coupland, D. (1980). 'Effect of simulated rain on retention, distribution, uptake, movement and activity of difenzoquat applied to *Avena fatua*', *Ann. Appl. Biol.*, **96**, 111–118.

Cheng, H. H. (ed.) (1990). *Pesticides in the Soil Environment: Processes, Impacts, and Modeling*. Soil Sci. Soc. Am. Book Ser. No. 2, Madison, WI, 530 pp.

Cohen, S. Z., Creeger, S. M., Carsel, R. F., and Enfield, C. G. (1984). 'Potential pesticide contamination of groundwater from agricultural uses', in *Treatment and Disposal of Pesticide Wastes* (eds Krueger, R. F. and Seiber, J. N.), pp. 297–325. ACS Symp. Ser. No. 259, Am. Chem. Soc., Washington, DC.

Coody, P. N., White, J. W., and Graney, R. L. (1990). 'A "small plot" approach to predicting pesticide runoff and aquatic exposure', *Proc. Soc. Environ. Toxicol. Chem., 11th Ann. Meet., Nov. 11–15*, Washington, DC.

Cooper, C. M., Dendy, F. E., McHenry, J. R., and Ritchie, J. C. (1987). 'Residual pesticide concentrations in Bear Creek, Mississippi, 1976 to 1979', *J. Environ. Qual.*, **16**, 69–72.

DeCoursey, D. G. (ed.) (1990). *Proceedingss of the International Symposium on Water Quality Modeling of Agricultural Non-point Sources*, USDA-ARS Rep. ARS-1, June 1990, 881 pp. USDA, Tifton, GA.

Donigian, A. G., Beyerlein, D. C., Davis, Jr, H. H., and Crawford, N. H. (1977). *Agricultural Runoff Management (ARM) Model Version II: Refinement and Testing.* US EPA Rep. No. EPA-600/3-77-098, 294 pp, US EPA, Washington, DC.

Donigian, A. G., and Crawford, N. H. (1976). *Modeling Pesticides and Nutrients on Agricultural Lands.* US EPA Rep. EPA-600/2-76-043, 317 pp, US EPA, Washington, DC.

Edwards, W. M., Triplett, Jr, C. G., and Rramer, R. M. (1980). 'A watershed study of glyphosate transport in runoff', *J. Environ. Qual.*, **9**, 661–665.

Emmerich, W. E., Woolhiser, D. A., and Shirley, E. D. (1989). 'Comparison of lumped and distributed models for chemical transport by surface runoff', *J. Environ. Qual.*, **18**, 120–126.

Ertel, M. (ed.) (1990). *Information Exchange on Models and Data Needs relating to the Impact of Agricultural Practices on Water Quality: Workshop Proceedings.* Avail. Office of Water Quality, U.S. Geol. Survey, Reston, Virginia (703) 648–5023.

Fawcett, R. S. (1991). 'Scratching the surface', *Agricultural Engineering*, **May 1991**, 16–17.

Felsot, A. S., Mitchell, J. K., and Renimer, A. L. (1990). 'Assessment of management practices for reducing pesticide runoff from sloping cropland in Illinois', *J. Environ. Qual.*, **19**, 539–545.

Frank, R., Braun, H. E., Holdrinet, M. V., Sirons, G. J., and Ripley, B. D. (1982). 'Agriculture and water quality in the Canadian Great Lakes Basin 5. Pesticide use in eleven agricultural watersheds and presence in stream waters—1975-1977', *J. Environ. Qual.*, **11**, 497–505.

Freeze, R. A., and Cherry, J. A. (1979). *Groundwater.* Prentice-Hall, Inc., Inglewood Cliffs, NJ, 604 pp.

Glenn, S., and Angle, J. S. (1987). 'Atrazine and simazine in runoff from conventional and no-till corn watersheds', *Agric. Ecosys. Environ.*, **18**, 273–280.

Glotfelty, D. E., Taylor, A. W., Isensee, A. R., Jersey, J., and Glenn, S. (1984). 'Atrazine and simazine movement to Wye River estuary', *J. Environ. Qual.*, **13**, 115–121.

Goss, D. W. (1992). 'Screening procedure for soils and pesticides relative to water quality impacts', *Weed Technol.*, **6**, 701–708.

Gustafson, D. I. (1989). 'Groundwater ubiquity score: a simple method for assessing pesticide leachability', *Environ. Toxicol. Chem.*, **8**, 339–357.

Hall, J. K. (1974). 'Erosional losses of s-triazine herbicides', *J. Environ. Qual.*, **3**, 174–180.

Hall, J. K., Hartwig, N. L., and Hoffman, L. D. (1983). 'Application mode and alternate cropping effects on atrazine losses from a hillside', *J. Environ. Qual.*, **12**, 336–340.

Hall, J. K., Hartwig, N. L., and Hoffman, L. D. (1984). 'Cyanazine losses in runoff from no-tillage corn in "living" and dead mulches vs. unmulched conventional tillage', *J. Environ. Qual.*, **13**, 105–110.

Hillel, D. (1980). *Applications of Soil Physics.* Academic Press, New York.

Hollis, J. M. (1990). 'Assessments of the vulnerability of aquifers and surface waters to contamination by pesticides', *Res. Rep. for UK MAFF*, Soil Survey and Land Research Centre, Silsoe Campus, Silsoe, Bedford MK45 4DT, England, 27 pp.

Hornsby, A. G. (1992). 'Site-specific pesticide recommendations: the final step in environmental impact prevention', *Weed Technol.*, **6**, 736–742.

Hubbard, R. R., Williams, R. G., Erdman, M. D., and Marti, L. R. (1989). 'Chemical transport from coastal plain soils under simulated rainfall: II. Movement of cyanazine, sulfometuron-methyl, and bromide', *Trans. ASAE*, **32**, 1250–1257.

Johanson, R. C., Imhoff, J. C., Kittle, J. L., and Donigian, A. S. (1984). *Hydrological Simulation Program-Fortran (HSPF): User's Manual for Release 8.0.* US EPA Rep. No. EPA-600/3-84-066, 767 pp, US EPA, Washington, DC.

Johnsen, T. N., Jr. (1980). 'Picloran in water and soil from a semiarid pinyon-juniper watershed', *J. Environ. Qual.*, **9**, 601–605.

Johnson, M. G., Kelso, J. R. M., and George, S. E. (1988). 'Loading of organochlorine contaminants and trace elements to two Ontario lake systems and their concentration in fish', *Can. J. Fish. Aquat. Sci.*, **45**, 170–178.

Jury, W. A., Focht, D. C., and Farmer, W. J. (1987). 'Evaluation of pesticide groundwater pollution potential from standard indices of soil chemical adsorption and degradation', *J. Environ. Qual.*, **16**, 422–428.

Kaiser, K. L. E., Lum, K. R., Comba, M. E., and Palabrica, V. S. (1990). 'Organic trace contaminants in St. Laurence river water and suspended sediments, 1985–1987', *Sci. Total Environ.* **97**, 23–40.

Klaine, S. J., Hinman, M. L., Winkelman, D. A., Sauser, R. R., Martin, J. R., and Moore, L. W. (1988). 'Characterization of agricultural non-point pollution: pesticides migration in a West Tennessee watershed', *Environ. Toxicol. Chem.*, **7**, 609–614.

Knisel, W. G. (ed.) (1980). *CREAMS: a Field Scale Model for Chemicals, Runoff, and Erosion from Agricultural Management Systems*. USDA-SEA Conservation Research Rep. No. 26, 643 pp. USDA, Tifton, GA.

Lavy, T. L., Mattice, J. D., and Kochenderfer, J. N. (1989). 'Hexazinone persistence and mobility of a steep forested watershed', *J. Environ. Qual.*, **18**, 507–514.

Leavesley, G. H., Bealey, D. B., Pionke, H. B., and Leonard, R. A. (1990). 'Modeling of agricultural nonpoint-source surface runoff and sediment yield—a review from the modeller's perspective' In *Proceedings of the International Symposium on Water Quality Modeling of Agricultural Non-Point Sources, June 19–23, 1988, Utah State University* (ed. D. G. DeCoursey), pp. 171–194. USDA-ARS Publication ARS-81, 421 pp. USDA, Tifton, GA.

Leitch, C., and Fagg, P. (1985). 'Clopyralid herbicide residues in streamwater after aerial spraying of *Pinus radiata* plantation', *N.Z. J. For. Sci.*, **15**, 195–206.

Leonard, R. A. (1990). 'Movement of pesticides into surface waters', in *Pesticides in the Soil Environment* (ed. H. H. Cheng), pp. 303–349. Soil Sci. Soc. Am. Book Series, No. 2, SSSA, Madison, WI, 530 pp.

Leonard, R. A., Knisel, W. G., and Still, D. A. (1987). 'GLEAMS: groundwater loading effects of agricultural management systems', *Trans ASAE*, **30**, 1403–1418.

Leonard, R. A., Langdale, G. W., and Fleming, W. G. (1979). 'Herbicide runoff from upland Piedmont Watersheds—data and implications for modeling pesticide transport', *J. Environ. Qual.*, **8**, 223–229.

Leonard, R. A., and Wauchope, R. D. (1980). 'The pesticide submodel', in *CREAMS: a Field Scale Model for Chemicals, Runoff, and Erosion from Agricultural Management Systems* (ed. W. G. Knisel), pp. 88–112. USDA-SEA Conservation Research Rep. No. 26, 643 pp. USDA, Tifton, GA.

Mayack, D. T., Bush, P. B., Neary, D. G., and Douglass, J. E. (1982). 'Impact of hexazinone on invertebrates after application to forested watersheds', *Arch. Environ. Contam. Toxicol.*, **11**, 209–217.

Mayer, J. R., and Elkins, N. R. (1990). 'Potential for agricultural pesticide runoff to a Puget Sound estuary: Padilla Bay, Washington', *Bull. Environ. Contam. Toxicol.*, **45**, 215–222.

Mayeux, H. S., Jr, Richardson, C. W., Bovey, C. W., Burnett, R. W., and Merkle, E. (1984). 'Dissipation of picloram in storm runoff', *J. Environ. Qual.*, **13**, 44–49.

McDowell, L. L., Willis, G. H., Murphree, C. B., Southwick, L. M., and Smith, S. (1981). 'Toxaphene and sediment yields in runoff from a Mississippi Delta watershed', *J. Environ. Qual.*, **10**, 120–125.

McDowell, L. L., Willis, G. H., Southwick, L. M., and Smith, S. (1984). 'Methyl parathion and EPN washoff from cotton plants by simulated rainfall', *Environ. Sci. Technol.*, **18**, 423–427.

McDowell, L. L., Willis, G. H., Southwick, L. M., and Smith, S. (1987). 'Fenvalerate washoff from cotton plants by rainfall', *Pestic. Sci.*, **21**, 83–92.

Morton, H. L., Johnsen, Jr, T. N., and Simanton, R. (1989). 'Movement of tebuthiuron applied to wet and dry rangeland soils', *Weed Sci.*, **37**, 117–122.

Neary, D. G., Bush, P. B., Douglas, J. E., and Todd, R. L. (1985). 'Picloram movement in an Appalachian hardwood forest watershed', *J. Environ. Qual.*, **14**, 585–592.

Nicholaichuk, W., and Grover, R. (1983). 'Loss of fall applied 2,4-D in spring runoff from a small agricultural watershed', *J. Environ. Qual.*, **12**, 412–414.

Ohyama, T., Jin, K., Katoh, Y., Chiba, Y., and Inoue, K. (1986). 1,3,5-Trichloro-2-(4-nitrophenoxy) benzene (CNP) in water, sediment, and shellfish of the Ishikari river', *Bull. Environ. Contam. Toxicol.*, **37**, 344–349.

Parsons, D. W., and Witt, J. M. (1989). *Pesticides in Groundwater in the United States of America: a Report of a 1988 Survey of State Lead Agencies*. Oregon State University Ext. Service Rep. No. EM 8406, August, 1989, 92 pp Corvallis, Oregon.

Penrose, D. L., and Lenat, D. R. (1982). 'Effects of apple orchard runoff of the aquatic macrofauna of a mountain stream', *Arch. Environ. Contam. Toxicol.*, **11**, 383–388.

Pereira, W. E., and Rostad, C. E. (1990). 'Occurrence, distribution, and transport of herbicides and their degradation products in the Lower Mississippi River and its tributaries', *Environ. Sci. Technol.*, **24**, 1400–1406.

Pereira, W. E., Rostad, C. E., and Leiker, T. J. (1990). 'Distribution of agrochemicals in the lower Mississippi river and its tributaries', *Sci. Total Environ.*, **97/98**, 41–53.

Peyton, R. L., and Sanders, T. G. (1990). 'Mixing in overland flow during rainfall', *J. Environ. Eng.*, **116**, 764–783.

Pfeuffer, R. J. (1991). *Pesticide Residues Monitoring in Sediment and Surface Water within the South Florida Water Management District*. Volume 2. TP 91-01. South Florida Water Management District, West Palm Beach, FL.

Richards, R. P., Kramer, J. W., Baker, D. B., and Krieger, K. A. (1987). 'Pesticides in rainwater in the northeastern United States', *Nature*, **327**, 129–131.

Rohde, W. A., Asmussen, L. E., Hauser, E. W., Wauchope, R. D., and Allison, H. D. (1980). 'Trifluralin movement in runoff from a small agricultural watershed', *J. Environ. Qual.*, **9**, 37–42.

Sauer, T. J., and Daniel, T. C. (1987). 'Effect of tillage system on runoff losses of surface-applied pesticides', *Soil Sci. Soc. Am. J.*, **51**, 410–415.

Shiraishi, H., Pula, F., Otsuki, A., and Iwakuma, T. (1988). 'Behavior of pesticides in Lake Kasumigaura, Japan', *Sci. Total Environ.*, **72**, 29–42.

Skaggs, R. W., Gilliam, J. W., Sheets, T. J., and Barnes, J. S. (1980). 'Effect of agricultural land development on drainage water in the North Carolina tidewater region', *Water Resource Res. Inst. Rep.* **159**, Univ. North Carolina, Raleigh, NC.

Smith, C. N., and Carsel, R. F. (1984). 'Foliar washoff of pesticides (FWOP): development and evaluation', *J. Environ. Sci. Health*, **B19(3)**, 323–342.

Smith, S., Reagan, T. E., Flynn, J. L., and Willis, G. H. (1983). 'Azinphos-methyl and fenvalerate runoff loss from a sugarcane-insect integrated pest management system', *J. Environ. Qual.*, **12**, 534–537.

Southwick, L. M., Willis, G. H., Bengtson, R. L., and Lormand, T. J. (1990). 'Effect of subsurface drainage on runoff losses of atrazine and metolachlor in Southern Louisiana', *Bull. Environ. Contam. Toxicol.*, **45**, 113–119.

Spencer, W. F., Cliath, M. M., Blair, J. W., and LeMert, R. A. (1985). *Transport of Pesticides from Irrigated Fields in Surface Runoff and Tile Drain Waters.* Agricultural Research Service Conservation Research Report No. 31, USDA, Washington, DC, 71 pp.

US EPA (1990). *National Survey of Pesticides in Drinking Water Wells: Phase I Report*. EPA Report No. 570/9-90-015, 98 pp. US EPA, Washington, DC.

Viessman, W., Jr, Lewis, G. L., and Knapp, J. W. (1989). *Introduction to Hydrology*, 3rd edn, Harper & Row Publishers, New York, NY, 780 pp.

Wallach, R., Jury, W. A., and Spencer, W. F. (1988). 'Transfer of chemicals from soil solution to surface runoff: a diffusion based soil model', *Soil Sci. Soc. Am. J.*, **52**, 612–618.

Wallach, R., Jury, W. A., and Spencer, W. F. (1989). 'The concept of convective mass transfer for prediction of surface-runoff pollution by soil surface applied chemicals', *Trans. ASAE*, **32**, 906–912.

Wallach, R., and van Genuchten, M. Th. (1990). 'A physically based model for predicting solute transfer from soil solution to rainfall-induced runoff water', *Water Resource Res.*, **26**, 2119–2126.

Wan, M. T. (1989). 'Levels of selected pesticides in farm ditches leading to rivers in the lower mainland of British Columbia', *J. Environ. Sci. Health: Part B: PFCAW*, **24**, 183–203.

Wauchope, R. D. (1978). 'The pesticide content of surface water drainage from agricultural fields: a review', *J. Environ. Qual.*, **7**, 459–472.

Wauchope, R. D. (1987). 'Effects of conservation tillage on pesticide loss with water', in *Effects of Conservation Tillage on Water Quality: Nitrate and Pesticides* (ed. T. J. Logan), pp. 205–215, Lewis Publishers, Chelsea, MI.

Wauchope, R. D. (1992). 'Environmental risk assessment of pesticides: improving model credibility', *Weed Technol.*, **6**, 753–759.

Wauchope, R. D., Buttler, T. M., Hornsby, A. G., Augustijn-Beckers, P. W. M., and Burt, J. P. (1992). 'The SCS/ARS/CES pesticide properties database for environmental decision-making', *Rev. Environ. Contam. Toxicol.*, **123**, 1–164.

Wauchope, R. D., Dowler, C. C., Sumner, H. R., Truman, C., Johnson, A. W., Chandler, L. D., Gascho, G. J., Davis, J. G., and Hook, J. E. (1993). 'Herbicide runoff measurements from very small plots: how realistic?' *Proc. Brighton Crop Prot. Conf.—Weeds 1993*, 1291–1298.

Wauchope, R. D., and Leonard, R. A. (1980). 'Maximum pesticide concentrations in agricultural runoff: a semiempirical prediction formula', *J. Environ. Qual.*, **9**, 665–672.

Wauchope, R. D., and Street, J. E. (1987a). 'Fate of a water-soluble herbicide spray on foliage. Part I. Spray efficiency: measurement of initial deposition and absorption', *Pestic. Sci.*, **19**, 243–252.

Wauchope, R. D., and Street, J. E. (1987b). 'Fate of a water-soluble herbicide spray on foliage. Part II. Absorption and dissipation of foliar MSMA deposits: mathematical modelling', *Pestic. Sci.*, **19**, 253–263.

Wauchope, R. D., Williams, R. G., and Marti, L. R. (1989). 'Runoff of sulfometuron-methyl and cyanazine from small plots: effects of formulation and grass cover', *J. Environ. Qual.*, **19**, 119–125.

Weaver, J. E., Hogmire, H. W., Brooks, J. L., and Sencindiver, J. C. (1990). 'Assessment of pesticide residues in surface and soil water from a commercial apple orchard', *Applied Agric. Res.*, **5**, 37–43.

Weber, J. B. (1977). 'The pesticide scorecard', *Environ. Sci. Technol.*, **11**, 756–761.

Wiese, A. F., Savage, K. E., Chandler, J. M., Liu, L. C., Jeffrey, L. S., Weber, J. B., and LaFleur, K. S. (1980). 'Loss of fluometuron in runoff water', *J. Environ. Qual.*, **9**, 1–5.

Willis, G. H., McDowell, L. L., Murphree, C. B., Southwick, L. M., and Smith, M. (1983). 'Pesticide concentration and yields in runoff from silty soils in the lower Mississippi Valley, USA', *J. Agric. Food Chem.*, **31**(6), 1171–1177.

Willis, G. H., McDowell, L. L., Smith, S., and Southwick, L. M. (1986). 'Permethrin washoff from cotton plants by simulated rainfall', *J. Environ. Qual.*, **15**, 116–120.

Willis, G. H., McDowell, L. L., Southwick, L. M., and Smith, S. (1987). 'Methoxychlor and endosulfan concentrations in unit-source runoff and in channel flow of a complex watershed', *Trans. ASAE*, **30**, 394–399.

World Wildlife Fund (1991). *Improving Aquatic Risk Assessment under FIFRA: Report on the Aquatic Effects Dialog Group.* Available from RESOLVE, Inc. Suite 500, 1250 24th St. NW, Washington, DC 20037.

Wynne, D. (1986). 'The potential impact of pesticides on the Kinneret and its watershed, over the period 1980–1984', *Environ. Pollut.*, **42**, 373–386. (order 1-31-91)

Zahnow, E. W., and Riggleman, J. D. (1980). 'Search for linuron residues in tributaries of the Chesapeake Bay, USA', *J. Agric. Food Chem.*, **28**, 974–978.

CHAPTER 6

The Volatilization of pesticide residues

A. W. Taylor

Environmental Behaviour of Agrochemicals
Edited by T. R. Roberts and P. C. Kearney   © 1995 John Wiley & Sons Ltd

University of Maryland, Agricultural Experiment State Document No 8633.

SUMMARY

Volatilization is one of the principal pathways by which pesticides are lost from target areas after application. The rate of disappearance by this pathway is often greater than or equal to that due to chemical degradation. Volatilization losses can sometimes exceed 90% of the application within 48 hours or less when residues of volatile pesticides are exposed on moist soil or plant surfaces.

The principal factors controlling the rate of volatilization are: (1) the vapor pressure of the pesticide; (2) the distribution of the residues; and (3) the moisture status of the soil or plant surface.

Volatilization is greatly reduced by incorporation into the soil, where the rate becomes dependent upon movement of the residues to the soil surface by diffusion or by convective transport by soil water (the wick effect). Both these mechanisms are very sensitive to soil moisture conditions. Convective flow is dependent upon continuous upward movement of moisture caused by evaporation at the surface: changes in moisture flow during the day result in marked diurnal variations in the pesticide volatilization rate. When there is no upward moisture flow the movement of residues to the surface of moist soil is controlled by diffusion.

Most pesticides are very strongly adsorbed by dry soil so that the volatilization of both incorporated and surface-exposed residues ceases almost completely from dry soil. This reduction in chemical activity is reflected in the reduction of the biological activity of herbicides in dry soils. Since soil temperatures only show marked rises as the soil becomes dry, the decrease in fugacity of the residues caused by the increased adsorption in dry soil negates any increase in volatilization due to rising temperature.

Although the speed at which residues can enter the atmosphere is to some degree offset by rapid dilution, re-deposition as a result of adsorption and re-concentration of the vapor in rain or fog droplets is a known pathway that may represent the first step in the bioaccumulation of those pesticides and their degradation products that are sufficiently stable in air or water. Where organisms that are sensitive to a particular chemical are present in the environmental phase where re-concentration occurs, the possibilities of chronic low-level toxic effects cannot be discounted.

INTRODUCTION

The importance of volatilization as a major pathway by which pesticide residues disappear from target areas was recognized in some of the earliest studies of residue disappearance (Harris and Lichtenstein, 1961; Lichtenstein and Schultz, 1961). This work showed that up to more than 30% of the residues of aldrin and other insecticides could be lost from glass surfaces and soils in flasks in the laboratory and also revealed the importance of soil moisture content on the volatilization rate. Earlier work which demonstrated the existence of pesticide vapor in the air over treated surfaces has been summarized by Spencer *et al*. (1973). None of these results could, however, be used to estimate volatilization losses in the field. The full significance of volatilization was recognized in the late 1960s when it was measured in a number of field experiments measuring the rates by which residues were dispersed into the environment by various pathways.

It is now clear that volatilization from target areas can be the principal factor limiting the efficacy of some pesticide materials. In unfavorable conditions volatilization losses may approach 80–90% of the amount applied within a few days after application, although such losses are sensitive to weather and microclimatic conditions. Volatilization rates depend upon complex interactions between the chemical properties of the individual compounds, the weather, the properties of the soil or plant tissue on which they are adsorbed, and—not least—the way in which they are applied and how the field or crop receiving them is managed. Under some circumstances rates of disappearance of residues by volatilization can exceed that due to chemical degradation.

The purpose of this chapter is to review the fundamental physics and chemistry of pesticide volatilization, to show the importance of these for our understanding of the process, and to review and summarize the available field data and their environmental and agronomic significance. More detailed reviews of several aspects of the subject have been presented elsewhere (Taylor and Glotfelty, 1988; Glotfelty and Schomburg, 1989; Taylor and Spencer, 1990).

BASIC PROCESSES

The dispersal of pesticide residues into the atmosphere involves two distinct physical processes. The first is the evaporation of the pesticide molecules from the residues on the soil or plant surfaces. The second is the dispersion of the resulting vapor into the overlying atmosphere by diffusion and turbulent mixing. Since the two processes are fundamentally different in character it is convenient to consider them separately but, as will be clear in the later section of the review, in an overall view they must be regarded as parts of an integrated whole.

In thermodynamic terms the evaporation of pesticide residues represents a phase change into vapor from the liquid or solid state: a review of the thermodynamics of this process has been presented by Glotfelty and Schomburg (1989). Under field conditions the vapor pressures of the pure compounds are affected by adsorption upon the soil or plant surfaces. The way that this adsorption changes with the pesticide concentration and moisture status of the surfaces is of major importance.

The second process, the dispersal of vapor into the atmosphere, is the same process that controls the transfer of water, carbon dioxide and other gases between soil and plant surfaces and the overlying air. The principal difference between these gases and pesticides is that the concentration of pesticide vapor in the air more than 2–3 m above the ground is virtually zero. Without any appreciable background concentration there is no downward 'return flow' — as there is with water or CO_2 — and we are concerned only with upward dispersal.

Mathematically, the upward vertical flux of vapor, $F\uparrow$, through any plane at height z above the surface is described by the general equation

$$F\uparrow = k_z(\mathrm{d}p/\mathrm{d}z) \tag{1}$$

where k_z is a diffusivity coefficient and $\mathrm{d}p/\mathrm{d}z$ the gradient of vapor pressure at the height z. The flux $F\uparrow$ has dimensions of $\mathrm{m.l^{-2}.t^{-1}}$.

The flow of air over plant and soil surfaces is invariably turbulent except in a very shallow layer with a depth of about 1 mm or less in which the flow may be regarded as laminar and controlled by viscosity (Montieth, 1973). The dispersal of pesticide vapor through this layer, which is illustrated in Figure 1, can be described in terms of the molecular diffusion coefficient of the vapor. Expressing equation (1) in terms of the depth of this laminar surface layer, we have

$$F\uparrow = D(p_s - p_d)/z_d \tag{2}$$

where p_s and p_d are the vapor pressures at the evaporating surface and at the upper surface of the laminar layer of effective depth z_d (Figure 1). This depth can be equated with the thickness of the thin surface layer of streamline flow through which the drag or frictional force of the surface is imposed on the airstream by Newtonian viscosity. Since there is no turbulent mixing in this layer this implies that the effective depth z_d is defined as the distance above the surface through which the vapor moves by molecular diffusion. The value of D may then be identified with the molecular diffusion coefficient of the particular compound. As pointed out by Hartley and Graham-Bryce (1980), equation (2) really only defines z_d as an 'effective thickness' that cannot actually be measured, but must be expected

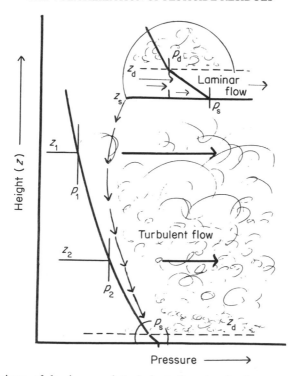

Figure 1   Regions of laminar and turbulent flow in the lower atmosphere and corresponding profiles of pesticide vapor pressure above the surface. (From Pesticides in the Soil Environment, SSSA Book Series No. 2, reprinted by permission of the Soil Science Society of America)

to vary with wind speed and surface roughness. The postulate of Newtonian flow implies that there is no single plane of wind shear, but a transition zone across which the flow becomes increasingly turbulent. Despite the usefulness of the interpretation summarized in equation (2), it must be recognized that the effective depth is physically indeterminate and the extent to which the diffusivity coefficient in it can be identified with the molecular diffusion coefficient measured in still air is also ambiguous. In an alternative view, the laminar layer can be regarded as the limiting distance above the surface to which the smallest eddies of the overlying turbulent flow can penetrate. In this picture the effective depth will, however, be equally indeterminate because it must fluctuate in response to the pressure variations and varying drag exerted by the turbulence in the overlying flow.

Outside this laminar surface zone the vapor is dispersed by mixing in the turbulent air flow in the atmosphere. The dispersal of vapor through any layer of finite thichness in the turbulent zone may be described by the equation

$$F \uparrow = k_z(p_1 - p_2)/(z_1 - z_2) \tag{3}$$

where k_z is the mean eddy diffusivity coefficient between the heights z_1 and z_2. As illustrated in Figure 1, the depth of the turbulent zone, often described as the 'atmospheric boundary layer', is several orders of magnitude greater than that of the laminar flow layer, being measured in meters rather than the millimetric depths of the latter. Dispersal of the pesticide vapor in the turbulent zone outside the laminar layer is relatively rapid except when the lower atmosphere is exceptionally stable, as in the evening or early morning hours.

The concept of separate regions of dilution by molecular diffusion and by turbulent mixing is important in the identification of the physical factors that impede the dispersion process under differing soil and microclimate conditions. As noted by Hartley and Graham-Bryce (1980), equation (1) can be rewritten as $F \uparrow = (f)[(p_1 - p_2)/R]$, where R is a resistance term such that $R = (z_1 - z_2)/k$. Stated in this way, the resistance terms are additive — as in electrical theory — and the appropriate forms of equations (2) and (3) may be regarded as describing the additive resistances (or impedances) to dispersive flow in the laminar and turbulent layers. If, as pointed out by Mackay (1984) and Mackay and Paterson (1982), the concentration or potential terms are converted to suitable units, the dispersion equations for all the environmental phases are additive and fluxes can be described in terms of similar potential gradients throughout the entire environmental system. This can be done by the use of the fugacity concept, originally posulated by Lewis (1901) and reintroduced by Mackay and Paterson (1981) and by Mackay (1991) as a fundamental principle in environmental chemistry. As will be shown below, this is particularly convenient in experimental work with volatile pesticides because their fugacities in the air are directly related to the fugacities of the residues, which can be described in terms of Henry's law coefficients and their adsorption isotherms on the solid phases in the soil.

VAPOR PRESSURES OF PESTICIDE RESIDUES

The background vapor pressures of all pesticides in the free atmosphere are extremely low, so that the overall gradient of vapor pressure or vapor density gradient is established by the value of the vapor pressure of the residues on the plant or soil surface, p_s. This is equivalent to a statement that the rate of dispersion will be a function of the fugacity of the residues. This fugacity, which is equivalent to the vapor pressure at the inner surface of the laminar flow layer illustrated in Figure 1, will only be that of the pure compound when the residues are present as a continuous exposed layer on the soil or plant surface whose fugacity is not reduced by chemical or physical adsorption reactions. Although under field conditions the effective vapor pressures (or fugacities) of actual residues will always be less than those of the pure compounds, it is useful to examine the magnitudes of the vapor pressures of some commonly used pesticides in order to establish the range of values to be expected.

Vapor pressures of pesticide chemicals

The measured vapor pressures (and vapor densities)* of a number of commonly used pesticides are presented in Table 1. These were selected to illustrate the range and scale of the values which extend over about six orders of magnitude from 2000 mPa for EPTC to 0.0085 mPa for simazine.

Table 1. Vapor pressures of pesticides

Compound	M.Wt. (g)	Temp. (°C)	Vap.Press (mPa)	Vap. Dens. (mg/m³)	Source[1]
			Figures in parentheses represent power of 10 factors.		
Alachlor	270	20	3.0	3.2 (−1)	a
Atrazine	216	20	3.0 (−2)	2.7 (−3)	a
Carbofuran	221	20	1.5	1.4 (−1)	a
Chlorpropham	214	20	1.5	1.3 (−1)	a
Chlopyrifos	351	20	1.5	2.2 (−1)	a
DDT	355	25	4.5 (−2)	6.4 (−3)	b
Diazinon	304	20	8.0	1.0	a
Dicamba	221	20	3.0	2.7 (−2)	a
Dieldrin	381	25	6.8 (−1)	1.0 (−1)	b
EPTC	189	20	2.0 (+3)	1.6 (+2)	a
Fenitrothion	277	20	4.0 (−1)	4.5 (−2)	a
Heptachlor	373	20	3.0 (+1)	4.6	a
Lindane	291	25	8.6	1.0	b
Malathion	330	20	1.0	1.4 (−1)	a
Methyl Parathion	263	25	2.4	2.5 (−1)	b
Parathion	291	25	1.3	1.5 (−1)	b
Picloram	241	20	6.0 (−2)	5.9 (−3)	a
Simazine	202	25	8.5 (−3)	7.0 (−4)	a
Triallate	305	25	2.6 (+1)	3.2	b
Trifluralin	335	20	1.5 (+1)	2.0	b

[1]Sources a. Suntio *et al.* (1988)
      b. Jury *et al.* (1983).

*The vapor pressure, p, and the vapor density, w/V, are related by the gas equation, $p = (w/V)(RT/M)$, where w is the mass and M the molecular weight of the pesticide, V is volume in liters, T is absolute temperature and R is the molecular gas constant. This equation reduces to $d = w/V = 0.12 \times p \cdot (M/T)$ where d is in $\mu g\,l^{-1}$ and p is in mPa. Both equations contain the implicit assumption that the pesticide vapor is completely dissociated.

The list is not comprehensive and other compounds may have values outside this range. The general scale is, however, very low in comparison with more familiar volatile organic chemicals such as alcohols or ethers which have vapor pressures close to one atmospere at normal temperatures. The pressure of a standard atmosphere (760 mmHg) is equal to 1×10^5 Pa, or about four orders of magnitude greater than the vapor pressure of EPTC. While the tabulated values may seem to suggest that the volatilization rates of the compounds would be very low, it must be remembered that the background vapor pressures of pesticides in the air are essentially zero so that the residues will evaporate as if they were exposed to a vacuum. This absence of any natural background concentration places pesticide vapor dispersion in marked contrast to that of other atmospheric components where a natural background concentration exists.

Comparison of the values in Table 1 with others in the literature may reveal some uncertainties about the 'true' value for some chemicals. These uncertainties arise from the methods used in their determination. Measurements of very low vapor pressures present considerable experimental difficulties. The methods used and the uncertainties they entail have been reviewed in detail by Plimmer (1976), Spencer and Cliath (1983) and Glotfelty and Schomburg (1989): they will, however, also be briefly described here because some of the same techniques have been used to measure the vapor pressures of residues in soil and water systems as well as those of the pure compounds.

The simplest method is the direct measurement of the pressure of the vapor in equilibrium with the pure solid or liquid chemical. This is most satisfactory where the vapor pressure is of the order of 130 Pa (= 1 mmHg) or more. This is much greater than that of most pesticides at room temperatures, so that elevated temperatures must often be used and the results extrapolated to the normal range by equations based upon the integrated Clapeyron–Clausius equation expressed in the form $\log P = A - B/(t\,°C + 273.17)$. Data obtained from such high temperature methods are sometimes uncertain due to errors in both the experimental measurement and in the extrapolation: where the extrapolation crosses the temperature of a phase change for the solid or liquid the error may be considerable.

The gas saturation method described by Spencer and Cliath (1969, 1983) is used to measure concentrations of pesticides in air equilibrated with either pure compounds or residues in soils. Since chromatographic analyses permit accurate determination of nano- and picogram quantities of pesticides, measurements can be made in the 10–40 °C range and no extrapolations are necessary.

Vapor pressures may also be calculated from gas/liquid chromatographic retention times. This method assumes that column retention times depend only on the vapor pressure, so that this can be calculated by comparison with the retention time of a standard compound of known vapor pressure. Successful measurements depend on a suitable choice of both column materials and standard reference compounds appropriate to the polarity and chemical character of the particular pesticide (Bidleman, 1984; Glotfelty and Schomburg, 1989). The advantages of the method are the speed at which data can be obtained and its applicability to mixtures and impure samples: a summary of the chemical principles of the method is presented by Glotfelty and Schomburg (1989).

The most consistent results appear to be obtained by the gas saturation and gas/liquid chromatographic methods. Where critical values are sought from the literature, those obtained by these means should be preferred provided that they are supported by an adequate presentation of the original data and description of the experimental conditions. Unfortunately this is far from always the case for many published values. Where data are required for critical purposes the results should be traced, if possible, to their origin.

Solubility and Henry's law

The water solubility of pesticides is important in relation to volatility because it is a major factor in controlling the distribution and movement of residues within the soil (Spencer, 1970; Jury *et al.*, 1983a, 1983b). The distribution between air and water is described by Henry's law, defined by the equation

$$K_{AW} = C_A/C_W \tag{4}$$

Where C_A is the vapor density and C_W is the solution concentration. Where these are both in the same units the Henry's law coefficient K_{AW} is a dimensionless number.*

Unfortunately, problems similar to those in the measurement of vapor pressure are found in measurements of pesticide solubilities and conflicting values are again found in the literature, particularly where the solubilities are very low. Difficulties sometimes also arise in the interpretation of the measured values. The value of C_W used in equation (4) should denote the concentration of pesticide in monomolecular solution. This may be considerably lower than the total concentration where the pesticide is surface active or tends to solubilize with the formation of micelles or where the solution contains foreign colloidal material on which the pesticide is adsorbed. Table 2 contains data on the water solubilities and Henry's law coefficients for the

---

*The ratio is not always stated in these terms. Air concentrations may be given in various pressure units, when the coefficient is represented by the symbol H which is not a dimensionless number. Care must be taken about dimensions in comparing various values of K_{AW} and H in the literature.

Table 2. Solubility and Henrys law coefficients

Compound	Temp. °C	Solubility (g/m³)	Coefficient $K_{AW} = C_A/C_W$	Source[1]
		Figures in parentheses represent power of 10 factors.		
Alachlor	20	1.3 (+2)	2.5 (−6)	a
Atrazine	20	3.0 (+1)	9.0 (−8)	a
Carbofuran	20	6.5 (+2)	2.1 (−7)	a
Chlorpropham	20	1.0 (+2)	1.3 (−6)	a
Chlorpyrifos	20	3.0 (−1)	7.3 (−4)	a
DDT (pp)	25	3.0 (−3)	2.1 (−3)	b
Diazinon	20	3.8 (+1)	2.6 (−5)	a
Dicamba	20	5.6 (+3)	5.0 (−8)	a
Dieldrin	25	1.5 (−1)	6.7 (−4)	b
EPTC	20	3.7 (+2)	4.2 (−4)	a
Fenitrothion	20	3.0 (+1)	1.5 (−6)	a
Heptachlor	20	1.0 (−1)	4.6 (−2)	a
Lindane	25	7.5 (−3)	1.3 (−4)	b
Malathion	20	1.5 (+2)	9.0 (−7)	a
Methyl Parathion	25	5.7 (+1)	4.4 (−6)	b
Parathion	25	2.4 (+1)	6.4 (−6)	b
Picloram	20	4.3 (+2)	1.3 (−8)	a
Simazine	20	5.0	1.4 (−7)	a
Triallate	25	4.0	7.9 (−4)	b
Trifluralin	25	3.0 (−1)	6.8 (−3)	b

[1]Sources a. Suntio *et al.* (1988)
    b. Jury *et al.* (1983).

pesticides listed in Table 1. Almost all these are calculated from vapor densities and concentrations in saturated solutions (C_A^s/C_W^s) where both are in equilibrium with the pure material. Very few measurements are available under unsaturated conditions, although experimental techniques to make these have been recently described (Suntio *et al.*, 1987; Fendinger and Glotfelty, 1988).

FUGACITY RELATIONSHIPS

The fugacity of a chemical component is a measure of the 'escaping tendency' of that component in any phase of a system (Mackay, 1991). When all the mobile components within a chemical system are at equilibrium, the fugacity of each will be uniform throughout the system. This fugacity, which is related to the chemical potential, has units of pressure and may be thought of as the vapor pressure of the pesticide in the atmosphere equilibrated with the phase (water, soil organic matter or soil minerals, sediment, plant or animal tissue) in which the pesticide is dissolved or on which it is adsorbed.

The fugacity (f_i) of the pesticide in any phase is linearly related to its concentration in that phase (C_i), so that

$$C_i = f_i Z_i \qquad (5)$$

where Z_i is the 'fugacity capacity' for the phase 'i'. The concentration C_i is in $mol\,m^{-3}$, and f_i in pascals (Pa) so that the units of Z_i are $mol\,m^{-3}\,Pa^{-1}$. Fugacity capacity is analogous to heat capacity: a phase with a high value of Z will take up a large amount of chemical for a small change in the fugacity, just as a substance of high heat capacity will absorb a large amount of heat for a small change in temperature. For a particular set of conditions each phase in the environment has a particular value of Z for each pesticide. If these values are known, the final equilibrium distribution of the pesticide between the phases can be calculated. When the initial non-equilibrium distribution is also known the fugacity may be used to calculate the redistribution between the air, water and solid phases. The values of Z for each phase are not independent of the state of the system, and may vary with temperature and with soil moisture content: the significance of such changes will be considered below.

Fugacity in air

In air, C is related to the fugacity through the ideal gas law, so that $C_a = n/V = p/RT = f_a Z_a$, and

$$f_a = C_a/Z_a = C_a RT \qquad (6)$$

and $Z_a = 1/RT$. This shows that the fugacity capacity of the air is the same for all chemicals, with a value of about $4 \times 10^{-4}\,mol\,m^{-3}\,Pa^{-1}$, varying only with temperature. Air is the only phase in which Z_a is the same for all chemicals.

Fugacity in water

In water solution equation (5) becomes $C_w = f_w Z_w$, or $Z_w = C_w/f_w = C_w/p$, where p (Pa) is the vapor pressure in the atmosphere equilibrated with the solution, and C_w is the concentration in water, again in $mol\,m^{-3}$. Thus, $Z_w = 1/H$ where H is the Henry's law coefficient, here expressed in terms of pressure (Pa) and molar concentrations. Since under ideal conditions fugacity and vapor pressure are identical, fugacity and concentration in water are related by the equation

$$f_w = C_w/Z_w = C_w H \qquad (7)$$

The fugacity of a pesticide in solution in water will thus reflect the Henry's law distribution between the vapor and the true molecular solution concentration in the water. This latter concentration may be less that the 'total concentration' where residues are associated with suspended organic materials or other solid particles or present as emulsions or micelles.

Fugacity and adsorption

Fugacities of adsorbed residues are defined in terms of their adsorption isotherms. Where the amount of adsorbate is a small fraction of the adsorption capacity, we may write the isotherm as $C_s = K_p C_w$, where, as above, C_w is the concentration in the water phase, C_s is that in the adsorbent (mol m^{-3}) and K_p is the soil/water partition coefficient.*

Since the fugacity of the adsorbed residues is defined by $f_s = C_s/Z_s$ and under the equilibrium conditions described by the partition coefficient this fugacity must be equal to that in the water phase, $f_s = f_w = C_s/Z_s = HC_w$, and $Z_s = C_s/(HC_w)$, so that $Z_s = K_p/H$ and

$$f_s = C_s(H/K_p) \tag{8}$$

This equation defines the fugacity capacity of the adsorbed phase in terms of the solid/water and air/water partition coefficients.

The interpretation of adsoption isotherms in terms of fugacity is illustrated in Figure 2, where the curve OM is an isotherm describing adsorption by moist soil in terms of fugacity and adsorbate concentration. Equation (8) shows that the slope of this line at any point is an inverse function of the fugacity capacity of the adsorbent. Changes in this slope must therefore reflect changes in the value of K_p as adsorption sites are filled. The isotherms presented in Figures 3 and 4 for desorption of DDT and lindane from moist Gila soil (Spencer and Cliath, 1970, 1972) are essentially rectilinear until the fugacity is close to that of the pure compounds, showing that Z_s remains esssentially constant over this range. Dieldrin and trifluralin isotherms on the same soil (Figures 5 and 6, Spencer and Cliath, 1969, 1974) are only rectilinear close to the origin: with increasing concentrations the decreasing slopes indicate that the fugacity capacity increases with pesticide loading.

The inflection in the isotherms at the point of intersection with the 'saturation isotherm' of the pure compound (Figures 3 and 5) does not

---

*Classically, partition coefficients are defined by $c_{ad} = x/m = k_p c$, where C_{ad} is in mass concentration units (x gram sorbate/m gram sorbent). In volumetric terms the sorbed concentration is $[(x \text{ gram})(d_B/m \text{ gram})]$, where d_B is the bulk density of the sorbent. Hence $C_s = c_{ad} \times d_B$ and $K_p = k_p \times d_B$. In molar units the numerical values of the coefficients are unchanged.

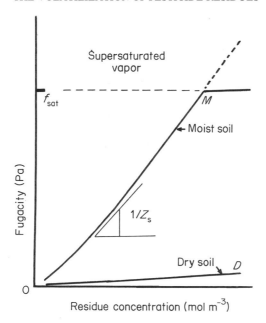

Figure 2 Isotherms relating the fugacity of pesticide vapor in air equilibrated with residues on moist and dry soil as a function of the pesticide residue concentration in the soil and the fugacity capacity of the soil (Z_s). (From *Pesticides in the Soil Environment*, SSSA Book Series No. 2, reprinted by permission of the Soil Science Society of America)

reflect any sudden change in the character of the adsorbing surface, but implies that points above this fugacity level are inaccessible because they represent supersaturation with the pure residues. The p,p'-DDT data in Figure 3 show that on the moist Gila soil at 30 °C this point is reached at a fugacity of 0.097 mPa: in contrast, adsorption of the o,p isomer, a very similar molecule, continues up to a fugacity of 0.74 mPa.

Adsorption by soil organic matter

Chiou and Shoup (1985) and Chiou *et al*. (1983) suggested that in moist soils the adsorption of organic compounds by soil organic matter represents partitioning of the adsorbate into this phase with the formation of a 'solid solution'. The partition between soil and water may thus be written in terms of the partition coefficient $K_{oc} = C_{oc}/C_w$, where C_{oc} is the (volumetric) concentration of pesticide in the organic phase. If the density of the soil organic matter is assumed to be the same as the bulk density of the soil, the soil organic matter content can be expressed as a volume fraction, and $C_{oc} = C_s/\phi$ where ϕ is the organic matter content (as a decimal fraction).

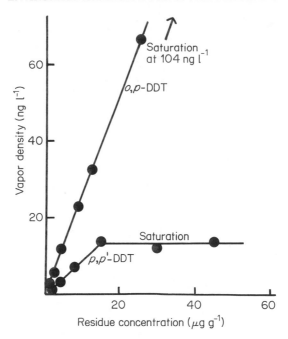

Figure 3 Vapor density of *o,p'* and *p,p'*-DDT in air equilibrated at 30 °C with residues adsorbed on Gila silt loam soil at a moisture content of 0.0394 kg kg^{-1}. (From *Pesticides in the Soil Environment*, SSSA Book Series No. 2, reprinted by permission of the Soil Science Society of America)

Also, since $K_{oc} = C_s/\phi C_w = K_p/\phi$ and the fugacity in the organic phase is given by $f = C_{oc}/Z_{oc}$, where Z_{oc}, the fugacity capacity of the organic phase is given by H/K_{oc}. Where the residue concentration is expressed in terms of the entire solid phase this becomes $f_s = (C_s H)/(\phi K_{oc})$, indicating that the slopes of the isotherms obtained by plotting fugacity as a function of C_s will reflect the partition of the pesticide between vapor, solution and soil organic matter.*

Few values of K_{oc} for pesticides are available in the literature, but they may be approximately estimated from values of the octanol/water distribution coefficients (K_{ow}) that have been measured for many compounds, using the regression equation $\log K_{oc} = (1.029 \log K_{ow}) - 0.18$ (Rao and Davidson, 1980). This approach should, however, be used with caution. A correlation based upon data for a wide range of compounds does not

---

*K_{oc} values are calculated in terms of soil organic carbon content. This is related to 'soil organic matter fraction' by the empirical equation (soil organic matter) = 1.73 × (organic carbon content). The conversion reflects the N,S,O content of the organic matter.

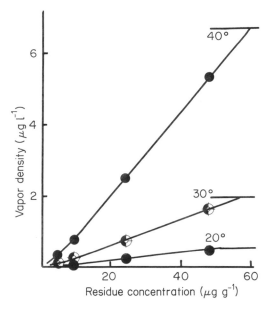

Figure 4 Vapor density of lindane in air equilibrated at 20, 30 and 40 °C with residues adsorbed on Gila silt loam soil at a moisture content of 0.0394 kg kg^{-1}. (From *Pesticides in the Soil Environment*, SSSA Book Series No. 2, reprinted by permission of the Soil Science Society of America)

provide a good basis for predicting the properties of individuals in a population with wide variances.

Fugacity of soil residues

The isotherms in Figures 3–6 show that the fugacities of all the compounds increase with residue concentrations up to the values corresponding to saturation with the pure compounds at 15 mg kg^{-1} for DDT, between 50 and 60 mg kg^{-1} for lindane, 25 mg kg^{-1} for dieldrin and 73 mg kg^{-1} for trifluralin. To relate these results to field conditions we may assume that a pesticide applied as a spray application at 1.25 kg ha^{-1} to the surface of a bare, uncultivated, moist soil remains in the top 1 mm layer of soil (with bulk density 1.25). The residue concentration will then be between 75 and 100 mg kg^{-1}. The data in Figures 3–6 therefore suggest that the fugacities of such surface residues will be close to those of the pure compounds. If, however, the pesticides are mixed into the soil by cultivation to a depth of about 75 mm, the residue concentrations will be reduced to 1.25 mg kg^{-1} or less and the fugacities will be reduced to 10% or less of those of the pure compounds with volatilization rates reduced in proportion.

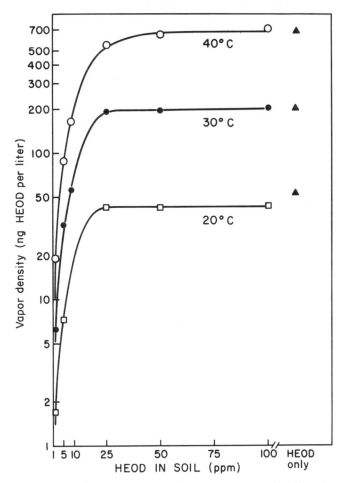

Figure 5   Vapor density of dieldrin (HEOD) in air equilibrated at 20, 30 and 40 °C with residues adsorbed on Gila silt loam soil at a moisture content of 0.10 kg kg^{-1}. (Spencer *et al*. SSSA Proceedings **33**, 509–511, 1969. Reprinted by permission of the Soil Science Society of America)

Table 3 presents the distribution of three typical pesticides between the air, water and solid phases of a soil matrix after incorporation to the 10 cm depth, calculated using the fugacity approach (Mackay, 1991). The large fugacity capacity of the solid phase dominates the distribution of all three pesticides. Owing to the low fugacity capacity of the air, only a very small fraction of the pesticide is present as vapor in the soil air, but this fraction is important because it acts as the point of supply to the lower boundary of the laminar layer between the soil and the overlying atmosphere as illustrated in

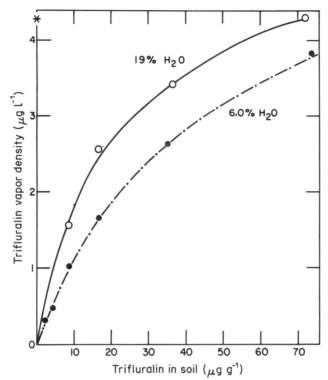

Figure 6 Vapor density of trifluralin in air equilibrated at 30 °C with residues adsorbed on Gila silt loam soil at a moisture contents of 0.06 and 0.19 kg kg⁻¹. (From Spencer and Cliath, *J. Agric. Food Chem* **22**, 987–981, 1974. Reprinted by permission of the American Chemical Society)

Figure 1. The values of the overall fugacity therefore represent the limiting values of p_s in equation (2) for the three pesticides. It is evident that in this hypothetical case these p_s values would in some degree reflect the initial rates of volatilization that would be expected for these compounds, with diazinon evaporating slightly faster than lindane, and both between 15 and 20 times faster than atrazine.

The relative transport of the pesticide through the soil matrix either as vapor or through the water phase and the replenishment of the fraction in the soil air from that in the water and solid phases are major factors controlling the volatilization rate of soil-incorporated pesticides. Jury *et al.*, (1983a,b; 1984a,b,c) developed a model describing diffusion, convective water flow, leaching and chemical degradation and gaseous diffusion into the laminar layer and their effects on the relative volatility, mobility and persistence of pesticides and other trace organic chemicals in soils. This

Table 3. Equilibrium distribution and fugacity capacity of selected pesticides in a model soil (after Glotfelty and Schomburg, 1989)

	Lindane	Diazinon	Atrazine
Equilbrium Distribution (% of total residues)			
Air	4.5×10^{-5}	6.3×10^{-5}	5.0×10^{-7}
Water	0.78	2.3	16
Soil Matrix	99	97	84
Fugacity Capacity (mol.m3.Pa$_1$)			
Air	4.0×10^{-4}	4.0×10^{-4}	4.0×10^{-4}
Water	7.1	15	3.4×10^3
Soil Matrix	460	310	8.9×10^3
Equilbrium Fugacity (Pa)			
	1.5×10^{-5}	2.1×10^{-5}	8.8×10^{-7}

Model soil contains: 25% by volume air,
25% by volume water,
2% organic carbon,
bulk density 1.25 kg/lit.
All pesticides assumed to be applied at 1 kg/ha, incorporated to 10 cm.

model assumes that the concentrations in the soil air and water are described by Henry's law, that the adsorption isotherms are linear, that degradation follows first-order rate kinetics and that there is a stagnant boundary layer at the soil–air interface through which the vapor must diffuse to reach the atmosphere. The model is based on a mass balance equation for the amount of pesticide contained in a specified volume of soil with a uniform initial concentration C_T (the total mass per unit weight of soil, in $\mu g\,cm^{-3}$),

$$(\partial C_T/\partial t) + (\partial J_S/\partial Z) + \mu C_T = 0 \qquad (9)$$

where $C_T = d_B C_S + \theta C_L + a C_G$. Here C_S is the adsorbed concentration ($\mu g\,g^{-1}$ soil) and d_B the dry soil bulk density ($g\,cm^{-3}$); θ is the volumetric water content ($cm^3\,cm^{-3}$) and C_L the concentration in soil solution ($\mu g\,cm^{-3}$ water); a is the volumetric air content ($cm^3\,cm^{-3}$) and C_G the vapor concentration ($\mu g\,cm^{-3}$ of soil air). Chemical transport through the soil J_s, ($\mu g\,cm^{-2}\,d^{-1}$) is described by a solute flux equation containing vapor and liquid diffusion and solute convection terms

$$J_S = -D_G(\partial C_G/\partial Z) - D_L(\partial C_L/\partial Z) + J_W C_L \qquad (10)$$

in which D_G and D_L are soil vapor and liquid diffusion coefficients ($cm^2\,d^{-1}$). Chemical degradation is described by a net first-order degradation coefficient $\mu(d^{-1})$. J_W is the soil water flux through the volume containing the pesticide.

The equations are solved analytically for an initial soil concentration averaged over an arbitrary finite depth interval $0 < Z < L$. The loss of pesticide to the atmosphere is presumed to be regulated by a stagnant boundary layer through which the vapor must move by diffusion. The description of the effects of resistances to vapor loss at the soil surface are introduced into the model by varying the values for C_A and the depth of the stagnant layer.

Basically the model is intended to classify pesticides and other chemicals according to their environmental behavior calculated from their physical and chemical characteristics such as vapor pressure, solubility, Henry's law and organic carbon partition coefficients and degradation rate. The model cannot be used to calculate the amounts of pesticides that will move from soil to air from specific places or land areas. A comparison of the predicted behavior of 20 pesticides showed that although the relative rates of volatilization depended on the chemical properties of the pesticides and the water evaporation rate, the volatilization patterns were mainly controlled by the Henry's law coefficient (K_{AW}) (Jury et al., 1984b). This coefficient (equation 4) can thus be used as a criterion for classifying pesticides depending on whether their volatilization is controlled within the soil or in the boundary layer above the soil surface. The results suggest that pesticides can be divided into three categories depending on whether the Henry's law coefficient (in dimensionless units) is greater than, approximately equal to, or less than 2.65×10^{-5}. For compounds with values below this the volatilization will be controlled in the atmospheric boundary layer: if the chemical moves to the surface more quickly than it evaporates it will accumulate at the surface, and the volatilization rate will tend to increase with time. By contrast, the volatilization rates of pesticides with higher Henry's law coefficients will decrease with time whether water is evaporating or not. The details of the pesticide classification based on the model and the supporting experimental evidence have been presented by Jury et al. (1983a, 1984a,b,c).

In applying this model to the interpretation of field data it is important to remember that it is assumed that the pesticide is initially uniformly distributed throughout the soil volume and that the soil is assumed to remain moist with a constant water content. The model does not therefore contain any function to predict the effect of soil drying upon volatilization rate.

SOIL PHYSICAL FACTORS

In addition to the chemical and physical characteristics of the pesticide compounds, physical conditions in the soil have major effects on the rates of volatilization of incorporated residues.

Temperature effects

Figures 4 and 5 suggest that where soil residue concentrations are high the effect of temperature on the fugacities will be similar to its effect on the fugacities of the pure compounds. At lower residue concentrations (where the relative vapor densities (C_A/C_A^s) are $\ll 1.0$), the temperature relations are more complex. In a detailed analysis of the lindane data of Figure 4, Spencer and Cliath (1970) showed that below the saturation level the vapor density increased with temperature more slowly than that of pure lindane, due to a combination of the effects of temperature on solubility, the Henry's law coefficient and the free energy of adsorption.

Soil moisture content

Retardation of volatilization in dry soils has been noted by a number of investigators (Fang *et al.*, 1961; Harris and Lichtenstein, 1961; Lichtenstein and Schultz, 1961; Deming, 1963; Gray and Weierich, 1965; Parochetti and Warren, 1966; Lichtenstein *et al.*, 1970). The increased adsorption of pesticides in dry soil has been discussed in detail by Spencer (1970) on the basis of the data presented in Figures 7 and 8. These results show that even where the concentrations of dieldrin and trifluralin in moist Gila soil were high enough to support vapor densities close to those of the pure compounds, reduction of the soil water content to less than 3 or 4% caused a large reduction in the equilibrium vapor densities. At water contents below 2% or less the relative vapor densities were below 10%, and at 1.6% water content dieldrin vapor density approached the lower levels of detection. Similar data have been presented for DDT and DDE isomers (Spencer and Cliath, 1972) and lindane (Spencer and Cliath, 1970).

In the Gila soil a water content of 2.8% corresponds to a monomolecular layer on the adsorption surfaces. At moisture contents below this, adsorption sites on which the pesticide molecules are more strongly bound become active so that their fugacity (and hence their vapor density) is greatly reduced. The effect is reversible in response to changes in water content (Spencer and Cliath, 1974; Spencer *et al.*, 1969). This effect of reduced soil moisture content on the fugacity of adsorbed residues is represented by the isotherm OD in Figure 2. The large decrease in slope is due to a large increase in Z_s, the fugacity capacity of the adsorbent, as the adsorbing surfaces become dry. This reflects a corresponding change in the value of K_p due to changes in the free energy and entropy relations of the adsorption reaction as the moisture is lost from the surface. Adsorption on dry mineral surfaces and its sensitivity to moisture has been reviewed in detail by Chiou *et al.* (1985). The importance of this for pesticide volatility is that the fugacity of the residues in a dry soil surface is reduced to such levels that

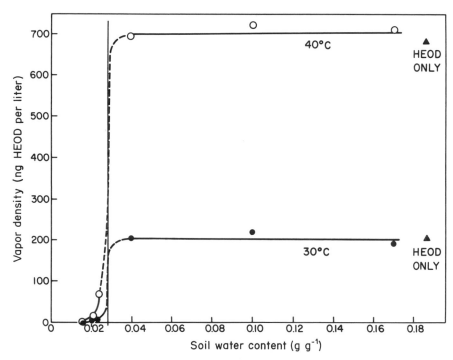

Figure 7 Vapor density of dieldrin (HEOD) in air equilibrated at 30 and 40 °C with Gila silt loam soil containing $10 \, mg \, kg^{-1}$ dieldrin and moisture contents up to $0.17 \, kg \, kg^{-1}$. (Spencer *et al.* SSSA Proceedings **33**, 509–511, 1989. Reprinted by permission of the Soil Science Society of America)

volatilization is almost completely suppressed. This, as will be seen below, is a major factor controlling the volatilization from soils in the field.

In practical terms it may be noted that in the Gila soil a moisture content of 3.9% is in equilibrium with air at a relative humidity of 94%. The moisture contents at which the equilibrium vapor densities (and hence the volatilization rates) of the residues are reduced to very low values are thus easily reached in surface solids drying into moderately dry air. This dependence of volatilization rate upon atmospheric relative humidity and soil moisture content has been observed for trifluralin in the laboratory by Spencer and Cliath (1974) and in the field by Harper *et al.* (1976) and Glotfelty *et al.* (1984). In the field, surface soil temperatures rise rapidly when the surface has dried to the point at which heat can no longer be removed by water evaporation: beyond this point pesticide adsorption will increase and volatilization will decrease rapidly. Despite the intuitive expectation that the pesticide volatilization should increase with temperature, no simple correlation between volatilization rates and soil temperature

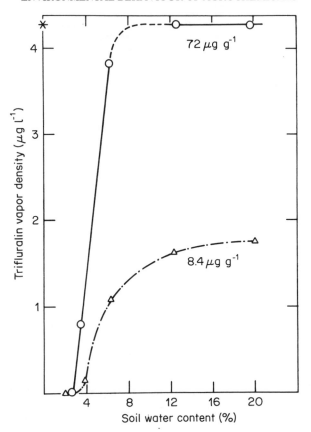

Figure 8 Vapor density of trifluralin in air equilibrated at 30 °C with Gila silt loam soil containing 8.4 and 72 mg kg^{-1} trifluralin and moisture contents up to 0.20 kg kg^{-1}. (From Spencer and Cliath, *J. Agric. Food Chem.* **22**, 987–981, 1974. Reprinted by permission of the American Chemical Society)

can actually be expected. Vaporization of residues on transpiring leaves, where the water supply is not depleted, may be expected to be independent of the dryness of the soil surface, and to respond to changes in the temperature of the leaf. This will not, however, rise as rapidly as the temperature of dry soil owing to the cooling effect of evapotranspiration.

This interrelation between temperature, soil moisture content and the fugacity (or activity) of pesticide residues is of major importance in controlling losses of residues of many pesticides from soil surfaces. The effect is also of particular importance in controlling the effectiveness of herbicides: the common field observation that herbicides applied to dry soils are not effective until they have been activated by moisture is clearly related to their desorption from the strong adsorption sites as soil moisture increases.

Early observations of the interaction between the rates of evaporation of water and DDT were interpreted as 'codistillation' (Acree *et al.*, 1963). When correctly used this term describes a bulk flow process that takes place where the combined vapor pressures of the two components being distilled together equal or exceed the ambient pressure. Since at normal temperatures the vapor pressure of water is always much less than one atmosphere and the pesticide vapor pressure is a very small fraction of this, 'codistillation' is a misnomer for pesticide evaporation. Experimental proof that the volatilization of trifluralin and dieldrin from soils is not directly dependent on water evaporation has been presented by Spencer and Cliath (1974) who showed that maximum pesticide volatilization rates were found at 100% relative humidity, when the water loss from the soil was, of course, zero. Any use of the term 'codistillation' as a description of pesticide volatilization is inappropriate and misleading and its use in this context should be discontinued.

RESIDUE DISTRIBUTION EFFECTS

For simplicity the previous discussion of the effects of pesticide and soil properties on volatilization has assumed that the pesticide is uniformly distributed throughout a defined volume of the soil profile. This condition is, however, never met under field conditions where the distribution is controlled by the way the pesticide is applied, the way the soil is subsequently managed and, in some degree, by the type of formulation used. Under field conditions irregularities in spatial distribution of residues are of primary importance because they determine the length of the pathway that the residues must travel by diffusion or bulk flow to the evaporating surface.

Surface residues

When pesticides are applied as water-based sprays of solutions, dispersed emulsions or wettable powders, the residues are initially deposited as a thin layer on the exposed porous soil surface so that the droplets will be adsorbed into the underlying soil or plant tissue by capillary action. The residues will then remain in a shallow surface layer. As the most exposed residues in the outer surface are depleted, the volatilization will become dependent upon the rate at which the residues diffuse upward to the free surface as a concentration gradient is established. The volatilization rate will then be controlled by the 'back-diffusion' of the residue. The way that the volatilization rates controlled by diffusive flow change with time have been analyzed by Mayer *et al.* (1974), who showed that the volatilization rate should decrease approximately in proportion to the square root of time, with the exact relationship depending upon the boundary conditions. This prediction was verified in laboratory measurements of volatilization rates of dieldrin (Mayer *et al.*, 1974) and lindane (Farmer *et al.*, 1972) residues mixed into the top 5 mm depth of moist soil: a similar laboratory test of the volatilization of incorporated triallate was reported by Jury *et al.* (1980). In the field validations of the predictions of such models are very difficult owing to the complex effects of residue distribution, sampling errors and the uncertainties about boundary conditions, but the type of relationship predicted was observed by Glotfelty *et al.* (1984), in an experiment measuring the volatilization of surface applied trifluralin, heptachlor, chlordane and lindane from moist soil in the field.

Since the increased adsorption in dry soils reduces the vapor pressures to such low values that volatilization effectively ceases whenever the topmost layer of the soil surface becomes dry, the predicted relationships can only be expected to be found over time periods where the soil remains moist. This reduction in fugacity by increased adsorption in dry soils can become the dominant factor controlling the volatilization loss. This effect also negates any temperature effect on the volatilization rate because marked increases in soil surface temperature only occur after water evaporation has ceased and the soil surface is dry. This effect of increased adsorption is strikingly shwon in the data obtained by Turner *et al.* (1978) on the volatilization of surface-applied chlorpropham.

The volatilization rates of surface-applied pesticides in the field may therefore be expected to change in a complex manner depending upon coverage, residue penetration, age and the moisture status of the surface. In specific field situations the effects of these physical factors will depend on the way in which the pesticides are applied and the soil is managed. Differences in volatilization rates due to differences in chemical properties may be completely obscured by the effects of management and weather unless different chemicals are compared under identical usage patterns at the same time and place.

Incorporated residues

Where pesticides are incorporated into the soil by cultivation, their volatilization is controlled by their movement through the body of the soil to the overlying air. Incorporated residues differ from surface applications in two respects. First, as noted above, the average concentrations in the soil are lower by an order of magnitude or more so that, as illustrated in Figures 3–6, the equilibrium vapor pressures will be much less than those of surface applications. Second, the average diffusion path length to the surface will be much longer so that in addition to the decrease in fugacity the resistance to flow will be much greater. Thus, except for the small fraction of exposed material remaining on the surface that is readily lost, the volatilization of incorporated residues will be greatly restricted in comparison to that of surface residues. The factors controlling their movement to the surface have been discussed in detail by Spencer *et al.* (1973), who contrasted the volatilization patterns to be expected when the flow to the surface is controlled by a simple diffusion gradient with those expected where pesticide is transported to the surface in the upward flow of soil water (convective flow).

Diffusion-controlled flow

After initial incorporation into a moist soil the pesticide residues will partition towards a uniform fugacity of the air, water and solid phases where the concentrations in the soil solution and the soil atmosphere are controlled by desorption from the solid phase. Where there is no upward evaporative water movement to the surface the diffusive flow of pesticide to the surface will become limiting. This situation is that described mathematically in the model developed by Jury *et al.* (1983a,b, 1984a,b,c). The final rate of diffusive flow will reflect the overall diffusion coefficient of the pesticide in the particular soil. Since this coefficient is a complex function of the diffusion pathways through the solid, liquid and vapor phases in the soil, and these will change with the water content, no single value of such an 'effective diffusion coefficient' can be expected to characterize the behavior of even a single pesticide in one soil. Predictions of diffusive flow must rely upon experimental values obtained by calibrations with the chosen combination of pesticide and soil. Such 'effective diffusion coefficients' may be expected to be sensitive to soil bulk density, soil water content and temperature. When the soil moisture content is reduced to a level where the fugacity becomes very low, diffusive movement will be almost wholly inhibited. These conditions may frequently be found in the field, when fallow surface soils—and particularly those of light texture—are exposed to intense drying by wind and sun. The consequent formation of a very dry surface layer will act as an essentially impermeable barrier to pesticide evaporation.

Convective flow and the wick effect

In a moist soil where water evaporation maintains a continuous upward movement of water through the soil pores to the open surface, the resulting upward transport of pesticide to the surface may prevent or distort the development of a diffusion gradient, and volatilization at the surface will be supported by the water movement. This flow mechanism has been described as the 'wick effect' by analogy with capillary liquid flow to the ends of exposed fibers (Hartley, 1969; Hartley and Graham-Bryce, 1980). Laboratory data for lindane and dieldrin show that the volatilization of these insecticides supported by convective flow in moist soil can be up to five times faster than that controlled by diffusion (Spencer and Cliath, 1973). Other experiments with trifluralin (Spencer and Cliath, 1974; Harper et al., 1976) showed much smaller differences due to its lower concentration in the soil water. Jury et al. (1980) also showed that for triallate the relative contribution from convective flow was much less in a silt loam than in a sandy loam soil due to the lower solution concentration in the silt loam, which also had a much higher organic matter content.

Under field conditions upward movement of water to the soil surface is not continuous and may cease during rainfall and become very slow at high relative humidities or when there is no insolation to evaporate the water. It may therefore vary greatly during daylight hours and will almost always be very small at night. Convective flow of pesticide to the surface will vary in the same way.

In the field, volatilization rates controlled by convective flow show a characteristic pattern that follows the daily cycle of insolation with a noonday maximum and much smaller rates in the morning and evening. Diffusion-controlled patterns tend to be much more uniform, often with noonday minima where the soil surface becomes dry and surface adsorption is enhanced. Both these characteristic patterns have been observed in some of the field observations discussed below. Patterns with noonday minima appear most frequently where residues are evaporating from light textured or sandy soils which have a lower capillary tension in the soil pores that makes the soil surface more susceptible to rapid drying by wind or sunshine.

VOLATILIZATION IN THE FIELD

Field measurements of volatilization are usually based upon flux rates calculated from profiles of vapor concentrations measured in the air over treated fields, supported by micrometerorological measurements of temperature, water vapor or wind profiles coupled with soil water loss or energy balance data. Such experiments require elaborate instrumentation and extensive analytical work due to the large number of samples that must be analyzed; more recent statistically based techniques have, however, offset the latter difficulty in some degree.

The results reflect the natural uncontrolled soil and weather conditions which can, of course, themselves show rapid changes. Where sampling periods are relatively short, the data can, however, show the effects of such changes on the volatilization rates very clearly.

Vapor flux methods

These methods are based on calculations of upward flux rates using meteorological theory that describes the upward dispersive flow in the turbulent boundary layer of the atmosphere up to about 2 m above the ground. Several approaches are used, differing slightly in their experimental requirements and underlying theory. Full descriptions of the latter lie outside the scope of this review, which will be confined to general descriptions of the ways in which their differences affect the design of field experiments and the interpretation of the results.

Vapor profile methods

These are based on measurements of the gradient of pesticide concentration with height in the air over the treated field area obtained by sampling the air at a series of heights (usually up to about 2 m) above the soil surface. Gradients of temperature, water vapor concentration, wind speed and other supporting meteorological variables must also be measured over the same time periods. Care must be taken to ensure that all the measurements are taken within the atmospheric boundary layer characteristic of the area treated with pesticide. Since the slope of the upper edge of this layer is often as low as 1%, this means that a mast of 2 m height must have a clear fetch of 200 m to the upwind edge of the treated zone to ensure that characteristic profiles are obtained. This approach therefore requires the use of large experimental plots up to 1 ha in size, located in relatively flat terrain with no obstacles to wind flow for some distance outside them.

The calculation of the pesticide flux is based upon equation (1). For pesticide vapor (Parmele *et al.*, 1972) this becomes

$$F \uparrow = k_{\mathrm{P}}(\mathrm{d}p/\mathrm{d}z) \tag{11}$$

The value of $\mathrm{d}p/\mathrm{d}z$ at the chosen height z is obtained from the profile of vapor concentrations, but the value of k_{P} is not known and must be measured indirectly. To this end it must be noted that equation (1) is one of a general set of equations of the form $F \uparrow_i = k_{Zi}(\mathrm{d}x_i/\mathrm{d}z)$ that describe the vertical transport of water vapor, heat, or wind momentum through the turbulent boundary layer. As a first approximation the values of k_{Zi} in each of the appropriate forms of the equation are assumed to be the same (the Similarity Principle). If independent measurements of the flux of a particular component $(F \uparrow_i)$ are made together with the corresponding gradient,

(dx_i/dz), the value of k_P may be calculated and used in equation (11) to calculate the pesticide flux ($F\uparrow$).

Of the three approaches—water vapor flux, energy balance (or heat transport), and/or aerodynamic (or momentum flux)—the first is probably subject to the fewest assumptions. The gradient of water vapor concentration is calculated from the humidity measured with wet and dry bulb thermometers mounted at chosen heights. The need for the companion measurement of water vapor flux ($F\uparrow_W$) imposes a considerable restriction on the possible use of this method because it requires direct measurements of evapotranspiration from the soil or crop surface. Direct measurement of this can only be made at locations equipped with recording lysimeters. Few measurements of pesticide volatilization using this approach have therefore been reported. The energy balance, or Bowen ratio method, described by Parmele *et al.* (1972), Lemon (1969) and Rose (1966), was originally developed to calculate the water vapor flux in the absence of lysimeter data. The flux of water vapor, $F\uparrow_W$, is calculated from the equation $F\uparrow_W = (R_n - G)/L(1 + B)$ where R_n is the net incident radiation, G the soil heat flux, L the latent heat of evaporation, and B is a function of the air temperature and water vapor density gradients. The value of k_W is then estimated from the water vapor flux equation $k_W = F\uparrow_W/(dq/dz)$. The principal disadvantage of the energy balance method is in the complexity of the supporting data which must include measurements of net radiation, soil heat flux and both temperature and humidity profiles. The method can present difficulties where the energy balance requires terms for advected energy as well as radiation and conduction: most typically this happens in the 'oasis situation' where warm dry air blowing into the experimental area causes water evaporation in excess of that measured from R_n and G. These conditions were encountered by Cliath *et al.* (1980) in measurements of EPTC volatilization from irrigated fields in southern California, when the water evaporation caused stable atmospheric inversions over the plot at midday, and also by Majewski *et al.* (1990) in other comparative methodological studies.

The aerodynamic method is based on measurement of the gradient of wind speed with height above the surface. The decrease in wind speed in the boundary layer close to the ground reflects the friction or drag exerted by the soil or crop surface as the momentum in the wind is transferred to the surface. This downward flux or diffusion of momentum is described by the equation $F\downarrow_M = -k_M(du/dz)$ which is similar to those for heat, water vapor and other gases: the negative sign is required because the momentum is transferred downwards. Here k_M is the eddy diffusivity coefficient for momentum transfer at the height z, and du/dz is the wind speed gradient. In principle it is clear that if we equate the diffusivity coefficients k_P and k_M, the pesticide flux $F\uparrow$ can be derived from measurements of pesticide vapor concentration and wind speed. The conditions under which this equality may be assumed have been discussed by Montieth (1973) and Rose (1966). The assumption is most likely to fail when the air becomes unstable due to strong surface heating or under radiative cooling when the stirring action of the wind is damped out by the increased stability. The final form of the equation used in the calculation of the pesticide flux using the aerodynamic method is

$$F\uparrow = \frac{v^2(C_1 - C_2)(U_2 - U_1)}{\phi^2[\ln(z_2 - z_0)/(z_1 - z_0)]^2} \tag{12}$$

where C_1, U_1 and C_2, U_2 are the vapor concentrations and wind speeds at the heights z_1 and z_2 above the surface, v is the von Karman constant, assumed to be 0.4. Following Parmele et al. (1972), the stability correction term may be defined as $\phi = (1 - bR_i)^{-0.25}$, where R_i is the Richardson number and b is a constant (close to 18) dependent upon the roughness of the surface. The relative simplicity of the aerodynamic method, which only requires measurement of wind profiles as supporting data, has made it the method used in a large number of the published field experiments, despite the possible ambiguities in the assumptions upon which it is based.

Theoretical profile and integrated horizonal flux methods

These are based upon fluxes calculated using theoretical descriptions of the dispersive flow. The 'theoretical profile shape' method is based on a two-dimensional trajectory simulation model which predicts the upward flow from measurements of wind speed and pesticide vapor concentration at a single height at the center of a circular treated plot with a radius of close to 50 m (Wilson et al., 1981). This permits measurements with much fewer pesticide analyses and requiring less supporting data. The method has been increasingly used in more recent field studies.

The 'integrated horizontal flux method' is a time-averaged mass balance technique in which the flux is calculated from the average wind speed and vapor concentration at a particular height and distance from the upwind edge of the treated plot (Denmead et al., 1977). Although the method gives results closely comparable to other approaches (Majewski et al., 1990), few results have been reported using the method.

Selection of methods

The choice of the method used in a particular experiment is often dictated by the resources available. The energy balance method is to be preferred when the resources are available—particularly where fluxes and evapo-transpiration rates are likely to be high—but it has the disadvantage that the calculated values of k_E may be indeterminate in the morning and evening or under cool conditions when the net energy input $(R_n - G)$ is small. By contrast, the aerodynamic method requires less equipment but is theoretically less reliable on hot calm days when wind speeds are low and thermal turbulence is dominant: in the morning and evening—when pesticide volatilization rates may sometimes be large—its value depends upon the wind speed gradient. On days when this is maintained outside daylight hours the aerodynamic method may give valuable results when other methods fail.

Where measurements that permit direct comparisons between the various methods have been made (Harper *et al.*, 1976, 1983; Glotfelty *et al.*, 1984; Grover *et al.*, 1985; Majewski *et al.*, 1989, 1991), the data suggest that the differences between them tend to vary with atmospheric stability and the character of the soil or crop surface. All gave comparable results over fallow soil in moderate wind conditions.

Sampling chambers

Measurements of volatilization using sampling chambers are occasionally reported in the literature. Fundamentally this method represents the measurement of the amount of pesticide that accumulates in a closed or partially closed vessel under negative pressure placed on or close to the soil surface. The volatilization rate is then calculated as the amount of pesticide released from the surface area under the sampler during the duration of the experiment. This does not represent a measure of flux, but is rather a measure of the vapor density close to the surface. This disadvantage of the method is twofold. First, the results only represent single point determinations which must be arbitrarily extrapolated to large field areas and second, the presence of the chamber itself is likely to affect the temperature and soil moisture conditions at the point of sampling so that they are not representative of the whole treated area.

No experimental data comparing data from sampling chambers with those of other methods are available.

Air sampling techniques

Determinations of pesticide concentrations in air are made with sampling devices which absorb pesticide vapor from a stream of air drawn through them at a known velocity. A range of materials may be used as the adsorbing substrate in the samplers. In many early experiments air was pulled through bubblers containing hexylene or ethylene glycol. This method has been discarded in favor of solid adsorbents because of the difficulties of transferring liquids in the field and because in all but the driest atmospheres the adsorption of water by the glycol caused changes in liquid volume and flow rate. Solid adsorbers such as polyurethane foam offer great convenience in transfer and storage of samples, but require elaborate cleaning and purification before use (Turner and Glotfelty, 1977). For many purposes powdered materials such as chromasorb or XAD resin may be used but they require more careful handling in transfer and filling operations and in controlling the density of packing in the sampler, which may affect resistance to air flow. The stability of the residues on the adsorbent during sampling and storage is of major importance: some adsorbents may adsorb significant amounts of water from humid air, which may cause the hydrolysis of some pesticide residues. The range of materials available has been reviewed by Bidleman (1985).

FIELD MEASUREMENTS

The earlier discussion shows that the properties of both pesticide and soil and the interactions between them are likely to be important in controlling the volatilization in the field. Predictions of field volatilization rates, however, require simplifying assumptions about the complex and often transient conditions of the field environment. Field measurements of volatilization, supported by data on usage and management, residue concentrations, soil moisture levels, surface temperatures and weather conditions are therefore essential to clarify these interactions between the chemical properties and soil factors. Without field data no real calibration of laboratory data or modelling projections can be made.

The considerable body of such data that has been published in recent years is summarized in Table 4. All the original reports contain much data that cannot be presented here and the individual experiments will therefore be reviewed only in so far as the results relate to the effects of chemical properties, residue distribution, soil conditions and management practices discussed above.

Preliminary inspection of the data reveals a wide range in the observed volatilization rates extending from losses of 70% within a few hours to small losses over weeks or months (Table 4, Expt no. 18, 21, 24, 41, and 16, 20, 35). A general review shows that the principal factors that are important in controlling volatilization are the fugacity of the pesticide residues, the way the pesticide is used—and particularly whether or it is incorporated into the soil or is surface applied—and the surface soil moisture status.

Surface applications

The results in Table 4 show a considerable range of volatilization rates for surface-applied residues, extending from 50% within 6–8 hours for heptachlor, lindane and trifluralin on moist soil (Glotfelty et al., 1984; Table 4, nos.

Table 4. Field measurements of volatilization

Compound Expt#. Usage	Surface	Season Place	% Fraction volat. in period	Source
Alachlor				
1. surface	fallow silt loam	May 1981 Mayland	26 in 24 d.	[c]
Atrazine				
2. surface	fallow silt loam	May 1981 Maryland	2.4 in 24 d.	[c]
3. surface	conventional & no-till	Apr. 1990 Maryland	Cnv: 4.4 in 26 d. N-t: 4.5 in 26 d.	[n]
Chlorpropham				
4. surface	soil under soybeans	May 1976 Maryland	49 in 50 d.	[m]
5. microencap	"	"	20 in 50 d.	[m]
Chlordane				
6. surface	moist fallow silt loam	Aug. 1975 Maryland	50 in 50 hrs.	[b]
7. surface	dry fallow sandy loam	Jun. 1978 Maryland	2 in 50 hrs.	[b]
Chlorpyrifos				
8. surface	bare, moist, drying soil	Sep. 1985 California	<1 in 4 d.	[g]
9. surface	conventional & no-till	Apr. 1990 Maryland	Cnv: 23 in 26 d. N-t: 76 in 26 d.	[n]
Dacthal				
10. surface	moist fallow silt loam	Aug. 1975 Maryland	2 in 34 hrs.	[b]
11. surface	moist (irrig) bare soil	Apr. 1987 California	15 in 21 d.	[h]
DDT				
12. incorp. and surface	moist fallow silt loam	Oct. 1968 Louisiana	44 in 120 d.	[p]
13. cotton field	foliage, 45% ground cover	Sep. 1976 Mississippi	21 in 5 d. 58 in 11 d.	[s]
Diazinon				
14. surface	bare, moist, drying soil	Sep. 1985 California	<0.2 in 4 d.	[g]
Dieldrin				
15. surface	moist soil fallow	Sep. 1969 Louisiana	20 in 50 d.	[q]
16. incorp. to 7.5 cm	soil under maize	May 1969 Ohio	4 in 170 d.	[i]
17. vegetation	short orchard grass	Jul. 1973 Maryland	90 in 30 d.	[j]
EPTC				
18. surface in irrigation water	soil under 25 cm alfalfa	May 1977 California	74 in 52 hrs.	[a]

Table 4. (*continued*)

Compound Expt#. Usage	Surface	Season Place	% Fraction volat. in period	Source
Fonofos				
19. surface	conventional & no-till	Apr. 1990 Maryland	Cnv: 44 in 26 d. N-t: 73 in 26 d.	(n)
Heptachlor				
20. incorp. to 7.5 cm	soil under maize	May 1969 Ohio	7 in 170 d.	(i)
21. surface	moist fallow silt loam	Aug. 1975 Maryland	50 in 6 hrs: 90 in 6 d.	(b)
22. surface	dry fallow sandy loam	Jun. 1978 Maryland	40 in 50 hrs.	(b)
23. vegetation	short orchard grass	Jul. 1973 Maryland	90 in 7 d.	(j)
Lindane				
24. surface	moist fallow silt loam	Jun. 1977 Maryland	50 in 6 hrs: 90 in 6 d.	(b)
25. surface	dry fallow sandy loam	Jun. 1978 Maryland	12 in 50 hrs.	(b)
26. surface	bare, moist, drying soil	Sep. 1985 California	>34 in 4 d.	(g)
MCPA				
27. open water	flooded rice field	May 1986 California	rapid degradation prevented measurement	(s)
Methyl Parathion				
28. open water	flooded rice field	May 1986 California	50 in 120 days	(s)
Molinate				
29. open water	flooded rice field	May 1986 california	50 in 11 days	(s)
Nitrapyrin				
30. surface	bare, moist, drying soil	Sep. 1985 California	>30 in 4 d.	(g)
Photodieldrin				
31. vegetation	short orchard grass	Jul. 1973 Maryland	5 per d.	(l)
Simazine				
32. surface	fallow silt loam	May 1981 Maryland	1.1 in 24 d.	(c)
Thiobencarb				
33. open water	flooded rice field	May 1986 California	50 in 11 days	(s)
Toxaphene				
34. cotton crop	foliage	Sep. 1974 Mississippi	25 in 5 d.	(r)

continued overleaf

Table 4. (*continued*)

Compound Expt#. Usage	Surface	Season Place	% Fraction volat. in period	Source
35. cotton field	foliage, 45% ground cover	Aug. 1976 Mississippi	33 in 11 d. 53 in 33 d.	(s)
36. surface	fallow silt loam	May 1981 Maryland	33 in 24 d.	(c)
Triallate				
37. incorp to 5 cm	bare soil	May 1983 Saskatchewan	17 in 67 d. 9 in 7 d.	(e)
Trifluralin				
38. incorp. to 2.5 cm	soybeans on sandy loam 0.5% o.m.	Jun. 1973 Georgia	22 in 120 d.	(f,o)
39. incorp. to 7.5 cm	soybeans on loam 4% o.m.	May 1973 New York	3.4 in 90 d.	(k)
40. surface	moist fallow silt loam	Aug. 1975 Maryland	50 in 7.5 hrs: 90 in 7 d.	(b)
41. surface	moist fallow silt loam	Jun. 1977 Maryland	87 in 50 hrs.	(b)
42. surface	dry fallow sandy loam	Jun. 1978 Maryland	25 in 50 hrs.	(b)
43. incorp. to 5 cm	bare soil	May 1983 Saskatchewan	24 in 67 d. 11 in 7 d.	(e)
2,4-D				
44. surface	20 cm high spring wheat	Jun. 1981 Saskatchewan	21 in 5 d.	(d)

Sources:
[a]Cliath *et al.* (1980)
[c]Glotfelty *et al.* (1989)
[e]Grover *et al.* (1988)
[g]Majewski *et al.* (1989)
[i]Taylor *et al.* (1976)
[k]Taylor: unpub. data.
[m]Turner *et al.* (1978)
[o]White *et al.* (1977)
[q]Willis *et al.* (1972)
[s]Willis *et al.* (1983)

[b]Glotfelty *et al.* (1984)
[d]Grover *et al.* (1985)
[f]Harper *et al.* (1976)
[h]Majewski *et al.* (1991)
[j]Taylor *et al.* (1977)
[l]Turner *et al.* (1977)
[n]Whang *et al.* (1993)
[p]Willis *et al.* (1971)
[r]Willis *et al.* (1980)
[t]Woodrow *et al.* (1990)

21, 22, 24, 25, 40, 41, 42) down to less than 2% per day for alachlor, and dacthal, and even lower values for atrazine and simazine (nos. 1, 10, 2, 3, 32). As may be expected from the earlier discussion, the most rapid losses are found where materials with the highest vapor pressures are exposed on the surface of moist soils. The lindane and trifluralin data presented in Figure 9 (Glotfelty, 1981) show the typical flux patterns for volatile compounds applied as spray applications to moist soil surfaces where fugacity is not reduced by strong adsorption on dry mineral surfaces.

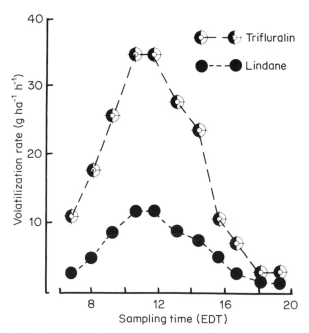

Figure 9 Diurnal variations in volatilization rates of trifluralin and lindane residues on the second day after surface applications of 2.80 and 1.10 kg ha^{-1} to moist silt loam soil at Beltsville, Maryland, June 1977 (Glotfelty, 1981)

Glotfelty *et al.* (1984) found that the disappearance of these residues over several days was best described by the equation $\log m = \log m_0 - kt^{-1/2}$, where m_0 was the initial residue concentration and m that at time t, measured in hours of daylight after the application. This equation can be interpreted as describing volatilization losses which continue during the daylight hours when the rate is controlled by upward movement of the pesticide to the surface through the shallow layer of soil penetrated by the spray. This result suggests that the rates of volatilization of surface residues from moist soils are controlled by solar energy input and by upward movement through the outermost soil layer penetrated by the formulation spray.

The effect of reduced soil moisture was shown in the loss patterns of surface-applied chlordane, heptachlor, trifluralin and lindane reported by (Glotfelty, 1981; Glotfelty *et al.*, 1984; Table 4: nos. 7, 22, 25, 42). In this experiment the volatilization rates of all four pesticides decreased quickly as the surface of the soil dried during the day. The low moisture retention capacity of the sandy soil allowed the top 5 mm to become very dry during the day, and the fugacity of the pesticide residues, which were all concentrated in this layer, were reduced to a very low value by the increased adsorption. The same effect was dramatically shown in the data presented by Majewski *et al*. (1989; Table 4: nos. 8, 14, 26, 30) for surface applications of chlorpyrifos, diazinon, lindane and nitrapyrin, where volatilization essentially ceased within a few hours after application to a drying California soil. Similar results were also reported by Turner *et al*. (1978; Table 4: no. 4) from an experiment in which chlorpropham was applied to a very dry bare soil: on the morning of the third day, when the soil moisture had been raised to 18% by an intervening period of rain, the volatilization exceeded the initial rate, but decreased again in the afternoon as the surface soil dried, despite a rise in temperature.

A characteristic feature of the inhibition of surface volatilization from drying soil is the appearance of a diurnal curve in which the highest rates are observed in the morning and evening, with a midday minimum. The increases in the evening are due to the moistening of the soil surface by dew formation, which persists through the night until its evaporation after sunrise. This effect is strikingly shown in the results presented by Majewski *et al*. (1989) and Turner *et al*. (1978). The effect of re-moistening of the soil is also shown by the alachlor data obtained by Glotfelty *et al*. (1989), presented in Figure 10. The volatilization reduction by dry soil is clearly

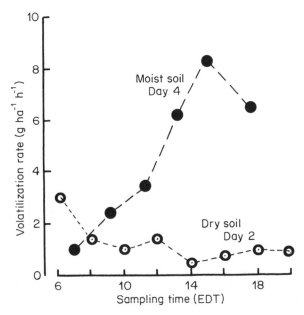

Figure 10   Diurnal variations in volatilization rate of alachlor residues from dry (day 2) and moist (day 4) silt loam soil after surface application of 2.24 kg ha^{-1}. Wye Research Center, Maryland, May 1981. (Glotfelty *et al.*, 1989)

evident in the volatilization pattern obtained on the second day after application, but on the fourth day this was transformed to that of free volatilization from moist soil by a light rainfall during the interim period.

Measurements of the volatilization of the octyl ester of 2,4-D by Grover *et al*. (1985); (Table 4, no. 44) after application to young 20-cm high wheat in Saskatchewan in May showed a more complex pattern. During the first two days, maximum volatilization occurred in the early afternoon despite very low surface soil moisture levels. This pattern was due to volatilization from the residues on the plant leaves, which had intercepted 52% of the application; there was little or no contribution from soil during this time. A morning maximum on the third day was, however, due to the desorption of the soil residues after rain. These data illustrate the complexity of volatilization patterns where surfaces with different characteristics are present at the same time. Other work with insecticides has shown that volatilization from transpiring plant leaves can take place freely during the day even when the underlying soil surface is dry (Taylor *et al*., 1977; Table 4, nos. 17, 23; Willis *et al*., 1983; Table 4, nos. 13, 35).

Comparative measurements of the volatilization of chlorpyrifos, fonofos and atrazine from soil under conventional and no-tillage management (Whang *et al*., 1993; Table 4, nos. 3, 9, 19) showed that volatilization of the fonofos and chlorphyrifos was between two and four times greater from the no-till surface than the conventional. This difference was due to the greater volatility of the 40% fraction of all the pesticides retained by the residues of plant material on the no-till surface during the application. Although the adsorption capacity of the plant residues was similar to that of the soil organic matter, the increased air exchange in the loose, open and spongy residues permitted much more rapid loss than from the soil. Due to its low vapor pressure, volatilization losses of atrazine were small from both surfaces.

Thus, the field observations all confirm the primary importance of the moisture content and management of the surface soil layer in controlling the volatilization of surface residues. The effect is clearly due to the reduction in the fugacity of the residues by the increased adsorption in dry soil.

Wind erosion

Data from two experiments (Glotfelty *et al*., 1984, 1989; Table 4, nos. 1, 2, 10, 32, 36) show that small but measurable amounts of some pesticides can enter the atmosphere by wind erosion of surface residues. The first experiment indicated that between 20 and 30% of the dacthal in the lowest part of the profiles was carried by particulates, whereas trifluralin, heptachlor and chlordane were entirely in the vapor form. The fraction of dacthal carried by the particulates increased with increasing wind speed and decreased with height: under calm conditions the amounts retained by both the filters and the adsorption samplers was the same as that obtained in laboratory measurements with vapor alone. The particle-borne fraction of the dacthal reflected wind erosion of the wettable powder formulation in which it was applied: all the other compounds were applied as water-based emulsions. It is likely that the dacthal residues remained concentrated on a thin easily removed layer of powder on the soil surface whereas the water-based spray was absorbed to a small but significant depth by the capillary tension in the soil pores.

In a second experiment, alachlor, atrazine and simazine fluxes over moist soil showed diurnal variations with mid-afternoon maxima, but when the soil dried the triazines showed a small mid-afternoon maximum when the alachlor volatilization was inhibited. Data from glass fiber samplers again suggested that theses maxima over dry soil were due to wind erosion of the triazine formulation particles. The actual amount of pesticide transported in this way is small and may be neglected in comparison to vapor dissipation for all but the least volatile materials. The greater retention of low volatility materials by mineral particles may, however, be of some significance in the long distance transport and deposition of pesticides of this type (Glotfelty, 1983).

Losses from plant leaves

Volatilization of residues from plant leaves appears to follow the same general patterns as that from moist soils except that the effects of changes in moisture content are less evident and the effects of decreasing residue coverage are more prominent. Initial losses are sometimes very rapid. Measurements of heptachlor and dieldrin fluxes after application to short orchard grass indicated that 46% of the heptachlor and 12% of the dieldrin applied were lost within 8 hours (Taylor *et al.*, 1977, Table 4, nos. 17, 22): the leaves intercepted 75–80% of the application. These heptachlor losses, which are comparable to those of EPTC from irrigation water under desert conditions found by Cliath *et al.* (1980, Table 4, no. 18), appear to be the highest reported in the literature up to this time.

Data for toxaphene losses from cotton leaves (Willis *et al*, 1980, 1983; Table 4, nos. 34, 35) where fluxes and residue disappearance decreased together suggested that the decrease in rate was controlled by changing residue distribution. Other measurements of the rates of disappearance of toxaphene from cotton plants in California by Seiber *et al.* (1979) were also consistent with 'first-order' losses, but no measurements of vapor fluxes were made.

A full understanding of the rates of volatilization from plant leaves requires more data than is presently available. The reduction in fugacity by adsorption on plant leaves is more complex than in soils, and the picture is further complicated by more complex metabolism kinetics and the possibilities of wash-off from leaves during rainfall or overhead irrigation.

Incorporated residues

The data summarized in Table 4 show that, as may perhaps be expected, volatilization rates are very much less when residues are incorporated into soils. This observation is, of course, consistent with the reasons for this practice, which is aimed to extend the persistence of compounds which would otherwise be greatly shortened by physical losses: as an example trifluralin, which is very sensitive to photodecomposition, is used in this way to extend its effectiveness as a herbicide.

Taylor *et al.* (1976, Table 4, nos. 15, 20) showed that incorporation to a depth of 7.5 cm reduced volatilization losses of dieldrin and heptachlor to less than 10% over a growing season of 100 days: the highest daily volatilization losses were $4.0\,g\,ha^{-1}\,day^{-1}$ of dieldrin and $5.0\,g\,ha^{-1}\,day^{-1}$ of heptachlor from warm moist soil 6 weeks after application. Marked midday maxima were observed in measured pesticide fluxes on all the days in which the surface soil remained moist and there was free surface water evaporation. These diurnal variations were shown to be closely correlated with the evaporation of soil water as measured in the lysimeter, confirming the importance of the wick effect in maintaining the supply of both pesticides at the soil surface.

The effects of soil moisture on the volatilization of incorporated trifluralin in Georgia were described by Harper *et al.* (1976) and White *et al.* (1977; Table 4, no. 38). In this experiment 95% of the trifluralin was contained in the top 2.5 cm layer of soil. The water content of this layer was measured throughout the experiment, so that volatilization rates could be directly related to the soil moisture content. Major differences were found in the diurnal volatilization patterns as the soil moisture changed. When the surface soil dried to a water content of less than $0.01\,cm^3\,cm^{-3}$ between 4 am and noon on the second day, the volatilization rate fell from 1.7 to below $0.15\,g\,ha^{-1}\,h^{-1}$ despite a fourfold increase in the atmospheric diffusivity coefficient, k_Z. This soil moisture level corresponds to less than the monomolecular layer level at which the fugacity of trifluralin would be reduced to a very low level by adsorption (Spencer and Cliath, 1974). These results clearly demonstrate the effect of the reduction of the fugacity in very dry soil. In another field experiment in New York state (A. W. Taylor, unpublished data) (Table 4, no. 39) where trifluralin was incorporated to the 7.5 cm depth in a gravelly loam, the results suggested that diffusion-controlled flow was the limiting mechanism throughout the growing season. Even though the soil was moist on all the days when fluxes were measured, no diurnal variations with marked midday maxima were observed and the rates were very uniform from early May to the end of July, with daily losses ranging between 0.22 and $0.4 \, g \, ha^{-1} \, day^{-1}$. Comparison of this result with the laboratory data obtained by Spencer and Cliath (1974) on the rate of trifluralin diffusion of trifluralin in Gila silt loam shows an acceptable level of agreement between the laboratory and field studies when allowance is made for the different application rates and soil conditions.

Other studies demonstrating the importance of moisture effects on the loss of incorporated residues have been presented by Willis et al. (1971, 1972, Table 4, nos. 12, 15) and by Grover et al. (1988; Table 4, nos. 37, 43). None of these experiments revealed meaningful correlations between flux rates and air or soil temperature. All these results confirm the dominant role of soil moisture in controlling the volatilization of soil-incorporated pesticides. Where soil moisture moves continuously to the surface in response to surface evaporation, the volatilization is controlled by the wick effect and reflects the rate at which the residue concentration at the soil surface is replenished by the upward convective movement of the pesticide in the flow of soil moisture. Marked diurnal variations with midday maxima are often observed as the rate of water evaporation changes with insolation or energy advection by the wind. When the soil remains moist but there is little water evaporation, volatilization becomes controlled by diffusive flow: both daily and long-term volatilization then tend to be slower and more uniform.

Where the surface soil dries to a moisture content less than 2–5%, depending upon soil texture, volatilization is reduced to very low values or ceases completely due to reduction of the pesticide vapor density by increased adsorption on the dry soil. Since air and surface soil temperatures may rise very steeply over dry soil that is not cooled by water evaporation, no meaningful relationship can be expected between volatilization and temperature or wind speed.

Rate-limiting factors

All these field experiments show that the volatilization rates of incorporated residues are controlled by soil conditions. Meteorological variables such as temperature, radiation input and wind speed are important insofar as they control the soil conditions, but it is the latter that is critical in controlling the supply of available pesticide at the surface, which in turn controls the volatilization rate.

Data from field experiments can also give much valuable information in identifying the rate-limiting steps in the overall process of residue dispersal into the atmosphere. With one or two exceptions (Taylor et al., 1977), the vapor concentrations at even the lowest heights over treated areas always represent relative vapor densities of 10% or less, even where volatilization is taking place from moist soils where the vapor pressures of the residues are close to those of the pure compounds. This indicates that there is rapid dilution of the vapor by turbulent mixing in the lower atmospheric boundary layer and that the vapor flux rate depends on the rate at which vapor enters the turbulent atmosphere from the underlying surface. This rapid turbulent dispersal of the vapor is exemplified by the value of the 'residence time' of the vapor in the air above the surface, calculated by dividing the amount of pesticide present in a volume of air by the flux through it. Values of 20–40 s or less for depths up to about 1 m are typical of turbulent mixing during spring or summer days (Harper *et al.*, 1976; Taylor *et al.*, 1977), with values increasing to between 2 and 10 min under stable atmospheric conditions in the morning and evening. Where these increases are not associated with accumulations of pesticide vapor in the air close to the ground, as would be expected if volatilization continued but atmospheric dispersal ceased, they must clearly reflect reduced vapor input from the soil or plant surfaces, again demonstrating the importance of the surface conditions as the rate-limiting factors in the volatilization process.

Volatilization from water surfaces

The volatilization of pesticide residues applied to open water surfaces for the control of insects or aquatic vegetation is of special environmental interest, emphasized by the great importance of rice culture in many parts of the world. Although the physical chemistry of the distribution of the residues between the air, water and sediments will not differ fundamentally from that in soils, the large uniform water surface will greatly modify the dynamics of the volatilization process. Measurements of the volatilization of herbicides and insecticides from irrigated and flooded fields of rice and other crops in California have demonstrated the importance of the effects of different types of formulations and the distribution of the pesticide between the water and underlying soil on the rate at which the pesticides equilibrate between the soil and water (Cliath *et al.*, 1980; Woodrow *et al.*, 1990; Ross and Slater, 1986: Table 4, nos. 18, 27, 28, 29, 33). Aerial spray applications of water-based emulsions were distributed throughout the water much more quickly than granular formulations which tended to sink to the underlying soil. The two sets of compounds showed different patterns of volatilization due to the different rates of mixing into the water. The rates of mixing were also influenced by the retention of residues by the submerged sediments. Changes of the volatilization rate of molinate appeared to be associated with temperature-dependent changes in the Henry's law coefficient of this herbicide, although the solution concentrations were far from saturation; this effect may be due to shifts in the equilibrium between two isomeric forms of this compound.

The theoretical approach to volatilization loss from water bodies has been discussed by Liss and Slater (1974) in terms of the 'two resistance-layer model' originally introduced by Whitman (1923). A more comprehensive interpretation of this approach, with an excellent bibliography, has been presented by Mackay (1984, 1991) and by Mackay *et al.* (1986) in terms of fugacity. Hartley and Graham-Bryce (1980) point out the importance of solar heating, wave action and wind currents in renewing the concentrations at the surface and reducing the importance of diffusion as the rate-limiting step.

The California data suggest that volatilization losses from water bodies is very sensitive to the management of both the pesticide and the cropping practice. In view of the great importance of pesticides in rice culture, it is clear that there is a major need for environmental information on the behavior of pesticides in such cropping systems.

Comparative experiments

The data listed in Table 4 show that volatilization rates in the field are highly variable and very sensitive to weather, soil conditions and usage patterns. While the influence of the soil chemistry and chemical properties of individual compounds are often evident in particular experiments, the variable nature of the field data makes it difficult to compare results from experiments done at different times and places. The effects of chemical properties are best evaluated in experiments where two or more compounds are applied together and sampled simultaneously in the same experiment. Even where two materials are compared in separate experiments against a third 'standard' compound used in each (e.g. B compared with C from experiments with AB and AC) the results are less satisfactory than direct comparisons unless the weather, usage patterns and soils are very similar.

Comparison of the fluxes of compounds sampled together in the same equipment are in effect comparisons of simultaneous vapor density profiles. Any errors introduced by ambiguities in flux calculations are thus discounted, so that the data give direct comparisons of the fugacities of the compounds in the air above the soil or plant surfaces. Such comparisons have been made for both incorporated and surface residues of several compounds (Table 4: soil surface nos. 1, 2, 32, 36; 8, 14, 26, 30; 7, 22, 25, 42; 6, 10, 21, 40; 24, 41; plant surfaces 17, 23; 13, 34; soil incorporated 16, 20; 37, 43). Interpretations of some of these experiments have been presented by Taylor and Spencer (1990). The triallate/trifluralin results obtained by Grover *et al.* (1988) are, however, of particular interest in showing how field data can demonstrate the importance of residue fugacities as opposed to those of the pure compounds. Over the 70-day period of the experiment the proportional loss of trifluralin was consistently about 1.3 times greater than that of triallate, although the vapor pressure of trifluralin is only about 58% that of triallate and trifluralin is also known to be more strongly adsorbed by the soil. These differences were more than offset by the lower water solubility and higher Henry's law coefficient of trifluralin, which resulted in a more rapid volatilization loss. Calculation of the expected ratio by the Jury model predicted a value of 1.5, close to that observed. This result shows how the consequences of differences in the adsorption and distribution of the residues of two pesticides in the soil can be demonstrated in a single field experiment; the differences between them could not be shown unambiguously by the comparison of two experiments at different times.

In general, the results obtained from experiments comparing the behavior of two or more pesticides under identical conditions show that such comparative experiments can give information about the chemical and environmental factors controlling volatilization that cannot be obtained in any other way. In view of the cost in time and resources required in field experiments on volatility it is strongly recommended that such an approach should always be considered in future experimental work in the field.

PREDICTION AND MODELING

Any comprehensive model describing the volatilization of pesticides from soil or plant surfaces must include several factors and describe the ways in which they interact with each other. These factors include:

(1) The chemical properties of the pesticide, including the vapor pressure, the Henry's law coefficient, the soil/water partition coefficients in both wet and dry soils and the degradation rate in soil.
(2) Management and usage factors, including the method and rate of application, the resulting residue concentration and distribution in the soil profile or the surface loading rate in $g\,m^{-2}$ of field soil or plant leaf area.
(3) The soil environment, including the distribution of soil moisture over the depth containing the pesticide, the rate of flow of soil moisture through this layer and the variations in this flow due to surface water evaporation on both a daily and seasonal basis.
(4) Weather patterns of rainfall, insolation and wind speed on a daily or shorter basis for the prediction of soil moisture changes due to rainfall and soil drying by evaporation.

No comprehensive model is currently available that will take into account the full complexity of these factors in the field situation, but some have been developed that provide descriptions of parts of the overall process. These models are concerned with mechanistic descriptions of the factors controlling the redistribution and movement of residues within soil profiles or from the soil into the atmosphere. The model developed by Jury et al. (1983a,b; 1984a,b,c), discussed above, which includes diffusion, convective flow, leaching, chemical degradation and gaseous diffusion at the atmospheric boundary layer is of great value in assessing the relative volatility, mobility and persistence of a range of pesticides and other trace organics in soils. Basically this model is intended to classify pesticides and other chemicals according to their environmental behavior calculated from their physical and chemical characteristics such as vapor pressure, solubility, Henry's law and organic carbon partition coefficients and degradation rate. The model indicates that pesticides can be divided into three categories depending on whether the Henry's law coefficient (equation 4) is greater than, approximately equal to, or less than 2.65×10^{-5}. The model has been most successfully applied in predictions of differences in volatilization rates of incorporated residues of several pesticides where the moisture content remains high long enough to allow the necessary diffusion gradients to be established and there is no inhibition of evaporation by soil drying.

AGRONOMIC AND ENVIRONMENTAL SIGNIFICANCE

Environmentally, the most important aspect of pesticide volatilization is the potential for the rapid injection of a large fraction of the applied residue into the atmosphere. The further possibility of the rapid transport of the resulting vapor over long distances makes this a distribution pathway of unique importance. The large distances over which the airborne residues may be carried is offset by the rapid dilution in the atmosphere. This, coupled with the possible degradation by photochemical and oxidative reactions reduces the risk of acute environmental impacts except in areas close to the source. These risks cannot, however, be entirely discounted, particularly where the chemicals may be reconcentrated by bioaccumulaton. The discovery of DDT and other stable organochlorine residues in regions as remote as the polar icecaps is a dramatic demonstration of this. The transport and re-deposition of a number of materials in this way has been discussed by a number of authors (Kurtz, 1990). Re-concentration of vapor by adsorption into rain and fog droplets, with possible re-deposition on vegetation has also been reported (Glotfelty et al., 1987).

In assessing the environmental impacts of herbicide use, it should be recognized that there are very direct benefits from their use in minimum tillage or no-tillage soil management systems, where there are very important long-term benefits in the reduction of soil erosion and improvements in water quality which could not be achieved without their use.

Agronomically, volatilization must be recognized as a major cause of pesticide disappearance from target areas, particularly when they are applied to the surfaces of soils or plants. The rate of this loss often exceeds that by chemical degradation. Recognition of this fact is of major importance for improved pesticide management. In many instances the effective life of both herbicides and insecticides is probably limited by their rapid disappearance from the target area. This often results in the need for repeated applications, particularly when weather and soil conditions favor volatilization loss, as in the moist warm conditions of the tropics. Owing to the volatile nature of many pesticide chemicals the possibilities for improved usage must center on preventive measures that avoid use patterns that favor volatilization losses.

The most rapid losses are those from residues on the surfaces of bare moist soils. Serious losses may occur in no-tillage cropping practices where herbicide use is essential. Where possible, application to dry soil is likely to be the best preventive measure. Although subsequent volatilization losses will certainly occur when the soil is moistened by rain, this loss is likely to be less than that occurring immediately after direct application to moist soil surfaces where there may be little tendency for the formulation to diffuse downward into the upper surface soil layer before volatilization begins. Where a choice of chemicals is possible the selection of less volatile materials is clearly desirable.

Where cultivation is consistent with the soil management practice, incorporation immediately after application—even to a very shallow depth—will greatly reduce the loss of even the more volatile compounds.

Losses from plant surfaces can also be very rapid, although residues under the canopy may be in some degree protected by the sheltering action of the leaf cover.

In some circumstances the volatilization of residues into still or stable air may produce vapor concentrations high enough to have adverse impacts on sensitive non-target species if the vapor is carried out of the target area in stable low-lying air or by nonturbulent katabatic flow. Instances have been reported where this has apparently happened even where residues were dispersed over considerable distances in turbulent air. Adsorption of vapor into fog particles which may be deposited on non-target plant species (Glotfelty et al., 1987) represents another pathway that may cause phytotoxic off-target effects of herbicides on sensitive species or contamination of other crops by insecticides, similar to the effects associated with spray drift (Farwell et al., 1976).

REFERENCES

Acree, F., Beroza, M., and Bowman, M. C. (1963). 'Codistillation of DDT with water', *J. Agric. Food Chem.*, **11**, 278–280.

Bidleman, T. F. (1984). 'Estimation of vapor pressure for non polar organic compounds by capillary gas chromatography', *Anal. Chem.*, **56**, 2490–2496.

Bidleman, T. F. (1985). 'High volume collection of organic vapors using solid adsorbents, in *Trace Analysis* (ed. J. F. Lawrence), Vol. 4, pp. 51–100, Academic Press, New York.

Chiou, C. T., Porter, P. E., and Schmedding, D. W. (1983). 'Partition equilibria of nonionic compounds between soil organic matter and water', *Environ. Sci. Technol.*, **17**, 227–231.

Chiou, C. T., and Shoup, T. D. (1985). 'Soil sorption of organic vapors and effects of humidity on sorptive mechanism and capacity', *Environ. Sci. Technol.*, **19**, 1196–1200.

Chiou, C. T., Shoup, T. D., and Porter, P. E. (1985). 'Mechanistic roles of soil humus and minerals in the sorption of nonionic organic compounds from aqueous and organic solutions', *Org. Geochem.*, **8**, 9–14.

Cliath, M. M., Spencer, W. F., Farmer, W. J., Shoup, T. D., and Grover, R. (1980). 'Volatilization of s-ethyl n,n,dipropylthiocarbamate from water and wet soil during and after flood irrigation of an alfalfa field', *J. Agric. Food Chem.*, **28**, 610–613.

Deming, J. M. (1963). 'Determination of volatility losses of C14-CDAA from soil samples', *Weeds*, **11**, 91–95.

Denmead, O. T., Simpson, J. R., and Freney, J. R. (1977). 'A direct measurement of ammonia emission after injection of anhydrous ammonia', *Soil Sci. Soc. Am. J.*, **41**, 1001–1004.

Fang, S. C., Theisen, P., and Freed, V. H. (1961). 'Effects of water evaporation, temperature, and rates of application on the retention of ethyl-N,N-di-n-propyl-thiocarbamate in various soils', *Weeds*, **9**, 569–574.

Farmer, W. J., Igue, K., Spencer, W. F., and Martin, J. P. (1972). 'Volatility of organochlorine insecticides from soil: I. Effect of concentration, temperature, air flow rate, and vapor pressure', *Soil Sci. Soc. Am. Proc.*, **36**, 443–447.

Farwell, S. O., Robinson, E., Powell, W. J., and Adams, D. F. (1976). 'Survey of airborne 2,4-D in South-Central Washington', *J. Air Poll. Control Assoc.*, **26**, 224–230.

Fendinger, N. J., and Glotfelty, D. E. (1988). 'A laboratory method for the experimental determination of air-water Henry's law constants for several pesticides', *Environ. Sci. Technol.*, **22**, 1289–1293.

Glotfelty, D. E. (1981). 'Atmospheric dispersion of pesticides from treated fields', *PhD Dissertation*, University of Maryland, College Park, MD, Univ. Library, Call No. 10879681.

Glotfelty, D. E., Leech, M. M., Jersey, J., and Taylor, A. W. (1989). 'Volatilization and wind erosion of soil-surface applied atrazine, simazine, alachlor, and toxaphene', *J. Agric. Food Chem.*, **37**, 546–555.

Glotfelty, D. E., and Schomburg, C. J. (1989). 'Volatilization of pesticides from soil', in *Reactions and Movement of Organic Chemicals in Soils* (eds B. L. Sawnhey and K. Brown), Soil Sci. Soc. Am. Special Publication Number 22, Chap. 7, pp. 181–207, Soil Sci. Soc. Am., Madison Wisconsin.

Glotfelty, D. E., Seiber, J. N., and Liljedahl, L. A. (1987). 'Pesticides in fog', *Nature*, **325**, 602–605.

Glotfelty, D. E., Taylor, A. W., Turner, B. C., and Zoller, W. H. (1984).

'Volatilization of surface-applied pesticides from fallow soil', *J. Agric. Food Chem.*, **32**, 638–643.

Gray, R. A., and Weierich, A. J. (1965). 'Factors affecting the vapor loss of EPTC from soils', *Weeds*, **13**, 141–146.

Grover, R., Shewchuck, R. R., Cessna, A. J., Smith, A. E., and Hunter, J. H. (1985). 'Fate of 2,4-D iso-octyl ester after application to a wheat field', *J. Environ. Qual.*, **14**, 203–210.

Grover, R., Smith, A. E., Shewchuck, S. R., Cessna, A. J., and Hunter, J. H. (1988). 'Fate of trifluralin and triallate applied as a mixture to a wheat field', *J. Environ. Qual.*, **17**, 543–550.

Harper, L. A., McDowell, L. L., Willis, G. H., Smith, S., and Southwick, L. M. (1983). 'Microlimate effects on toxaphene and DDT volatilization from cotton plants', *Agron. J.*, **75**, 295–302.

Harper, L. A., White, A. W., Bruce, R. R., Thomas, A. W., and Leonard, R. A. (1976). 'Soil and microclimate effects on trifluralin volatilization', *J. Environ. Qual.*, **5**, 236–242.

Harris, C. F., and Lichtenstein, E. P. (1961). 'Factors affecting the volatilization of insecticidal reidues from soils', *J. Econ. Entomol.*, **54**, 1038–1045.

Hartley, G. S. (1969). 'Evaporation of pesticides', in *Pesticidal Formulations Research: Physical and Colloidal Chemical Aspects* (ed R. F. Gould), Adv. Chem, Series No. 86, pp. 115–134, Am. Chem. Soc., Washington, DC.

Hartley, G. S., and Graham-Bryce, I. J. (1980). 'Behavior of pesticides in air', in *Physical Principles of Pesticide Behavior*, **Vol. 1**, pp 110–203, Academic Press, New York.

Jury, W. A., Grover, R., Spencer, W. F., and Farmer, W. J. (1980). 'Modeling vapor losses of soil-incorporated triallate', *Soil Sci. Soc. Am. J.*, **44**, 445–450.

Jury, W. A., Farmer, W. J., and Spencer, W. F. (1984a). 'Behavior assessment model from trace organics in soil: II. Chemical classification and parameter sensitivity', *J. Environ. Qual.*, **13**, 567–572.

Jury, W. A., Spencer, W. F., and Farmer, W. J. (1983a). 'Use of models for assessing relative volatility, mobility, and persistence of pesticides and other trace organics in soil systems', in *Hazard Assessment of Chemicals; Current Developments*, (ed J. Saxena), **Vol. 2**, pp. 1–43, Academic Press, New York.

Jury, W. A., Spencer, W. F., and Farmer, W. J. (1983b). 'Behavior assessment model for trace organics in soil: I. Model description', *J. Environ. Qual.*, **12**, 558–564.

Jury, W. A., Spencer, W. F., and Farmer, W. J. (1984b). 'Behavior assessment model for trace organics in soil: III. Application of screening model', *J. Environ. Qual.*, **13**, 573–579.

Jury, W. A., Spencer, W. F., and Farmer, W. J. (1984c). 'Behavior assessment model for trace organics in soil: IV. Review of experimental evidence', *J. Environ. Qual.*, **13**, 580–585.

Kurtz, D. A. (ed.) (1990). *Long Range Transport of Pesticides*. Lewis Publishers, Chelsea, Michigan.

Lemon, E. R. (1969). 'Gaseous exchange in crop stands', in *Physiological Aspects of Crop Yield* (eds J. D. Eastin, F. A. Haskins, C. Y. Sullivan, and C. H. M. Van Bavel), pp. 117–142, Am. Soc. Agron., Madison, Wisconsin.

Lewis, G. N. (1901). 'The law of physico-chemical change', *Proc. Am. Acad.*, **37**, 49–96.

Lichtenstein, E. P., and Schultz, K. R. (1961). 'Effect of soil cultivation, soil surface and water on the persistence of insecticidial residues in soils', *J. Econ. Entomol.*, **54**, 517–522.

Lichtenstein, E. P., Schultz, K. R., Fuhremann, T. W., and Liang, T. T. (1970). 'Degradation of aldrin and heptachlor in field soils during a ten-year period: translocation into crops', *J. Agric. Food Chem.*, **18**, 100–106.

Liss, P. S., and Slater, P. G. (1974). 'Flux of gases across the air-sea interface', *Nature*, **247**, 181–184.

Mackay, D. (1984). 'Air/water exchange coefficients', in *Environmental Exposure from Chemicals* (eds W. Brock Neely and G. E. Blau), Vol. 1. pp 91–108, CRC Press, Boca Raton, FL.

Mackay, D. (1991). *Multimedia Environmental Models: the Fugacity Approach.* Lewis Publishers, Chelsea, Michigan.

Mackay, D., and Paterson, S. (1981). 'Calculating fugacity', *Environ. Sci. Technol*, **5**, 1006–1014.

Mackay, D., and Paterson, S. (1982). 'Fugacity revisited', *Environ. Sci. Technol.*, **16**, 645A–660A.

Mackay, D., Paterson, S., and Schroeder, W. H. (1986). 'Model describing the rates of transfer processes of organic chemicals between atmosphere and water', *Environ. Sci. Technol.*, **20**, 810–816.

Majewski, M. S., Glotfelty, D. E., Kyaw Tha Paw U., and Seiber, J. N. (1990). 'A field comparison of several methods for measuring pesticide evaporation rates from soil', *Environ. Sci. Technol.*, **24**, 1490–1496.

Majewski, M. S., Glotfelty, D. E., and Seiber, J. N. (1989). 'A comparison of the aerodynamic and the theoretical-profile-shape methods for measuring pesticide evaporation from soil', *Atmos. Environ.*, **23**, 929–938.

Majewski, M. S., McChesney, M. M., and Seiber, J. N. (1991). 'A field comparison of two methods for measuring DCPA soil evaporation rates', *Environ. Chem. Toxicol.*, **10**, 301–311.

Mayer, R., Letey, J., and Farmer, W. J. (1974). 'Models for predicting volatilization of soil-incorporated pesticides', *Soil Sci. Soc. Am. Proc.*, **38**, 563–568.

Montieth, J. L. (1973). *Principle of Environmental Physics*, Chaps., 2, 6, 7, 12. American Elsevier, New York.

Parmele, H., Lemon, E. R., and Taylor, A. W. (1972). 'Micrometeorological measurement of pesticide vapor flux from bare soil and corn under field conditions', *Water, Air, Soil Pollut.*, **1**, 433–451.

Parochetti, J. V., and Warren, G. F. (1966). 'Vapor losses of IPC and CIPC', *Weeds.*, **14**, 281–285.

Plimmer, J. R. (1976). 'Volatility', in *Herbicides: Chemistry, Degradation and Mode of Action*, (ed P. C. Kearney), Vol. 2, Chap. 19, pp. 891–934. Dekker, NY.

Rao, P. S. C., and Davidson, J. M. (1980). 'Estimation of pesticide retention and transformation parameters required in nonpoint source pollution models', in *Environmental Impact of Nonpoint Source Pollution* (eds M. R. Overcash and J. M. Davidson), pp. 23–67, Ann Arbor Sci. Pub., Ann Arbor, Michigan.

Rose, C. W. (1966). *Agricultural Physics*, Chap. 2, 3, Pergamon Press, New York.

Ross, L. J., and Sava, R. J. (1986). 'Fate of thiobencarb and molinate in rice fields', *J. Environ. Qual.*, **15**, 220–225.

Seiber, J. N., Madden, S. C., McChesney, M. M., and Winterlin, W. L. (1979). 'Toxaphene dissipation from treated cotton field measurements: component residual behavior on leaves and in air, soil, and sediments determined by capillary gas chromatogaphy', *J. Agric. Food Chem.*, **27**, 284–290.

Spencer, W. F. (1970). 'Distribution of pesticides between soil, water, and air', in *Pesticides in the Soil: Ecology, Degradation, and Movement*, pp 120–128, Michigan State University, East Lansing, MI.

Spencer, W. F., and Cliath, M. M. (1969). 'Vapor density of dieldrin', *Environ. Sci. Technol.*, **3**, 670–674.

Spencer, W. F., and Cliath, M. M. (1970). 'Desorption of lindane from soil as related to vapor density', *Soil Sci. Soc. Am. Proc.*, **34**, 574–578.

Spencer, W. F., and Cliath, M. M. (1972). 'Volatility of DDT and related compounds', *J. Agric. Food Chem.*, **20**, 645–649.

Spencer, W. F., and Cliath, M. M. (1973). 'Pesticide volatilization as related to water loss from soil', *J. Environ. Qual.*, **2**, 284–289.

Spencer, W. F., and Cliath, M. M. (1974). 'Factors affecting vapor loss of trifluralin from soil', *J. Agric. Food Chem.*, **22**, 987–991.

Spencer, W. F., and Cliath, M. M. (1983). 'Measurement of pesticide vapor pressures', *Residue Rev.*, **85**, 57–71.

Spencer, W. F., Cliath, M. M., and Farmer, W. J. (1969). 'Vapor density of soil-applied dieldrin as related to soil-water content, temperature, and dieldrin concentration', *Soil Sci. Soc. Am. Proc.*, **33**, 509–511.

Spencer, W. F., Farmer, W. J., and Cliath, M. M. (1973). 'Pesticide volatilization', *Residue Rev.*, **49**, 1–45.

Suntio, L. R., Shiu, W. Y., Mackay, D., Seiber, J. N., and Glotfelty, D. (1987). 'A critical review of Henry's law constants for pesticides', *Rev. Environ. Contam. Toxicol.*, **103**, 1–58.

Taylor, A. W., and Glotfelty, D. E. (1988). 'Evaporation from soils and crops', in *Environmental Chemistry of Herbicides* (ed. R. Grover), Chap. 4, pp. 89–129, CRC Press, Boca Raton, FL.

Taylor, A. W., Glotfelty, D. E., Glass, B. L., Freeman, H. P., and Edwards, W. M. (1976). 'The volatilization of dieldrin and heptachlor from a maize field', *J. Agric. Food Chem.*, **24**, 625–630.

Taylor, A. W., Glotfelty, D. E., Turner, B. C., Silver, R. E., Freeman, H. P., and Weiss, A. (1977). 'Volatilization of dieldrin and heptachlor residues from field vegetation', *J. Agric. Food Chem.*, **25**, 542–548.

Taylor, A. W., and Spencer, W. F. (1990). 'Volatilization and vapor transport processes', in *Pesticides in the Soil Environment: Processes, Impacts and Modeling*. (ed. H. H. Cheng). Soil Science Society of America Book Series 2, Chap. 7, pp. 214–269, Soil Sci. Soc. Am., Madison, WI.

Turner, B. C., and Glotfelty, D. E. (1977). 'Field air sampling of pesticide vapors with polyurethane foam', *Anal. Chem.*, **49**, 7–10.

Turner, B. C., Glotfelty, D. E., and Taylor, A. W. (1977). 'Photodieldrin formation and volatilization from grass', *J. Agric. Food Chem.*, **25**, 548–550.

Turner, B. C., Glotfelty, D. E., Taylor, A. W., and Watson, D. R. (1978). 'Volatilization of microencapsulated and conventionally applied chlorpropham in the field', *Agron. J.*, **70**, 933–937.

Whang, J. M., Schomburg, C. J., Glotfelty, D. E., and Taylor, A. W. (1993). 'Volatilization of fonofos, chlorpyrifos and atrazine from conventional and no-till surface soils in the field', *J. Environ. Qual.*, **22**, 173–180.

White, A. W., Harper, L. A., Leonard, R. A., and Turnbull, J. W. (1977). 'Trifluralin volatilization from a soybean field., *J. Environ. Qual.*, **6**, 105–110.

Whitman, W. G. (1923). 'Preliminary experimental confirmation of the two-film theory of gas adsorption', *Chem. Metall. Eng.*, **29**, 146.

Willis, G. H., McDowell, L. L., Harper, L. A., Southwick, L. M., and Smith, S. (1983). 'Seasonal disappearance and volatilization of toxaphene and DDT from a cotton field', *J. Environ. Qual.*, **12**, 80–85.

Willis, G. H., McDowell, L. L., Smith, S., Southwick, L. M., and Lemon, E. R.

(1980). 'Toxaphene volatilization from a mature cotton canopy', *Agron. J.*, **72**, 627–631.

Willis, G. H., Parr, J. F., and Smith, S. (1971). 'Volatilization of soil-applied DDT and DDD from flooded and nonflooded plots', *Pestic. Monit. J.*, **4**, 204–208.

Willis, G. H., Parr, J. F., Smith, S., and Carroll, B. R. (1972). 'Volatilization of dieldrin from fallow soil as affected by different soil water regimes', *J. Environ. Qual.*, **1**, 193–196.

Wilson, J. D., Thurtell, G. W., and Kidd, G. E. (1981). *Boundary Layer Meteorol.*, **21**, 295–313.

Woodrow, J. E., McChesney, M. M., and Seiber, J. N. (1990). 'Modeling the volatilization of pesticides and their distribution in the atmosphere', in *Long Range Transport of Pesticides* (ed. D. A. Kurtz). pp. 61–81 Lewis Publishers Inc., Chelsea Michigan.

CHAPTER 7

Biosensors for detection of pesticides

M. E. Eldefrawi, A. T. Eldefrawi and K. R. Rogers

INTRODUCTION

There is a critical and growing need for rapid and cost-effective on-site methods of detection and monitoring of pesticides and their toxic metabolites in plants, food, feed or in superfund sites, hazardous waste facilities and ground water. The magnitude of the problem is evident when current cost of analysis is calculated. For ten million wells at a cost of $100 per analysis for a single pesticide, excluding its metabolites, the cost would be a billion dollars.

There is also a need for rapid monitoring of pesticide levels in the serum and urine of industrial and farm workers for possible exposure. Biosensors

Environmental Behaviour of Agrochemicals
Edited by T. R. Roberts and P. C. Kearney   © 1995 John Wiley & Sons Ltd

are well positioned to serve this need because of their speed, simplicity, specificity, sensitivity and, in some cases, portability. According to the Superfund Records of Decision database, pesticides are major contaminants in a number of Superfund sites targeted for remediation. In many sites, these pesticides have been identified only by chemical class. As a result, biosensors that respond to a particular class of pesticides may be useful in ground water monitoring systems required in the remediation and management of many of these sites. The majority of biosensor research and development has been directed towards the clinical market, e.g. the glucose biosensor to monitor blood glucose in diabetics. However, driven by the need for environmental surveillance and for detection of contaminants, including pesticides in various matrices, a great deal of research effort is being directed toward these problems.

The field of biosensors is immense, with an estimated global market of $350 million by 1996 and $750 million by the year 2000 (Frost & Sullivan Inc., no. T057). As a reflection of the amount of activity in this field there is a Gordon Conference devoted to biosensors every 1–1/2 years, several international meetings every year, and several symposia on biosensors in each semiannual meeting of the American Chemical Society in the last few years, as well as two journals devoted to biosensors (e.g. *Biosensors and Bioelectronics* and *Sensors and Actuators*). This chapter will concentrate on biosensors in environmental monitoring for pesticides and is divided into two parts. In the first part, biosensors will be defined as to their constituents, and their uses explored. In the second part, the focus will be on important attributes of biosensors for use in detection and environmental monitoring for pesticides. Issues discussed include: their specificity, selectivity, dynamic range, response time, reproducibility, lifetime and stability, as well as the effect of protein immobilization on the specificity and reusability of the biosensor.

DEFINITIONS, ATTRIBUTES AND USES OF BIOSENSORS

A biosensor is an analytical device (Schultz, 1991) composed of a biological sensing element (for its specific recognition capabilities) in close proximity to a signal transducer, which together relate the concentration of an analyte to a measurable electrical signal (Figure 1).

Biological sensing elements

There are two major mechanisms of biological recognition used in biosensors:

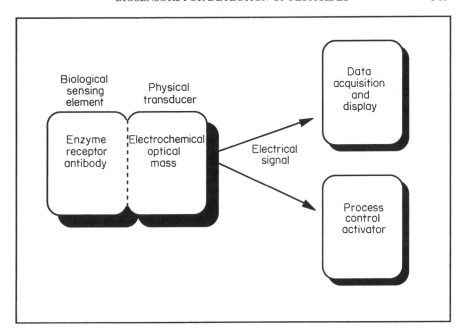

Figure 1 The biological sensing element converts an analyte concentration into a change in the electrical, optical or surface-bound mass properties of the biosensor system. The physical transducer then converts this change in property into an electrical signal which may be quantified and displayed in a variety of formats. The biosensor may also be interfaced to a process control actuator which can be used to instantaneously modulate or control a critical parameter up-line or down-line of the biosensor in the process stream

(1) A catalytic reaction mediated by an enzyme, where a product of biocatalysis is measured.

(2) A stoichiometric reaction that results from the binding of an analyte to a receptor, antibody or gene, where the one-to-one binding event is measured.

Biosensors that depend on a whole cell as the biological sensing element, utilize either or both of the above two mechanisms.

Enzymes have historically been used as the first biological sensing elements primarily because of their ability to amplify a signal by catalytically converting substrates into products which can be measured optically or electrochemically (Figure 2). The major advantages in using enzymes are sensitivity of detection, their reversible nature, and the extensive research data available on their immobilization, stabilization, and storage properties (Guilbault *et al.*, 1991). The best described enzyme biosensors are those

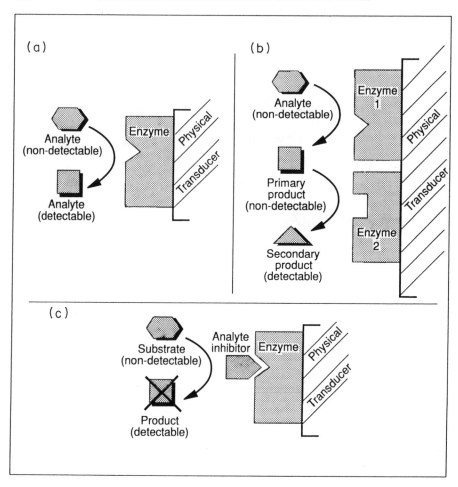

Figure 2 An enzyme may function as biological sensing element for an analyte of interest by several mechanisms. (a) The biocatalyst may change a non-detectable (by the system) analyte into a detectable product. (b) Enzymes may also be used in series to convert an analyte into a non-detectable primary product which is converted to a detectable secondary product either electrochemically at the sensor surface or by means of a second enzyme. (c) An analyte may be detected by its ability to inhibit an enzyme-catalytic reaction. Formats a and b can run continuously and reversibly (given the presence of adequate analyte). Although format c must be regenerated, it is inherently more sensitive than formats a or b due to enzyme amplification of inhibitor binding

using glucose oxidase for detecting glucose in blood (Wilson and Turner, 1992). Enzymes are also capable of functioning in certain organic media (Saini and Turner, 1991), which allows their use for detection of compounds in hydrophobic media such as phenolics in olive oil, cholesterol in butter, or alcohol in fuel. It is also possible to engineer miniaturized multi-enzyme biosensors for *in vivo* applications (Urban *et al.*, 1992). Most commercially available biosensors are enzyme-based electrochemical devices developed for clinical purposes and are durable, easy to use, disposable, or inexpensive to mass produce.

Biosensors that use receptors and antibodies as sensing elements typically measure analytes indirectly through the binding of an optically or enzymatically labeled probe (i.e. a chemical derivative of the analyte which competes with the compound of interest for a limited number of binding sites) (Figure 3). These biosensor configurations can be used to measure specific compounds or groups of closely related compounds with detection limits and reversibility characteristics being determined by the affinity of the antibody or receptor for several specific analytes. The nicotinic acetylcholine receptor (nAChR), which has been used in several biosensor configurations (Rogers *et al.*, 1989, 1991b,d; 1992b), is the direct target for the insecticide nicotine and the recently introduced nitromethylenes. The major inhibitory receptor in mammalian and insect brains (i.e. the γ-aminobutyric acid (GABA) receptor) is the primary target for the cyclodiene insecticide endosulfan as well as possibly a secondary target for pyrethroids. With the exception of anticholinesterase (antiChE) insecticide biosensors, antibody-based biosensors (i.e. immunosensors) dominate the literature with respect to potential for detection of pesticides. This is primarily because of their high specificity and sensitivity for a variety of analytes. Examples of immunoassays (see Kaufman and Clower, 1991) have been developed for a wide range of insecticides (e.g. organophosphates (OP) and chlorinated hydrocarbons), herbicides (e.g. thiocarbamates, phenoxyaliphatic acids, bipyridiliums and triazines), fungicides (e.g. benzimidazoles, acylalanines and triazoles), and rodenticides (e.g. coumarins).

More complex biological sensing elements such as microorganisms have also been used in biosensors (Gaisford *et al.*, 1991). An example of this type of biosensor is a device which utilizes *Nitrobacter* sp. to detect NO_2. Another is the biosensor that uses genetically engineered *E. coli* to detect the herbicides meturon, isoproturon, propanil and ioxynil at concentrations as low as 50 ppb. In addition to having the lux gene incorporated, this bacterium shows better transport of toxic substances across the cell membrane than the parent *E. coli*, thereby allowing inhibition of cellular respiration to be monitored via bioluminescence (Lee *et al.*, 1992). There are also whole organ biosensors such as the one utilizing intact crab antennae to detect salinity and trimethylamine *N*-oxide (Barker *et al.*, 1990). A biosensor that is finding increasing use in research is the silicon microphysiometer, which detects receptor activators or inhibitors via changes they produce in the physiologic state of cultured living cells. These metabolic changes are monitored by the rate at which the cells excrete acidic products of metabolism (Owicki and Parce, 1991).

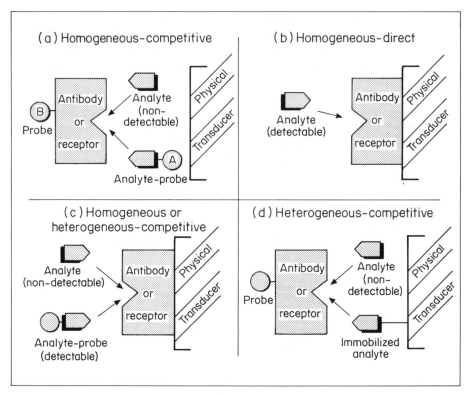

Figure 3 Antibody- or receptor-based biosensors typically operate using several general formats. These may be homologous (in which case the unbound analyte probe need not be physically separated from the antibody–analyte probe complex, usually in a washing step). (a) In this assay format, labeled and unlabeled analyte compete for a limited amount of soluble antibody or receptor. The binding of analyte probe to the antibody is detected as a proximity-induced change in optical properties of one or both probes. (b) In this homogeneous assay, the binding of analyte to the immobilized antibody or receptor is directly detected as a change in mass at the sensor surface or by a measurable optical property of the analyte. (c) For this assay, a limited amount of immobilized antibody or receptor competes for labeled and unlabeled analyte. If the physical transducer can differentiate between bound and free analyte probe, then the assay is homogeneous. (d) In this heterogeneous assay, antibody binds to either free or immobilized analyte. After a separation step, the amount of probe-labeled antibody or receptor bound to the surface is inversely proportional to the analyte concentration

Physical transducers

Biosensors are generally classified by their physical transducers which fall into four main classes: electrochemical, optical, mass and thermal. Any type of biological sensing element may be interfaced with any of these classes of transducers provided an appropriate reaction product or probe can be devised and measured. However, the recognition of the analyte by the biological sensing element must result in a change in a physical or chemical property of the system which is recognized and measured by the transducer.

Electrochemical transducers were incorporated into the first biosensors and constitute the largest area of current research and commercial activity. Of these transducers, the most common are potentiometric (measuring voltage), amperometric (measuring current) and conductimetric (measuring conductivity). An example of an amperometric sensor (Figure 4a) is an oxygen-consuming enzyme coupled to an oxygen measuring electrode. An example of a potentiometric sensor (Figure 4b), is an enzyme which produces or consumes protons coupled to a pH electrode. There are also integrated multibiosensors for simultaneous amperometric and potentiometric measurements (Kimura *et al.*, 1988).

Extensions of potentiometric measurement systems include the chemical-

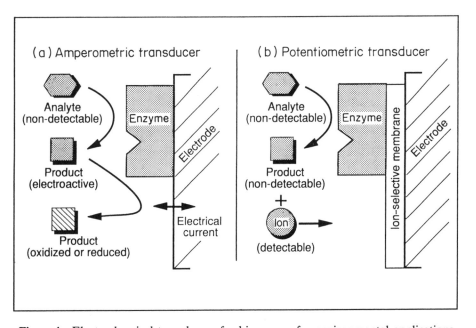

Figure 4 Electrochemical transducers for biosensors for environmental applications typically operate using two basic mechanisms. (a) For amperometric sensors, an electrically non-reactive analyte is converted by an enzyme (immobilized on the sensor surface) to an electrically reactive product, which is then oxidized or reduced at the electrode surface, resulting in a measurable flow of current that is proportional to analyte concentration. (b) For potentiometric sensors, a non-detectable analyte is converted, by an immobilized enzyme, to a non-detectable product and an ion (e.g., H^+, F^-, I^-, etc). The concentration of ion which is proportional to the analyte concentration is then detected as a change in electrical surface potential by movement of the ion into an ion-selective membrane at the electrode surface

selective and ion-selective field effect transistor (CHEMFET and ISFET, respectively). The former combines solid state integrated circuits with ion-selective electrodes (Blackburn, 1989). In the ISFETs, the gate of the semiconductor device is replaced by an ion-insulating layer which is covered with the selective reagent (Krull *et al.*, 1991). These devices can combine the selectivity of the sensing element with a miniature device of low noise and capability of on-chip signal amplification and processing.

Fiberoptic biosensors have typically been used to transmit light between the optical instrument and the biological sensing element housed in some type of a reaction chamber or immobilized on the distal end of the fiber. Association of an analyte with the bioselective layer triggers a concentration-dependent change in the optical property being measured (Figure 5a). The latter mode is ideal for remote monitoring whether in a hostile environment or inaccessible areas such as river and lake sediments, ground water or waste dumps. Another mechanism by which a fiberoptic transducer can be used to measure the binding of a fluorescent probe (in this case bound to the side of the fiber) relies on the evanescent wave effect (Figure 5b). When light is directed through the wave guide (i.e. the fiber), the associated electromagnetic field does not abruptly switch to zero at the interface between the fiber and the buffer, but decays exponentially away from the fiber. This evanescent zone is proportional to the wavelength of excitation light and has a limiting value of one wavelength. Thus the fluorescent dye bound to the sensing protein, that is immobilized on the surface of the fiber, is within the evanescent zone and when excited by the evanescent wave part of the emitted fluorescence enters the fiber and is transmitted to a detector. Although this configuration requires that the analyte flow through the instrument, we have observed that the speed and sensitivity characteristics for this format are typically much better than for the end-of-fiber format (Rogers *et al.*, 1992a). Nevertheless, in comparing the two excitation modes on a hard-clad silica fiber at different launching conditions, Weber and Schultz (1992) observed that the evanescent signal was similar to that of the end face configuration in spectroscopic efficiency and ultimate sensitivities (Weber and Schultz, 1992).

The mass-sensitive transducers such as the piezoelectric crystals respond primarily to bulk surface weight. Thus, the larger the size of the analyte relative to the immobilized protein, the more sensitive is the biosensor. The surface acoustic wave (SAW) devices appear to respond additionally to characteristics such as surface charge and viscosity (Clarke *et al.*, 1989). Thus, changes in these characteristics as a result of binding of the analyte to the protein-sensing element are detected. The piezoelectric crystal or quartz crystal microbalance (QCM) biosensor operates by means of a shearing action in which the frequency of displacement is proportional to the mass bound to the surface, much like a weight attached to the end of a spring

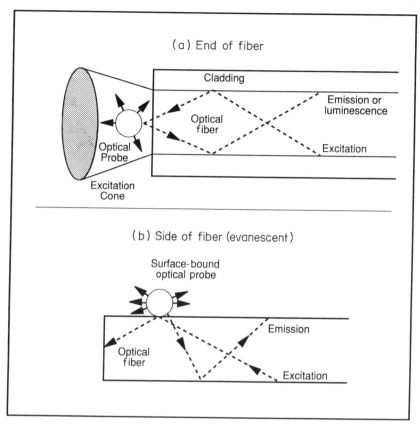

Figure 5 Fiberoptic transducers typically function by means of two mechanisms. (a) The optical fiber may act as 'light pipe' to transfer light between the spectrometer and the reaction medium. Light (excitation and emission frequencies) are then measured in a cone extending from the end of the fiber. A variety of catalytic or binding formats may then be used to measure an analyte using fluorescence, absorbance or luminescence properties of the analyte or analyte probe. (b) The optical fiber may also be used to measure an analyte or analyte probe which binds to the side of an unclad fiber by means of the evanescent wave. Thus, the light leaves the fiber just far enough (i.e. =1000 Å) to excite a fluorescent probe bound to a protein immobilized on the fiber surface. A portion of the emitted fluorescence is then trapped and propagated back up the fiber where it can be detected

(Figure 6a). In the case of the SAW sensor, the frequency of the electrically generated surface wave is also proportional to the mass bound to the crystal between the electrodes (see Figure 6b). Most of the reported work using these devices has focused on the use of chemically selective coatings. However, a new class of piezoelectric biosensors has used antibodies, antigens or enzymes to coat the crystals. Antibody-coated sensors were used

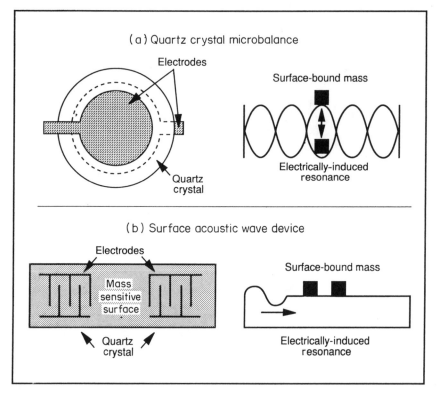

Figure 6 Mass-sensitive biosensors typically operate using one of two types of devices. (a) For the quartz microbalance (QCM), an applied voltage causes the crystal to deform. Upon removal of the potential, the crystal vibrates according to its natural frequency which will change as a result of mass propagation to the crystal surface. (b) For the surface acoustic wave (SAW) devices, the delay in propagation of the wave is sensitive to the surface bound mass

to measure formaldehyde (Guilbault *et al.*, 1991), parathion (Ngeh-Ngwainbi *et al.*, 1986) and atrazine (Guilbault *et al.*, 1992).

Another transducer, termed surface plasmon resonance (SPR) spectroscopy, enables measurements to be made of the refractive index of thin dielectric layers adsorbed onto metallic films such as gold, silver, or aluminum. By immobilizing on the metal surface one member of a binding pair (e.g. antibody), these sensors can be made highly specific (Morgan and Taylor, 1992). Their primary attractive feature is the fact that the method does not require the use of labeled analyte. However, because of the need to produce a significant surface refractive index change before a signal is detected, this transducer shows a low sensitivity for small molecules (Attridge *et al.*, 1991). Thermal detectors (e.g. thermistors and pyroelectric devices) use heat generated by a specific reaction, such as that by an enzyme, as a source of analytical information (Mosbach, 1991).

Uses of biosensors in pesticide detection

Several biosensors have been developed to detect OP and carbamate antiChE insecticides, using AChE or butyrylcholinesterase as the biological sensing element. In one biosensor, acetic acid which results from acetylcholine hydrolysis acidifies the medium, thereby changing the optical properties of a dye (fluorescein), that is detected using a fiberoptic system (Rogers *et al.*, 1991a, 1992a). In another biosensor, the production of protons results in a change in surface potential measured using a light addressable potentiometric sensor (LAPS) (Rogers *et al.*, 1991c; Fernando *et al.*, 1993). In another strategy, a primary hydrolysis product (choline) is acted on by a second enzyme (choline oxidase) which results in the production of hydrogen peroxide, that is then measured using an amperometric electrode (Bernabei *et al.*, 1991). Another amperometric biosensor was constructed as a disposable strip containing a cobalt pthalocyanine-modified composite electrode and a cross-linked cholinesterase layer (Skladal, 1992; Skladal and Mascini, 1992). This sensor was able to detect paraoxon, dichlorvos and carbaryl at 0.3–11 nM concentrations in a few seconds to 11 min. Butyrylcholinesterase has also been used as the biological sensing element in a potentiometric biosensor to detect OP and carbamate antiChEs (Kumaran and Tran-Minh, 1992). Characteristics of these biosensors, such as operating format, sensitivity and dynamic range, make them potentially useful for continuous and remote monitoring of ground water and agricultural run-off.

Immunosensors used in environmental detection utilize optical, surface mass and electrochemical transducers in a wide variety of signal transduction formats. Because of the small size of most environmental pollutants as compared to biological macromolecules (e.g. $\leqslant 1000$ Da vs $\geqslant 50\,000$ Da, respectively), signal transduction formats developed for many clinical analytes are not easily transferable to environmental applications. Initial reports of immunosensors for environmental applications have focused largely on parathion and atrazine and this has been most likely due to the availability of well characterized antibodies. Because of the widespread use of atrazine, it is of particular interest as a ground water contaminant, primarily in the midwest agricultural belt. Parathion, although no longer approved for extensive use, still poses an environmental hazard due to storage and disposal concerns as well as from its limited use. Fiberoptic immunosensors have been reported for triazines (Bier *et al.*, 1992) and parathion (Anis *et al.*, 1992). In both cases, the analyte was measured using a competitive immunoassay format. For these assay protocols, a pesticide derivative is first immobilized onto the side of an optic fiber, then primary antipesticide antibodies in the medium, that are not bound to the analyte of interest, will bind to the antigen-coated fiber. The antibodies that are bound to the fiber are then detected by a second fluorescent-tagged antibody which binds to the primary antibody. The amount of antipesticide antibodies, and consequently fluorescent-tagged secondary antibodies which bind to the optic fiber, is inversely proportional to the amount of analyte.

There is growing interest in continuous flow immunoassays such as one developed for 2,4-dinitrophenol that does not require incubation periods near introduction of additional reagents following sample loading into the system (Kusterbeck *et al.*, 1990). This method is highly specific and sensitive with detection limits for 2,4-dinitrophenol of 143 nM or 14 pmol 100 ml^{-1}. It is easy to perform with only little user input and no radiolabel is necessary. A flow injection immunoanalysis has also been developed based on a sequential/competitive enzyme immunoassay, where the hapten (atrazine and propazine of the triazine herbicides) competes with the corresponding enzyme-labeled hapten (i.e. modified atrazine conjugated to peroxidase) for a limited number of anti-atrazine antibodies (Kramer and Schmid, 1991).

Piezoelectric immunosensors have been reported for both atrazine (Guilbault *et al.*, 1991) and parathion. Since these devices can measure changes in surface mass of the order of picograms, binding of atrazine or parathion to antibodies (which are immobilized on the surface of the device) can be directly measured electronically as a change in the resonance frequency at which these devices operate. Because these devices are rugged, simple and relatively inexpensive to manufacture, they appear to be good candidates for commercialization.

BIOSENSOR CHARACTERISTICS

Protein immobilization

Ideally, immobilization of the receptor, enzyme or antibody protein on the surface of a polymer, metal, membrane or glass should not affect the stability, affinity or sensitivity of the protein. Many enzymes and immunoglobulins (IgGs) lose 30% or more of their activities by immobilization, possibly because of limitations on their conformational motility due to multipoint binding or stearic hindrance of the active site by the binding surface (Wu and Walters, 1988). Although covalent immobilization of the biological sensing element to the physical transducer is generally used, the primary thermodynamic driving force for physical adsorption of the protein to the metal or glass surfaces is via hydrophobic and surface charge interactions. This property has been exploited in non-covalent immobilization of proteins such as the nAChR (Rogers *et al.*, 1991b) or antibodies

(Anis *et al.*, 1992) to quartz fibers. Excess binding sites on the glass are blocked by subsequent incubation with proteins such as casein or serum albumin.

Not all proteins, however, are able to form stable layers. Some may denature and many nonspecific adsorption or exchange interactions may occur. Consequently, procedures have evolved for covalently binding proteins to transducer surfaces. These methods generally involve derivatization of the surface of the sensor with thiols or silanes followed by modification of the distal moiety of the adlayer to accept a sensing ligand by direct covalent linkage or via a secondary binding system such as by biotin–streptavidin. Streptavidin (a tetravalant protein obtained from *Streptomyces avidinii*) possesses four very high affinity biotin binding sites arranged in pairs on opposite faces of the molecule (Hendrickson *et al.*, 1989) and is frequently used as a linking molecule to biotinylated proteins. Immobilization of AChE by the biotin–streptavidin method (Figure 7) was used in the light addressable potentiometric sensor (LAPS) for detection of

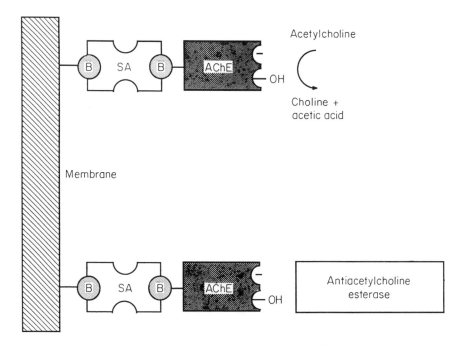

Figure 7   Schematic illustration of the chemistry of immobilization of AChE on the cellulose nitrate membrane of the LAPS and the reaction of AChE. Acetic acid, the product of hydrolysis of ACh, is detected by the sensor. AntiChEs inhibit formation of acetic acid. B, biotin; SA, streptavidin. (With permission from Fernando *et al.*, 1993)

antiChEs (Fernando *et al.*, 1993). The enzyme activity was unaffected by the immobilization.

For most enzymes and antibodies, protein function or affinities do not appear to be substantially altered by immobilization. However, after adsorption of the nAChR onto a quartz fiber, its affinities for agonists (e.g. ACh and succinylcholine [suxamethonium]) were reduced without changes occurring in its affinities for antagonists Rogers *et al.*, 1991b). This is an indication of a change in receptor conformation as a result of its immobilization. The problem can be overcome by using another strategy reported for a LAP sensor configuration, where the receptor and ligand bind in solution before the complex is immobilized to the membrane (Rogers *et al.*, 1992b). For this biosensor, α-bungarotoxin (α-BGT), the highly toxic polypeptide in the venom of the banded krait, *Bungarus* sp., which binds semi-irreversibly to the nAChR, was first biotinylated with DNP–biotin–NHS. The nAChR, which was labeled with carboxyfluorescein (CF), then bound to the biotinylated α-BGT in solution and the complex was immobilized to a biotinylated nitrocellulose membrane via a streptavidin bridge. The number of receptor complexes immobilized to the membrane were then probed using anti-CF IgG conjugated to urease (Figure 8). Urease hydrolyzed urea,

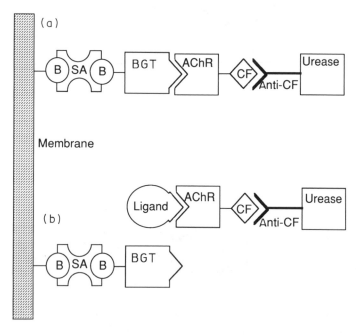

Figure 8 Immobilization and detection schematic for LAPS-nAChR biosensor. B, biotin; SA, streptavidin; B-BGT, biotinylated α-bungarotoxin; CF-AChR, carboxyfluorescein-labeled nAChR; anti-CF urease, anti-carboxyfluorescein–IgG–urease conjugate. (With permission from Rogers *et al.*, 1992a)

thereby releasing NH_3 that generated the potentiometric changes. Accordingly, the presence of a nicotinic ligand in a solution competed with the receptor's binding to the immobilized α-BGT, resulting in a decreased signal.

Immobilization of AChE by co-cross-linking, at the surface of a pH electrode, with albumin and glutaraldehyde in a membrane filter and trapping in a polyacrylamide gel, showed little effect on the enzyme activity and produced an effective AChE biosensor for detection of antiChEs (Stein and Schwedt, 1993).

Lifetime and stability of the biosensor

It is very important for the biosensor to have a long lifetime, whether 'shelf' or 'in-use'. One of the major problems restricting the expanded use of enzyme sensors is protein stabilization required for sensors that can withstand prolonged storage and perform well for long periods. Although lyophilized glucose oxidase is stable for at least 2 years at $0\,°C$ and 8 years at $-15\,°C$, its stability in solution is dependent on pH (Wilson and Turner, 1992). Enzyme and receptor conformational changes are necessary for their activities. Thus, it is important to maintain a delicate balance of movement in the protein to ensure activity but not enough to allow unfolding, escape or loss of activity. Enzymes have been stabilized by methods including: cross-linking their subunits, physical entrapment in a polymer or gel matrix, or immobilization via a spacer onto an inert body, or by freeze drying often in presence of stabilizers such as polyalcohols and various sugars (e.g. polyethylene glycols, trehalose and cellobiose, sucrose and xylitols). These stabilizers strengthen the structural relationship between the protein and its immediate environment, thereby preserving its three-dimensional structure. A combination of positively charged polymer, DEAE-dextran, and the non-reducing sugar alcohol lactitol has been shown to stabilize the activities of alcohol oxidase, peroxidase and L-glutamine oxidase, used in the fabrication of alcohol and L-glutamate biosensors. This treatment substantially increased the storage stability of both the enzyme activity and biosensor function. Dried alcohol oxidase lost only 15–20% of activity after one year at $37\,°C$ and alcohol oxidase/peroxidase electrodes using these stabilizers lost 33% of their activities by 35 days at $37\,°C$ (Gibson et al., 1992).

Most receptors attain their optimum conformation when bound to membranes. After purification they are not devoid of bound detergents, which are acquired during their solubilization. During purification the nAChR binds 0.113 mg of Triton X-100 per mg protein (Edelstein et al., 1975). The solubilized nAChR is stable for months during storage at $-80\,°C$, but deteriorates by over 50% in 7 days if stored at ambient temperature. Immobilization of this receptor, however, can increase its stability. For example, storage of the receptor immobilized onto the glass fiber in phosphate buffer at 4 °C results in <10% loss of activity after 3 days, 25% in 10 days and 55% in 30 days (Rogers et al., 1991d). In this case, it appears that immobilization of the receptor may have afforded some protection.

Speed of detection

A major advantage of biosensors over classical analytical methods is their simplicity and speed of detection. Whether the sensing element on the optical fiber is an enzyme such as AChE (Rogers et al., 1991a), an antibody (Anis et al., 1992) or a receptor (Rogers et al., 1991d), detection can be made in seconds to minutes if the initial rates of changes in fluorescence are measured. Also, in many cases, such as in water samples and sometimes urine, there is no need for sample preparation before the assay, which speeds the process and eliminates the use of organic solvents. For biosensors using the LAPS transducer, there is the added advantage that eight samples can be assayed simultaneously. Also, the use of basal activity of each compartment acts as its own control, thus giving precise measurements, and minimizes variabilities. The LAPS biosensor, however, is not yet adaptable for field use.

Sensitivity of detection

The sensitivity of the biosensor for a particular analyte depends on the affinity of the immobilized protein for the analyte, the amount of protein immobilized, and the sensitivity of the physical transducer. For a biosensor that utilizes a stoichiometric reaction to evoke a response from the transducer, a greater density (up to a limit) of the receptor or antibody immobilized to the physical transducer would increase detection limits for the analyte. This is not the case, however, for the irreversible inhibition of an enzyme by the stoichiometric binding of an inhibitor (i.e. AChE inhibited by an antiChE, such as paraoxon). Because inhibition of a percentage of the total number of immobilized enzyme molecules is the parameter being measured, in this case, time of exposure would be the most important determinant of sensitivity. For an enzyme-based biosensor which is irreversibly inhibited, increasing the incubation time should increase sensitivity.

Given differing incubation times and protein immobilization protocols, it is remarkable that four different biosensors using AChE for detection of antiChEs yielded similar sensitivities for the compounds measured. Examples include the following: an optic fiber biosensor, using pH-dependent change in fluorescent signal of fluorescein isothiocyanate-tagged AChE on quartz fibers, utilizing evanescent wave guide mode (Rogers et al., 1991a); a fiberoptic biosensor utilizing fluorescein-tagged dextran as a pH sensor together with AChE immobilized on a thin polyacrylamide gel layer (Hobel and Polster, 1992); a LAP biosensor using AChE immobilized on a cellulose nitrate membrane (Fernando *et al.*, 1993); and an electrode biosensor with co-immobilized AChE and choline oxidase (Marty *et al.*, 1992).

Selectivity of detection

The selectivity of a biosensor required for a specific application depends on whether one specific chemical or chemical class is to be determined. The ability of the biosensor to measure one chemical component in the presence of closely related chemicals is a characteristic that reflects its selectivity. For example, the AChE fiberoptic biosensor can detect paraoxon at <100 ppb while its inactive parent insecticide parathion was hardly detected at 1000 ppm (Rogers *et al.*, 1991a; Eldefrawi *et al.*, 1992). Conversely, an immunosensor for parathion using parathion antibodies could detect parathion at ppb concentrations, whereas paraoxon was detectable at 10–100-fold higher concentrations (Figure 9) (Anis *et al.*, 1992). Receptor-based biosensors are typically selective for groups of analytes that represent the receptor's agonists and antagonists. Although enzyme-based biosensors are highly selective for detection of the enzyme's substrate and related compounds, they can also be selective for groups of inhibitors such as antiChEs (Table 1). Antibody-based biosensors can also be selective for groups of compounds. Using an optic fiber biosensor with immobilized anti-imazethapyr antibodies as the sensing element, and fluorescein-labeled imazethapyr as the probe in a competition assay, imidazolinone compounds (imazapyr, imazaquin and imazmethabenz methyl) were detected, while primisulfuron, sethoxidim and chlorimuron ethyl had minimal effects (Figure 10) (Anis *et al.*, 1993; Wong *et al.*, 1993). Our preliminary results using an optic fiber immunosensor demonstrated a dose response with a dynamic range of 1–1000 ppb for aroclor 1248. Aroclors 1221, 1232, 1016, 1242, 1248, 1254, 1260, 1262, and 1268 show varying levels of cross-reactivity.

Another issue relevant to the application of biosensors to 'real world' scenarios is the effect of non-specific protein denaturants and sample matrices. In certain cases it is possible to distinguish between specific inhibitors such as an OP antiChE and non-specific inhibitors such as $HgCl_2$, both of which inhibit AChE. In this case, however, inhibition of the enzyme by the OP is reversible (using oximes) whereas that of $HgCl_2$ is not.

Many biosensors that depend on changes in fluorescence to detect the analyte can do so in the presence of colored contaminants or particulate matter in the sample. This eliminates the need for sample pretreatment. However, it must be emphasized that biosensors are limited by the sensitivity of their biological sensing elements to known as well as unknown

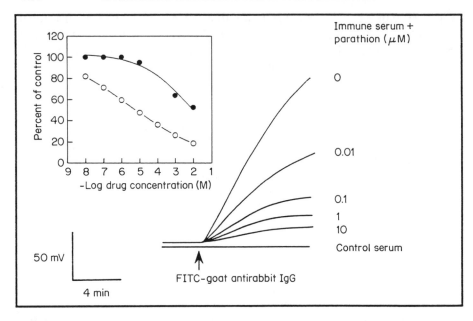

Figure 9 Inhibition by different parathion concentrations of the optical signal generated by binding of the complex of fluorescein isothiocyanate (FITC)-goat antirabbit IgG with rabbit antiparathion IgG (diluted 1/500) to the glass fiber-immobilized casein parathion. Control serum represents the signal generated by a fiber with immobilized casein–parathion but incubated with serum from a control rabbit rather than from the parathion immunized rabbit. *Inset:* The dose–effect of the presence of parathion (●) or paraoxon (○) in the medium, which competes for the fluorescent-labeled complex and prevents its binding, thereby quenching the fluorescence. The 100% control level is the rate recorded in absence of parathion or paraoxon. Symbols are means of triplicate measurements from three experiments with standard errors of <5%. (With permission from Marcel Dekker, Inc, New York, USA, Anis *et al.*, 1992)

contaminants present in samples. Nevertheless, in the case of the highly specific immunosensors, because of the speed of detection, matrix materials which may present a problem in a normal ELISA with its prolonged incubation time may cause minimal interference in the biosensor. Such an effect was found with the herbicide imazethapyr, which was extracted from different soils, and detected using an immunosensor (Anis *et al.*, 1993). In another example, phenolic biosensors could measure phenolics in the presence of high concentrations of hexane, chloroform, ethanol and acetonitrile (Hall *et al.*, 1988; Campanella *et al.*, 1992). Potential advantages of immunosensors over ELISA are speed of detection, reusability and the possible use of *in situ* applications.

Table 1. Comparative sensitivity of AChE to OP and carbamate antiChEs determined by LAPS, fiberoptic sensor and colorimetric assays

Compound	LAPS biosensor assay IC_{50} (M)	Fiberoptic biosensor assay IC_{50} (M)	Colorimetric assay IC_{50} (M)
Echothiophate	3.1×10^{-8}	3.8×10^{-8}	3.5×10^{-8}
Bendiocarb	9.0×10^{-8}	2.2×10^{-6}	1.7×10^{-7}
Methomyl	1.8×10^{-7}	9.0×10^{-6}	2.1×10^{-7}
Paraoxon	1.5×10^{-7}	3.7×10^{-7}	4.0×10^{-7}
Dicrotophos	1.9×10^{-4}	3.3×10^{-4}	1.1×10^{-4}
Dichlorvos	8.8×10^{-7}	–	8.1×10^{-7}
Monocrotophos	2.4×10^{-4}	–	3.1×10^{-5}
Phosdrin	2.9×10^{-7}	–	3.0×10^{-7}
Adlicarb	7.0×10^{-6}	–	1.2×10^{-6}
TEPP	1.3×10^{-7}	–	1.3×10^{-7}
Diazinon	8.0×10^{-4}	–	–

With permission from Fernando *et al.*, 1993.

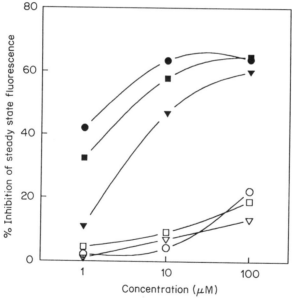

Figure 10 Concentration-dependent displacement of fluorescein hydrazino methylene imazapyr (FHMI) from the optic fiber with immobilized imazethapyr antibodies by imidazolinone (closed symbols) and non-imidazolinone (open symbols) compounds. Three concentrations of each compound were used for the displacement of the fluorescence signal from steady state bound levels. Each symbol is the mean of three measurements (sd <10%). Symbols: chlorimuron ethyl (○), primisulfuron (▽), sethoxydim (□), imazmethabenz methyl (▼), imazapyr (■), imazaquin (●). (With permission from Anis *et al.*, 1993)

Reusability

An important characteristic for a biosensor is reusability to allow for repeated or continuous monitoring. Receptor-, enzyme-, and antibody-based biosensors can be reversible. In some cases, however, the biosensor may be reactivated even if it has been irreversibly inhibited such as the case with AChE biosensor and OPs. Reactivation, in this case, was accomplished using the activator 2-pralidoxime (Figure 11).

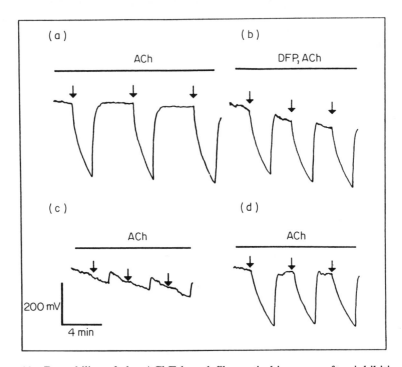

Figure 11 Reusability of the AChE-based fiberoptic biosensor after inhibition by diisopropylfluorophosphate (DFP). (a) The biosensor response to ACh (1 mM) by changing fluorescence of the immobilized fluorescein isothiocayanate-tagged AChE on the quartz fiber. The pH-dependent fluorescence was quenched by the protons resulting from acetylcholine hydrolysis. Arrows indicate when interruption in the flow of the perfusate occurs, thereby allowing accumulation of the protons. The flow is resumed after 2 min. The immobilized enzyme was then exposed to DFP for 10 min (0.1 mM) in the presence of ACh (1 mM), which competed for the catalytic site, and assayed for activity by interrupting the flow of the perfusate (b). The biosensor was subsequently exposed to 0.1 mM DFP alone for 10 min followed by addition of ACh and assay of activity, which was inhibited by its phosphorylation and the fluorescence was quenched. (c). The DFP-inhibited biosensor was then exposed to 2-pralidoxime (0.1 mM) for 10 min in the presence of ACh (1 mM) and assayed for reactivation (d). Note the speed of detection in seconds to minutes as shown on the time scale. (With permission from Academic Press, Florida, USA, from Rogers *et al.*, 1991a)

The very high affinity that some antibodies have for their specific antigens may preclude reversibility of the reaction. However, a strategy can be used to allow for the reuse of immunosensors as well. Since in order to make a toxicant (like the herbicide imazethapyr) antigenic, it is covalently linked to bovine serum albumin, a variety of antibodies are produced by the immunized rabbit. A few of these are directed against the imazethapyr epitope, most are directed against the large albumin protein and some are against both. These properties are reflected as different affinities of the antibodies for imazethapyr. The lower the affinity of the antibody for the analyte the more reversible is the reaction and thus the more reusable the biosensor. This was shown with the immunosensor for imazethapyr where the reaction between the herbicide and the antibodies was reversible by washing with buffer (Figure 12) (Anis *et al.*, 1993). Another strategy to reactivate immunosensors is to remove the bound antigen from the immobilized polyclonal antibody by lowering the pH with glycine-HCl buffer (pH 2.5) as was demonstrated with a piezoelectric immunosensor for atrazine (Guilbault *et al.*, 1992).

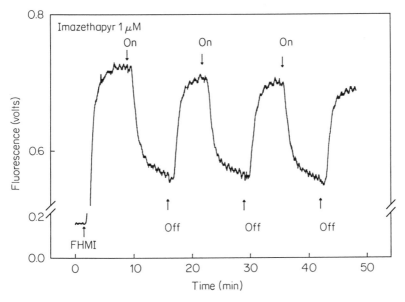

Figure 12 Reusability of the optic fiber immunosensor for imazethapyr, where imazethapyr antibodies were immobilized on the glass. A 25 mM fluorescein hydrazino methylene imazapyr (FHMI) in phosphate buffer casein was perfused to reach a steady state of binding (=5 min) then at the point indicated by 'on', 1 μM imazethapyr was introduced, which reduced the amount of FHMI bound and accordingly the fluorescence, the perfusion solution was switched back to 25 nM FHMI alone (indicated by 'off'). (With permission from Anis *et al.*, 1993)

CONCLUDING REMARKS

It is evident that there is a clear niche for biosensors in pesticide detection and that there are many advantages in their use, whether to complement or to replace traditional methods. Although biosensors have come a long way, these devices are still in their infancy when considering their full potential. There is no single ideal type of biosensor for all pesticides. Two major factors determine the choice of biosensor characteristics. The first is the nature of the chemical to be detected. High sensitivity is important for detection of minute amounts of very toxic compounds; however, lower sensitivity is acceptable for less toxic compounds, with higher environmental action levels. Although assay speed is an important requirement for a very toxic nerve gas, it is not as critical for many applications such as detection of pesticides. The second factor is the medium which is to be tested and the requirements of the analysis. Thus, for a single pesticide, the choice of biosensor depends on whether the analyte is to be detected in a river, well, soil or in human urine, whether the pesticide alone or its metabolites must be detected, and whether the need is for continuous monitoring such as for water quality (where a reversible sensor is needed) or for spot analysis and whether qualitative or quantitative data are required so as to meet regulatory requirements. All these factors will impact on the selection of the biological sensing element, the transducer, and the assay format.

Laboratory prototype biosensors have been demonstrated for detection of pesticide residues in foods, meat, water, soil, blood, and urine. Immunosensors appear to be the most suitable for detection of pesticides because of the well established database for immunoassay of these compounds. Further, under field conditions, portable, robust and reliable biosensors could be extremely useful for complimenting classical laboratory analysis. Low power requirements and absence of replaceable parts is recommended. The light-emitting-source colorimeter, which has been used for field measurement of plasma and erythrocyte AChE (Magnotti *et al.*, 1988) is a step in the right direction. Ideally, a field-operable biosensor should have a one-step assay protocol that requires no sample manipulation, reagent addition or separation step. The coated piezoelectric immunosensor can easily be automated or combined with flow injection systems, extending its capability for continuous and repeated assay (Suleiman and Guilbault, 1991).

In the last two years, numerous commercial immunoassays kits, with sensitivities of 1–100 ppb, have been developed for determination of pesticide residues in water. They include competition ELISA in tubes for atrazine, aldicarb, chlordane, carbofuran, 2,4-D; competition ELISA in microwells for alachlor, triazine and benomyl; immunoassay on card for paraquat. There are also colorimetric tests using cholinesterase inhibition on card to detect OP and carbamate antiChEs (see Kaufman and Clower,

1991). These kits are well suited for field use, although some are at best semi-quantitative or detect only the insecticide but not its metabolites. Guidelines need to be developed by the concerned government agencies (i.e. the US Environmental Protection Agency, Food and Drug Administration and Department of Agriculture) for evaluating the test kits and setting acceptable standards.

Antibodies collected from animals, immunized with a pesticide coupled to a large carrier protein, are variable with only a limited number of clones recognizing the pesticide. The kinds of antibodies in the immunized animals vary between individuals and even between different bleedings of the same animal. Thus, polyclonal antibodies may not be the most reliable source for antibodies to be used in biosensors. Monoclonal antibodies, which are produced from hybridoma cells in culture, cells produced by fusing a β lymphocyte with an immortal myeloma cell, are homogeneous and can be obtained in large amounts. The novel biotechnology for production of single chain variable fragments of the antibodies can provide larger amounts of a wider variety of monoclonal antibodies in much shorter time than by any other means (see Ward, 1992). The basis of this antibody engineering technology is the construction of an antibody gene fragment library (consisting of heavy and light chain fragments) from mRNA extracted from the spleen of a pesticide-immunized mouse. The gene fragments, connected by DNA linkers, are then separately expressed in *E. coli*, and several clones are isolated and expanded to yield gram quantities of monoclonal antibodies with the desired degrees of affinities for the analytes.

There is definitely a need for broad-spectrum devices capable of detecting chemical changes or toxicity, as may occur in major pollution incidents, where the exact nature of the threat is not known. A most interesting development is the multi-analyte microspot immunoassay, that confines a small number of antibody molecules onto a very small area in the form of a microspot of $\approx 100~\mu m^2$. Their occupancy by an analyte can then be measured, using very highly specific antibody probes (Ekins and Chu, 1991). This maximizes the signal-to-noise ratio in determining antibody occupancy and makes theoretically possible an array of hundreds of thousands of such microspots with different analytes on a very small chip using laser scanning technique, analogous to that used in compact disc technology. This would allow simultaneous detection of thousands of analytes in a single sample, thereby providing unparalleled scanning capabilities.

The potential success of biosensors is significantly increased in situations where no other analytical techniques exist. The use of biosensors is greatly needed in space stations and shuttles for biomedical experimentation and monitoring toxicants in water and air recycling processes. Biosensors save on space, power and time, in addition to facilitating digitized data transfer (Bonting, 1992). There are also biosensors that measure picograms of DNA, by utilizing as the biological sensing element high affinity DNA-binding proteins (Kung *et al.*, 1990). As for situations where there are competing analytical methods, an important determining factor for success is cost-effectiveness. This is enhanced by reduction in production cost, use of cheap disposable kits, the development of regenerable biosensors, savings in time, elimination of the need for sample preparation and procedure simplicity, which would not require highly trained personnel. A large enough market is also necessary to attract industry to scale up production and reduce cost.

ACKNOWLEDGEMENTS

The understandings of this research were funded by US Army contract No. DAAA15-89-C-0007 to M. Eldefrawi. Some of the information in this chapter was funded by the US Environmental Protection Agency under cooperative agreement No. CR820460-0, with the University of Maryland at Baltimore. This document has been subject to EPA's Agency review and approved for publication.

REFERENCES

Anis, N., Eldefrawi, M. E., and Wong, R. B. A. (1993). 'A reusable fiber optic immunosensor for rapid detection of imazethapyr herbicide', *J. Agric. Food Chem.*, **41**, 843–848.

Anis, N. A., Wright, J., Rogers, K. R., Thompson, R. G., Valdes, J. J., and Eldefrawi, M. E. (1992). 'A fiber-optic immunosensor for detecting parathion', *Anal. Lett.*, **25**, 627–635.

Attridge, J. W., Daniels, P. B., Deacon, J. K., Robinson, G. A., and Davidson, G. P. (1991). 'Sensitivity enhancement of optical immunosensors by use of a surface plasmon resonance fluoroimmunoassay', *Biosensors & Bioelectronics*, **6**, 201–214.

Barker, T. Q., Buch, R. M., and Rechnitz, G. A. (1990). 'Intact chemoreceptor-based biosensors', *Biotechnol. Prog.*, **6**, 498–503.

Bernabei, M., Cremiscini, M., and Palleschi, G. (1991). Determination of organo-phosphorus and carbamate pesticides with a choline and acetylcholine electro-chemical biosensor'. *Anal. Lett.*, **24**, 1317–1331.

Bier, F. F., Stocklein, W., Bocher, M., Bilitewski, V., and Schmid, R. D. (1992). 'Use of fiber optic immunosensor for the detection of pesticides', *Sensors & Actuators B*, **7**, 509–512.

Blackburn, G. F. (1989). 'Chemically sensitive field-effect transistors', in *Biosensors: Fundamentals and Application* (eds A. P. F. Turner, I. Karube and G. S. Wilson), pp. 481–530, Oxford University Press, New York.

Bonting, S. L. (1992). 'Utilization of biosensors and chemical sensors for space applications', *Biosensors & Bioelectronics*, **7**, 535–548.

Campanella, L., Sammartino, M. P., and Tomasetti, M. (1992). 'New enzyme sensor for phenol determination in non-aqueous and aqueous medium', *Sensors & Actuators B*, **7**, 383–388.

Clarke, D. J., Blake-Coleman, B. C., and Calder, M. R. (1989). 'Principles and potential of piezo-electric transducers and acoustical techniques', in *Biosensors: Fundamentals and Application* (eds A. P. F. Turner, I. Karube and G. S. Wilson), pp. 551–571, Oxford University Press, New York.

Edelstein, S. J., Beyer, W. B., Eldefrawi, A. T., and Eldefrawi, M. E. (1975). 'Molecular weight of the acetylcholine receptors of electric organs and the effect of Triton X-100', *J. Biol. Chem.*, **250**, 6101–6106.

Ekins, R. P., and Chu, F. W. (1991). 'Multianalyte microspot immuno-assay-microanalytical "compact disk" of the future', *Clin. Chem.*, **37/11**, 1955–1967.

Eldefrawi, M. E., Rogers, K., and Eldefrawi, A. (1992). 'Optical biosensor for detection of anticholinesterases', in *Pesticides and the Future: Toxicological Studies of Risk and Benefits*, (eds E. Hodgson, N. Motoyuma and R. M. Roe), pp. 329–334, N. Carolina State Univ. Publ., Raleigh, NC.

Fernando, J. C., Rogers, K. R., Anis, N. A., Valdes, J. J., Thompson, R. G., Eldefrawi, A. T., and Eldefrawi, M. E. (1993). 'Rapid detection of anticholine-sterase insecticides by a reusable light addressable potentiometric biosensor', *J. Agric. Food Chem.*, **41**, 511–516.

Frost & Sullivan Inc., 'Advanced Sensors' report (#T057), New York, New York.

Gaisford, W. C., Richardson, N. J., Haggett, B. G. D., and Rawson, D. M. (1991). 'Microbial biosensors for environmental monitoring', *Biochem. Soc. Trans.*, **19**, 15–18.

Gibson, T. D., Hulbert, J. N., Parker, S. M., Woodward, J. R., and Rawson, D. M. (1992). 'Extended shelf life of enzyme-based biosensors using a novel stabilization system', *Biosensors & Bioelectronics*, **7**, 701–708.

Guilbault, G. G., Hock, B., and Schmid, R. (1992). 'A piezoelectric immunobiosensor for atrazine in drinking water', *Biosensors & Bioelectronics*, **7**, 411–419.

Guilbault, G. G., Kauffman, J. M., and Patriarche, G. J. (1991). 'Immobilized enzyme electrodes as biosensors', *Bioproc. Technol.*, **14**, 209–262.

Hall, G. F., Best, D. J., and Turner, A. F. P. (1988). 'Amperometric enzyme electrode for the determination of phenols in chloroform', *Enzyme Microb. Technol.*, **10**, 543–546.

Hendrickson, W. A., Pahler, A., Smith, J. L., Satow, Y., Merritt, E. A., and Phizackerley, R. P. (1989). 'Crystal structure of core streptavidin determined from multiwavelength anomalous diffraction of synchrotron radiation', *Proc. Natl Acad. Sci. USA*, **86**, 2190–2194.

Hobel, W., and Polster, J. (1992). 'Fiber optic biosensor based on acetylcholine esterase', *Fresenius J. Anal. Chem.*, **343**, 101–102.

Kaufman, B. M., and Clower, M. Jr (1991). 'Immunoassay of pesticides', *J. Assoc. Off. Anal. Chem.*, **74**, 239–247.

Kimura, J., Murakami, T., and Kuriyama, T. (1988). 'An integrated multibiosensor for simultaneous amperometric and potentiometric measurement', *Sensors & Actuators*, **15**, 435–443.

Kramer, P., and Schmid, R. (1991). 'Flow injection immunoanalysis (FIIA)—a new immunoassay format for the determination of pesticides in water', *Biosensors & Bioelectronics*, **6**, 239–243.

Krull, U. J., Brown, R. S., Vandenberg, E. T., and Heckl, W. M. (1991). 'Determination of the physical structure of biological materials at biosensor interfaces by techniques of measuring magnification from microscopic to molecular scale', *J. Electron Microsc. Tech.*, **18**, 212–222.

Kumaran, S., and Tran-Minh, C. (1992). 'Insecticide determination with enzyme electrodes using different enzyme immobilization techniques', *Electroanalysis*, **4**, 949–954.

Kung, V. T., Panfilli, P. R., Sheldon, E. L., King, R. S., Nagainis, P. A., Gomez, B. Jr, Ross, D. A., Briggs, J., and Zuk, R. F. (1990). Picogram quantitation of total DNA-binding proteins in a silicon sensor-based system', *Anal. Biochem.*, **187**, 220–227.

Kusterbeck, A. W., Wemhoff, G. A., Charles, P. T., Yeager, D. A., Bredehorst, R., Vogel, C.-W., and Ligler, F. S. (1990). 'A continuous flow immunoassay for rapid and sensitive detection of small molecules', *J. Immunol. Methods*, **135**, 191–197.

Lee, S., Sode, K., Nakanishi, K., Marty, J.-L., Tamiyama, E., and Karube, I. (1992). 'A novel microbial sensor using luminous bacteria', *Biosensors & Bioelectronics*, **7**, 273–277.

Magnotti, R. A. Jr, Dowling, K., Eberly, J. P., and McConnell, R. S. (1988). 'Field measurement of plasma and erythrocyte cholinesterases', *Clin. Chim. Acta*, **315**, 315–332.

Marty, J. L., Sode, K., and Karube, I. (1992). 'Biosensor for detection of organophosphate and carbamate insecticides', *Electroanalysis*, **4**, 249–252.

Morgan, H., and Taylor, D. M. (1992). 'A surface plasmon resonance immunosensor based on the streptavidin-biotin complex', *Biosensors & Bioelectronics*, **7**, 405–410.

Mosbach, K. (1991). 'Thermal biosensors', *Biosensors & Bioelectronics*, **6**, 179–182.

Ngeh-Ngwainbi, J., Foley, P. H., Kuan, S. S., and Guilbault (1986). 'Parathion antibodies on piezoelectric crystals', *J. Am. Chem. Soc.*, **108**, 5444–5447.

Owicki, J. C., and Parce, J. W. (1991). 'Biosensors based on the energy metabolism of living cells: the physical chemistry and cell biology of extracellular acidification', *Biosensors & Bioelectronics*, **6**, 1–22.

Rogers, K. R., Anis, N. A., Valdes, J. J., and Eldefrawi, M. E. (1992a). 'Fiber-optic biosensors based on total internal reflection fluorescence', in *Biosensor Design and Application* (eds P. R. Mathewson and J. W. Finley), Am. Chem. Soc. Symp. Ser. 511, pp. 165–172, American Chemical Society, Washington, DC.

Rogers, K. R., Cao, C. J., Valdes, J. J., Eldefrawi, A. T., and Eldefrawi, M. E. (1991a). 'Acetylcholinesterase fiber-optic biosensor for the detection of anticholinesterases', *Fund. Appl. Toxicol.*, **16**, 810–820.

Rogers, K. R., Eldefrawi, M. E., Menking, D. E., Thompson, R. G., and Valdes, J. J. (1991b). 'Pharmacological specificity of a nicotinic acetylcholine receptor optical sensor', *Biosensors & Bioelectronics*, **6**, 507–516.

Rogers, K. R., Fernando, J. C., Thompson, R. G., Valdes, J. J., and Eldefrawi, M. E. (1992b). 'Detection of nicotinic receptor ligands with a light addressable potentiometric sensor', *Anal. Biochem.*, **202**, 111–116.

Rogers, K. R., Foley, M., Alter, S., Koga, P., and Eldefrawi, M. E. (1991c). 'Light addressable potentiometric biosensor for the detection of anticholinesterases', *Anal. Lett.*, **25**, 627–635.

Rogers, K. R., Valdes, J. J., and Eldefrawi, M. E. (1989). 'Acetylcholine receptor fiber optic evanescent fluorosensor', *Anal. Biochem.*, **182**, 353–359.

Rogers, K. R., Valdes, J. J., and Eldefrawi, M. E. (1991d). 'Effects of receptor concentration, media pH and storage on nicotinic receptor-transmitted signal in a fiber-optic biosensor', *Biosensors & Bioelectronics*, **6**, 1–8.

Saini, S., and Turner, A. P. F. (1991). 'Biosensors in organic phases', *Biochem. Soc. Soc. Trans.*, **19**, 28–31.

Schultz, J. S. (1991). 'Biosensors', *Scient. Amer.*, **265**, 64–69.

Skladal, P. (1992). 'Detection of organophosphate and carbamate pesticides using disposable biosensors based on chemically modified electrodes and immobilized cholinesterase', *Anal. Chim. Acta*, **269**, 281–287.

Skladal, P., and Mascini, M. (1992). 'Sensitive detection of pesticides using amperometric sensors based on cobalt phthalocyanamine-modified composite electrodes and immobilized electrodes and cholinesterases'. *Biosensors & Bioelectronics*, **7**, 335–343.

Stein, K., and Schwedt, G. (1993). 'Comparison of immobilization methods for the development of an acetylcholinesterase biosensor', *Anal. Chim. Acta*, **272**, 73–81.

Suleiman, A. A., and Guilbault, G. G. (1991). 'Piezoelectric (PZ) immunosensors and their applications'. *Anal. Lett.*, **24**, 1283–1292.

Urban, G., Jobst, G., Keplinger, F., Aschauer, E., Tilado, O., Fasching, R., and Kohl, F. (1992). 'Miniaturized multi-enzyme biosensors integrated with pH sensors on flexible polymer carriers for in vivo application', *Biosensors & Bioelectronics*, **7**, 733–739.

Ward, E. S. (1992). 'Antibody engineering: the use of *Escherichia coli* as an expression system', *The FASEB J.*, **6**, 2422–2427.

Weber, A., and Schultz, J. S. (1992). 'Fiber-optic fluorimetry in biosensors: comparison between evanescent wave operation and distal-face generation of fluorescent light', *Biosensors & Bioelectronics*, **7**, 193–197.

Wilson, R., and Turner, A. P. F. (1992). 'Glucose oxidase: an ideal enzyme', *Biosensors & Bioelectronics*, **7**, 165–185.

Wong, R. B., Anis, N., and Eldefrawi, M. E. (1993). 'Reusable fiber-optic-based immunosensor for rapid detection of imazethapyr herbicide', *Anal. Chem. Acta*, **279**, 141–147.

Wu, D., and Walters, R. R. (1988). 'Protein immobilization on silica supports: a ligand density study', *J. Chromatogr.*, **458**, 169–174.

CHAPTER 8

Techniques and procedures for the determination of pesticides in water

C. V. Eadsforth and A. P. Woodbridge

Environmental Behaviour of Agrochemicals
Edited by T. R. Roberts and P. C. Kearney   © 1995 John Wiley & Sons Ltd

The need to evaluate the risk to the environment from the use of chemicals has been a significant part of the regulations to control pesticides for many years in the UK and elsewhere. There has been an increased awareness and concern from the public and regulatory authorities regarding the potential for pesticides to contaminate air, soil and water sources. This pressure has resulted in the evaluation of different analytical methods and detection techniques in an effort to lower detection limits and improve confirmation procedures for pesticides in water.

THE NEED FOR ANALYTICAL METHODS

Companies involved in developing and marketing crop protection products to control weeds, pests or diseases or to improve the yields or quality of crops need to provide extensive packages of environmental fate and toxicology data to comply with the registration requirements. This includes assessments using laboratory studies and a range of outdoor studies ranging from small plot field experiments through to commercial scale applications (Figure 1). In order to quantify the fate and effects of such products and their metabolites, extensive residue determinations of each product in water (as well as in soil, crops and other matrices) are required. To do this appropriate analytical methods will need to have been developed and validated in advance of setting up the laboratory and field studies. Once the compound is marketed commercially, a variety of other bodies including industry associations, academia, national laboratories or institutes, as well as

Figure 1   Environmental studies needing analytical methods during the development of a new crop protection product

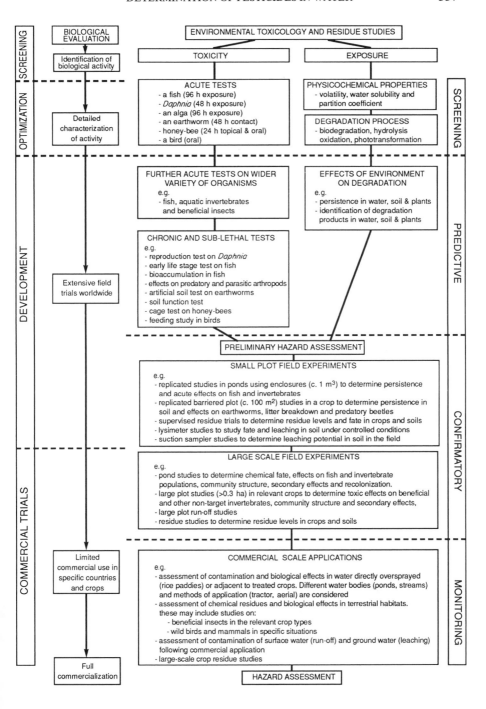

water companies/associations and river authorities, whose role it is to monitor for pesticides in effluents, surface and ground waters in their catchments, will require suitable methods for analysing for pesticides and their metabolites.

SCOPE OF METHODS

Generally there are two types of method for water that may be available, namely, so-called 'single-compound' methods and 'multiresidue' methods.

Single compound methods

Single compound methods are developed by agrochemical companies for individual research and development compounds which are subsequently to be marketed commercially.

At an early stage in the development of a pesticide a method for determining the parent compound in water with a limit of determination suitable for regulatory requirements (e.g. $0.1 \, \mu g \, l^{-1}$ or better for the EEC) is developed by the manufacturer. This method may initially be used and adapted for supporting physicochemical determinations (e.g. water solubility, rate of hydrolysis and partition coefficient) and early acute and chronic ecotoxicity tests (e.g. analysis of dosing levels for daphnia, fish and algae). Soil metabolism studies will identify the major soil degradation products and subsequently the method will be extended or further analytical methods will be developed for determining these compounds in water. Parent and metabolite methods may then be used to determine the fate of these products in subsequent laboratory and field studies (Table 1). Generally, methods are developed which should be applicable to a range of different types of water samples which arise from the laboratory and field studies. These range from relatively clean samples of ground or natural waters, rain water, and even pure laboratory water where low limits of determination are required, to water from other sources which could potentially be contaminated with natural organic compounds that might interfere in the analysis, e.g. surface water following run-off, pore water and piezometer samples, pond waters and samples from laboratory ecotoxicity studies. If the same low limit of determination is also required for these latter samples then invariably a greater degree of clean-up will be required in the analysis.

These single compound methods are normally available from the pesticide manufacturers or are published by them. They are generated with sufficient performance data, including recovery data to the required limit of determination, for submission along with other analytical methods (for crops, soil,

Table 1 Laboratory and field studies which require water analysis

---

Physicochemical parameters
  Water solubility
  Partition coefficient
  Hydrolysis/photolysis studies

Environmental fate
  Fish bioaccumulation
  Mobility in soil (via pore water and leachate analyses)
  Run-off
  Spray drift
  Ground water monitoring
  Water treatment

Environmental effects
  Aquatic ecotoxicity (laboratory)
  Mesocosm (field)

---

air, tissues, etc.) to regulatory authorities responsible for pesticide registration.

Multiresidue methods

Methods are also developed which are intended mainly for the simultaneous determination of residues of several compounds, often as a class (e.g. triazines, organochlorines or organophosphates) in one and the same procedure. These multiresidue methods are necessary for use in monitoring programmes as it would be unrealistically time consuming and costly to use a separate method for each pesticide and would limit the amount of monitoring possible. They tend to be developed for commercially marketed products by those bodies involved with monitoring programmes and also by interested governmental groups (e.g. US EPA, UK DoE and German DFG) which include scientists involved in monitoring programmes, often with the participation of the pesticide manufacturers. Such groups draw on available procedures to develop methods for use nationally as discussed later in this chapter.

The EC Drinking Water Directive states that the maximum admissible concentration for an individual pesticide in drinking water is $0.1 \mu g l^{-1}$. Consequently, analytical methods for use in the UK and Europe are required to have high sensitivity and should have a detection limit lower than this if possible, of around $0.01–0.02 \mu g l^{-1}$, in order to monitor compliance. Elsewhere higher limits are adopted in some cases, as indicated in this chapter in the section entitled 'National residue methods'.

As a large number of pesticides may need to be included in multiresidue methods, they may not be tested as thoroughly as the single compound methods until they are used for monitoring purposes. A number of monitoring studies where multiresidue methods have been used are listed at the end of this chapter.

Previously, monitoring studies concentrated on the identification of parent compounds, though more recently there has been interest in identifying, quantifying and monitoring the fate of the major degradation products of pesticides in the environment. Multicompound methods have been developed which include the determination of the parent compound and selected degradation products in a common HPLC or GC method. Steinheimer (1993) developed a method based on bonded solid phase (cyclohexyl) extraction followed by gradient HPLC with detection by multiple wavelength monitoring (diode array) for atrazine and its principal degradation products (Figure 2) to assist with field research of water quality in the Midwest of the USA. A method for N-methyl carbamates in river and saline waters based on HPLC with post-column fluorometric detection or thermospray HPLC-MS which includes aldicarb and its sulphoxide and sulphone metabolites has recently been published by the Standing Committee of Analysts (SCA) in the UK (SCA, 1994b).

Other pesticides such as cyanazine, are readily degraded in soil into a large number of products, many of which (e.g. chloro- and hydroxy-acids, **3**, **7**, **4** and **8**; Figure 3) are considerably more polar than the parent compound. In order to monitor successfully for these compounds in water as well as the parent cyanazine (**1**) and its amide metabolite (**2**), a range of different analytical approaches has proved effective:

(1) cyanazine: extraction from water with dichloromethane, cartridge clean-up (NH$_2$) followed by GC-MS.
(2) cyanazine amide: extraction from water/ammonia with ethyl acetate, cartridge clean-up (NH$_2$) followed by GC-MS.
(3) chloro-acids: extraction from acidified water with ethyl acetate, cartridge clean-up (hydrophobic SAX) followed by HPLC-UV.
(4) hydroxy-acids: extraction from water using a solid phase extraction cartridge (SCX), followed by HPLC-UV.

SAMPLING

Sampling of waters containing pesticide residues

The purpose of water sampling is to select appropriate locations and times for sampling to provide the required information on water quality, then to acquire a representative sample of the water at each location and time of sampling, and finally to ensure that the concentration of the determinands in

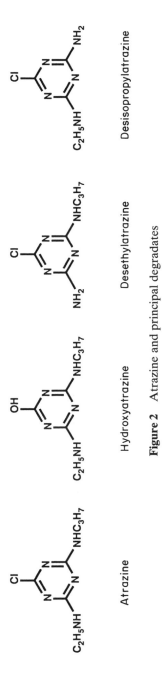

Figure 2 Atrazine and principal degradates

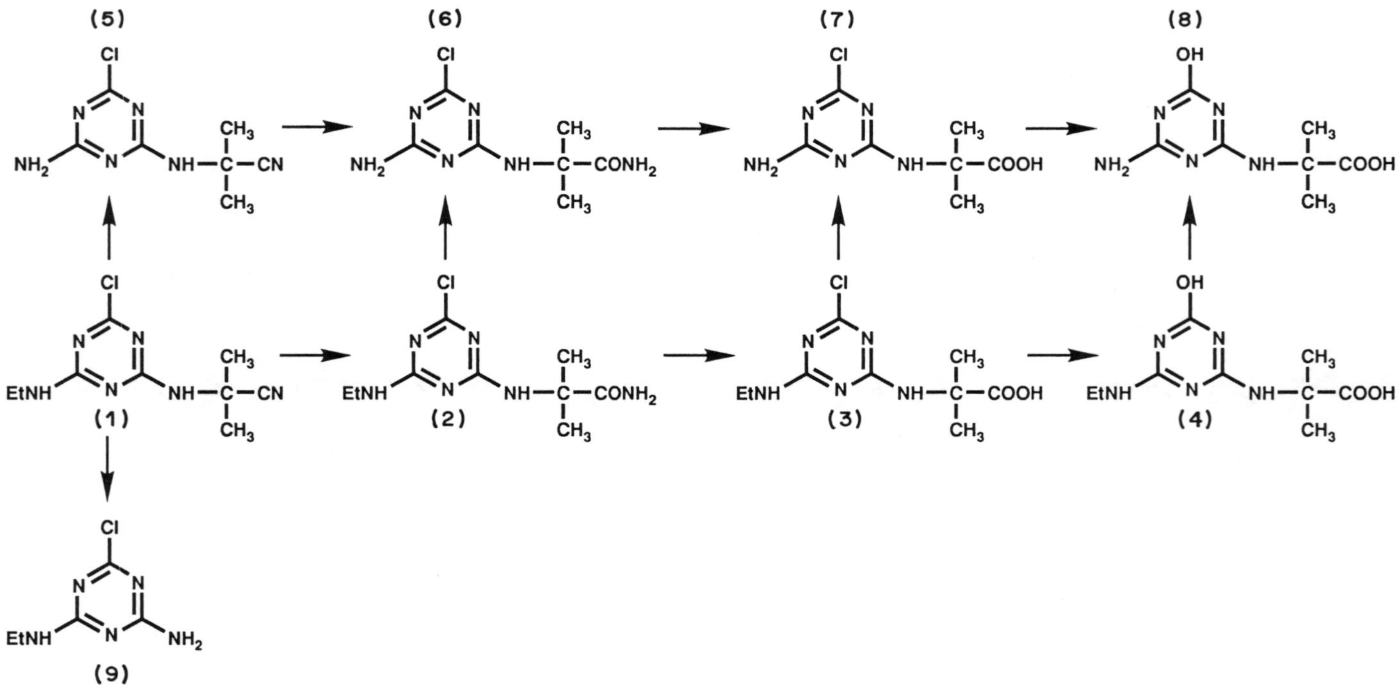

Figure 3 The breakdown of the herbicide cyanazine (1) in soils (Beynon *et al.*, 1972)

the samples does not change during the period between sampling and analysis. Generally samples of water that require analysis for trace concentrations of pesticides are normally transported to a specialist laboratory, though *in situ* and on-site extraction/analysis is now becoming more feasible (Buchmiller, 1989; Goerlitz and Franks, 1989).

More detailed information on sampling of particular types of water have been published:

• *Surface water*	Mohnke *et al*. (1986); DFG (1987); Kratochvil and Peak (1989); GIFAP (1990); Clark *et al*. (1991)
Integrated depth	Eadsforth *et al*. (1991)
surface film	Eadsforth *et al*. (1991)
run-off	Wauchope (1978); Harris *et al*. (1991); Waite *et al*. (1992)
• *Rain/fog*	Glotfelty *et al*. (1987); Clark *et al*. (1991); Nations and Hallberg (1992)
• *Ground water*	Pettyjohn *et al*. (1981); Kratochvil and Peak (1989); GIFAP (1990); Simmons (1991)
pore water	Clark *et al*. (1990); Williamson and Carter (1991); Banton *et al*. (1992)
piezometers	Clark (1988)
• *Water supplies*	DoE (1990)
• *Sea water*	Kratochvil and Peak (1989)

Accurate determination of pesticides in water samples at trace concentrations requires considerable effort and cost and it is vital that proper attention is paid to ensure a representative sample is taken and that losses of compounds (via volatilization, degradation or adsorption to the container) or the introduction of interfering compounds (e.g. from poor selection of container or other sampling materials) are prevented. The analyst should be aware of these factors and be intimately involved in the selection and generation of any sampling protocols.

Avoiding contamination and loss of pesticides during sampling

The importance, when sampling water for trace organic compounds, of considering the materials from which pumps, samplers, discharge lines, storage containers and well casings are constructed has been highlighted (Pettyjohn *et al*., 1981). Guidelines for the use of PTFE, PVC and stainless steel in samplers have been suggested (Parker, 1992).

The choice of the right container material for the collection and storage of water samples containing pesticides is important. Glass containers fitted with ground glass stoppers or PTFE-lined screw caps are best suited to the collection of water samples, though containers made of stainless steel or aluminium have been used.

Water samples should not normally be taken in plastic containers or come into contact with plastic tubing during sampling, since these materials can not only absorb pesticides, but can also release other organic compounds (e.g. plasticizers) that may interfere with the subsequent analysis.

House and Ou (1992) have investigated the affinity of pesticides to glass, Teflon and various filtration membranes at low ($< 0.25\ \mu g\,l^{-1}$) concentrations. Adsorption of pesticides onto containers is important for compounds with high octanol/water partition coefficients (e.g. organochlorine pesticides and pyrethroids). Consequently, triazines are not adsorbed to glass, although Topp and Smith (1992) have found that atrazine and metolachlor adsorb tightly to silicon rubber and selected polymers. Smith (1993) found small but quantifiable losses of pesticides (metribuzin, chlorpyrifos and metolachlor) to a programmable water/suspended sediment sampler. Under certain conditions when the tubing aged and became more adsorbent, losses could be as high as 40–50%. Other contaminants from nylon cord (as part of the bailer system) and latex gloves have been observed in ground water samples taken from a Superfund programme (Canova and Muthig, 1991).

Checks to ensure that none of the equipment or procedures introduces contamination which may interfere with the determination of pesticide residues and give rise to invalid false positive results have been outlined (Eadsforth et al., 1991). To ensure that such contamination is avoided, all water sampling equipment used should be made of inert materials (glass and PTFE) and be thoroughly washed before use, rinsed with solvent (e.g. acetone) and, if necessary, the solvent wash analysed to the required limit of determination.

As part of the sampling protocol used by the authors for an environmental monitoring study, some 15% of the sample bottles taken to the field are filled with laboratory distilled water checked to be interference-free by analysis. About one-third of these are never opened and act as trip blanks. Water in each of the remainder is transferred to empty sample bottles at selected field sampling locations to simulate field sampling. These samples act as field blanks. These blank samples are transported, stored and analysed side by side with field water samples.

APPROACH TO DEVELOPING METHODS

The objective is to develop methods which enable the efficient extraction, clean-up and determination of pesticide residues in water samples.

In addition to sampling, all residue method development studies including those for water must contain elements of some or all of the stages in the sequence shown in Figure 4. It is important that each of these steps be studied for its efficiency and relevance to a particular method and to use the appropriate control samples (or a range of control waters/samples) since extractability and clean-up can be influenced by components of the sample matrix (e.g. dissolved salts, pH, dissolved organics) (Huber *et al.*, 1992).

Sample preservation and storage

Once a sample of water has been taken and appropriately labelled, the whole sample should preferably be transported to the laboratory for analysis to maintain its integrity. Considerable care needs to be taken with the transport and storage of water samples, especially when the identity of the pesticide(s) is unknown, and any possibility of adventitious contact with any pesticide must be avoided. It is preferable that samples be extracted as soon as possible after collection to prevent loss of pesticides by hydrolysis, microbiological degradation or other changes. Sample preservation involves the physical and/or chemical treatment of samples to minimize any such changes in the concentrations of one or more of the determinands in the sample in the period between sampling and analysis. In the case of pesticides, this invariably involves cooling the samples, normally to 4 °C, or more rarely freezing. Storage in metal containers at −25 °C has been proposed (DFG, 1987) to prevent hydrolysis, but not generally adopted. The US EPA have recommended the addition of mercuric chloride as well as cooling to inhibit microbiological activity (US EPA, 1988a) but a

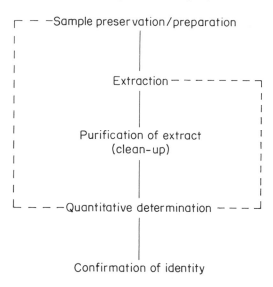

Figure 4 Key steps in a residue analytical method

disadvantage of this procedure is the toxic nature of mercury. Less stringent conditions are more appropriate if the storage stability data or the identity of the pesticide(s) is known prior to sampling. It is preferable not to add preservatives to samples unless it is necessary and then, only if it has been established that they will not interfere with the determination of the analytes.

Reference materials

Unlabelled compounds

These should be of suitable known purity to be used as certified analytical references. It is recommended that standards should be greater than 95% (m/m) pure if possible.

Radiolabelled compounds

These should be shown to be radiochemically pure by a combination of radio-TLC, GC and HPLC and be of known specific activity with, if possible, a minimum value of $10 \, nCi \, \mu g^{-1}$ for [^{14}C]. If mixtures of isomers are present, the initial isomer ratio should be determined.

Solutions

Dilute solutions ($\leqslant 1 \, \mu g \, ml^{-1}$) are prepared by weighing, followed by serial dilution from a stock solution (normally $\geqslant 100 \, \mu g \, ml^{-1}$). The stability of reference solutions must be checked under normal use conditions to ensure that the concentration of dissolved reference material is as expected and does not decline with time due to thermal or photochemical instability, reaction with the solvent, insolubility, adsorption onto the surface of the glass or some combination of these parameters. Normal storage conditions for reference solutions are at ambient temperature or $4\,°C$, although if compounds are light sensitive it may also be necessary to store solutions in the dark. Experiments should indicate solvents which are suitable for initial and serial dilution and should provide information on the shelf life up to a period of about 6 months. Reference materials and their concentrated solutions should always be kept separate from water samples stored for analysis.

Extraction

The efficiency of the chosen extraction procedure, liquid–liquid or solid–liquid (solid phase), must be checked by appropriate extractability experiments. These experiments should, if possible, lead to a procedure which ensures that the residue is efficiently extracted from a number of different spiked waters, with a range of differing pH, dissolved salts and organic matter (including solids) contents.

Unless water samples have high solids contents which necessitates separation of solids (by filtration or centrifugation) for separate analysis, the entire contents, rather than an aliquot, of the sample container should be taken for extraction. With liquid–liquid extraction, samples may be extracted directly in the sample container or using a separator. If a separator is used, the empty sample container should be thoroughly rinsed with extraction solvent to remove any adsorbed pesticide and the washings used for the extraction or combined with the sample extract. With solid phase extraction (SPE) rinsing of the container may not be possible because of the small volumes of organic solvent used to elute the cartridge. In this case, addition of a small volume (e.g. of the order of 0.5–1% v/v) of an organic solvent modifier such as isopropanol (IPA) or methanol to the sample in the sample container may prove helpful to desorb any pesticides adsorbed onto solid surfaces, before the whole sample or an aliquot is used for extraction. Unless solids are separated, SPE is more suited to samples with low solids contents to avoid the problem of cartridges becoming blocked. Whichever procedure is used to prepare the sample and introduce it onto the extraction cartridge, its validity must be checked as part of the method development programme.

Purification of extracts

Where the determination step is insufficiently selective and/or the quantity of co-extracted materials is too great, extracts need some degree of clean-up before determination of the residue can be undertaken.

A wide range of different purification procedures are available. Bulk separations are often based on solvent partitions with acid–base separations incorporated where appropriate. Chromatographic procedures are normally used to give more specific purification but are most important for the more difficult separation of the residue from co-extractives of a similar nature.

Whichever clean-up method is selected, the study must enable the parameters of the system used to be clearly defined. The selected procedure must also be examined in the presence of typical extracts so that any influence of co-extractives is observed.

Quantitative determination

Several alternative approaches are normally possible for the determination of the pesticide residue. These usually encompass the techniques of gas chromatography (GC), high performance liquid chromatography (HPLC) and thin-layer chromatography (TLC), with a choice of different detection systems. Chemical derivatization can be performed to enhance sensitivity and selectivity (and volatility in the case of GC). A number of other methods (e.g. immunoassay and NMR) have also been employed. The choice of technique and detection system depends on the balance between sensitivity, selectivity and the degree of clean-up required to allow the limit of determination of the method to be achieved for the range of water types to be encountered.

The importance of selecting appropriate determination techniques must be considered during any development of clean-up procedures. It may be necessary to combine a particular clean-up with a particular determination in order to achieve acceptable determination of the residue.

As with clean-up, the study must enable the necessary system parameters to be defined and, normally, more than one determination technique should be examined. The effect of co-extractives in real samples must also be checked.

Confirmation

A confirmation stage is most important for those cases, such as environmental monitoring samples, where no untreated water of the same type is available or knowledge of the pesticides which may be present is not available. This stage must, as far as possible, be selected to complement the determination stage by using a procedure of separation or detection (or both) based as far as possible on a different principle to the clean-up and determination steps.

Gas chromatography-mass spectrometry (GC-MS) with multiple ion monitoring is one of the best choices since it gives information on the identity of the residue and is highly sensitive, though HPLC, TLC and derivative formation methods can also be used to complement a GC procedure. Use of the same technique but under different conditions, such as GC with two different stationary phases or HPLC employing normal and reverse-phase separations, is confirmatory but the result is less conclusive.

However, care must still be taken in interpreting results. For example, Koskinen et al. (1992) have shown that atrazine and tris-(2-chloroethyl) phosphate, a flame retardant and plasticizer in many products, have the same GC retention time on capillary dimethylsiloxane and diphenyldimethylsiloxane columns. Both chemicals have similar responses to electron capture and NP detectors. Desethylatrazine and pynachlor have the same retention times on non-polar capillary columns, with pynachlor also fragmenting into the three ions that are used routinely for desethylatrazine confirmation by GC-MS.

In addition to these difficulties with trace level confirmation, potential interferences can arise from impure solvents and reagents or artefacts which may arise from derivatizing reactions and all such sources need to be checked. Poor recoveries and additional peaks in GC chromatograms have been observed with an old sample of dichloromethane used for herbicide extractions (Singmaster, 1991). Artefacts which have formed during sample preparation in US EPA methods 625 and 8270 for determination of semi-volatile organics have been identified by Chen et al. (1993) using GC-MS. These arose from a number of reactions, including oxidation of phenolic surrogates and phenolic target analytes, halogenation or nitration of phenolic surrogates, reaction of cyclohexene (preservative) in methylene chloride with halogens, auto-oxidation of cyclohexene and aldol condensation of acetone.

Method performance

When each stage in the method has been developed, the overall performance of the analytical procedure must be assessed with the matrix types likely to be encountered in practice. Information on limits of determination and recovery and blank values are then obtained.

With each set of performance data, several comparable waters should be analysed that have not been treated either with the compound(s) being analysed or with a related compound that might interfere with the analysis (control samples). A complete residue analysis carried out with such material gives the background control value; this includes any contribution from the reagent blank value. Unexpectedly high sample blank values often may be due to contaminated reagents, adsorbents, etc. For example, additional gas chromatographic peaks observed during the analysis of 2,4-D and Silvex in water (as methyl derivatives) were ethyl derivatives formed from ethanol impurities in the extraction solvents (Kongovi, 1981). In such cases, the reagent blank value should be checked by performing a complete analysis without the analytical material (i.e. water) present and the source of interference identified and eliminated before performance testing is initiated.

In the authors' laboratory a typical protocol used for performance testing a single compound method has included five replicate spike determinations at each of three concentration levels, which include the limit of determination, and five replicate unspiked control determinations. Normally four natural waters would be used in these experiments. Where possible, recoveries are expected to be better than 70% and preferably should be within the range 80–110%. Relative standard deviations for replicates have typically been in the range 5–15% for concentrations in the range 0.02–0.5 μgl^{-1}.

More detailed performance testing is necessary where methods are to be used for water quality compliance testing. For internal quality control an estimate of the within-laboratory standard deviation of individual analytical results for blanks, standard solutions, samples and spiked samples having concentrations over the range of interest (including the prescribed concentration), over at least five batches on five separate days is required (DoE, 1990; DWI, 1993). Each estimate of total standard deviation should have at least 10 degrees of freedom. Recovery of added spikes is assessed during these experiments and recoveries between 90 and 110% are considered acceptable (DWI, 1993), although such high recoveries may not be achievable for low concentrations of some pesticides. A continuing check on analytical performance during routine analysis of samples is also required.

CURRENT TECHNIQUES

Stability of pesticides in samples and extracts

As it is often impossible to analyse samples immediately following collection and sample extraction, it is essential to demonstrate that pesticides of interest are stable in both aqueous and organic solvent (extract) materials for all environmental studies.

Stability data on pesticides and their metabolites should normally be available to aid laboratory management of the analytical programme before samples are taken in the field. This can be obtained in laboratory studies which closely simulate the storage of field treated samples before analysis (RT, 4 °C or if necessary, deep frozen). Each water sample should be stored individually in a sample container, which should contain sufficient sample for a single analysis. The compound is normally added in a small volume (<1 ml) of a suitable, water-miscible organic solvent, such as acetone, using a syringe. Immediately after addition, the bottles are closed and stored for the appropriate length of time. Sufficient samples, under identical treatment, should be stored to enable determinations to be carried out (in duplicate or triplicate) at suitable (e.g. weekly) intervals.

At the time of each analysis, samples should (if necessary) be thawed and during the extraction of the water sample the bottle should be washed out as normal with extraction solvent and the washings added to the extract to ensure a quantitative recovery. The analysis should be carried out using a recommended method for the compound. Where the analytical method is still under development, radiolabelled [^{14}C] compounds should be used if available. Also, use of radiolabelled compounds can, with advantage, shorten the analysis time and can provide information on the nature and location of the material present if at any stage the recovery is non-quantitative.

A number of studies have been carried out indicating that certain pesticides (e.g. simazine, lindane, 1,3-dichloropropene and propanil) will eventually degrade in ground water (Loch and Verdam, 1989; Pestemer et al., 1989; Cavalier et al., 1991). Degradation appears to be dependent on the concentration of the pesticide, microbial activity and temperature (Cavalier et al., 1991) with some persistent compounds (e.g. alachlor, metolachlor, 2,4-D and dichlorprop) having half-lives of 3–5 years in water at 15–22 °C. Valkirs et al. (1990) have also shown that the concentration of tributyl-tin in sea water falls by 50% after frozen storage during a 2-year period.

The US EPA have carried out a comprehensive study on the stability of 147 pesticides and related compounds spiked into well water samples, which had been biologically inhibited using mercuric chloride or monochloroacetic acid and stored at 4 °C for 14 days (Munch and Frebis, 1992). Analyte stability was demonstrated for 121 compounds, though for another 26 up to 100% loss was observed during that same period. Analytes generally remained stable in stored organic extracts. A disadvantage of mercuric chloride is the toxic nature of the mercury.

An approach to improving the reliability of environmental water sampling procedures, particularly where there may be uncertainty of the stability of pesticides of interest, is to extract water samples onto SPE material in the field and store the SPE material. Senseman et al. (1993) have shown that pesticides have equivalent or greater stability on SPE discs compared to storage in water at 4 °C, with freezing the disc after pesticide extraction being the most favourable storage option.

Though not widely used and then normally for single compound studies, internal standards (e.g. deuterated forms) may also be added in the field to aid quality control in monitoring the integrity of low pesticide concentrations in samples during transport, storage and analysis. In work with cyanazine, penta-deuterated (d_5) cyanazine ($—NHCD_2CD_3$ form) has been used as the internal standard (Eadsforth et al., 1991). At the end of each sampling day, d_5-cyanazine was added to field water samples by syringe to give a concentration of $1 \, \mu g \, l^{-1}$. Cyanazine and d_5-cyanazine were then quantified in the same water extract by positive ion chemical ionization (CI) GC-MS using the characteristic M + 1 ions at m/z 241/243 (cyanazine, ^{35}Cl and ^{37}Cl isotopes) and at m/z 246/248 for the d_5 analogue.

Extraction of pesticides from water

The determination of pesticide residues in water samples is generally performed using gas chromatography (GC) or high performance liquid chromatography (HPLC). The detection limits of GC and HPLC detectors normally make large concentration factors essential. EC regulations state that the maximum admissible concentration of a single pesticide in drinking water is $0.1\,\mu g\,l^{-1}$ and the maximum total concentration of all pesticides in drinking water should not exceed $0.5\,\mu g\,l^{-1}$. In order to meet these requirements the detection limit for any one pesticide should be less than $0.05\,\mu g\,l^{-1}$ and preferably be lower and a significant volume of sample, generally about 1 litre, is taken for extraction. In many analytical procedures considerable sample work-up is needed before the final analysis can be performed. Sometimes various interfering compounds must be removed and/or the compounds of interest must be enriched before detection is possible. Frequently and especially in trace pesticide analysis, both these operations are needed. Several reviews of methods for sample extraction have been made (Hunt and Wilson, 1986; Poole et al., 1990). The most popular methods for removing pesticides from water include liquid–liquid extraction (LLE) or liquid–solid extraction, more commonly referred to as solid phase extraction (SPE).

Liquid–liquid extraction

Classically, water samples are extracted with organic solvents at appropriate pH values. These methods are still regularly used with dichloromethane, diethyl ether, chloroform, hexane, petroleum ether and toluene being the more common solvents of choice. However, their high solvent usage, slow speed and labour-intensive nature make them less suited to the modern analytical laboratory. Organic solvents must be disposed of safely and with minimum environmental impact. Ease of use and suitability for automation are increasingly important aspects in modern analysis and are probably major reasons for the attempts at replacement of classical liquid–liquid extraction by solid phase extraction.

Because of the long established use of liquid–liquid extraction it will not be discussed extensively in this paper. However, organochlorine insecticides, pyrethroids, carbamates, phenyl urea and acidic herbicides can be efficiently extracted by liquid–liquid extraction using hexane, dichloromethane or diethyl ether. Liquid–liquid extraction with methylene chloride is suitable for a range of nitrogen- and phosphorus-containing pesticides in finished drinking waters (Edgell et al., 1991) and for triazine and acetanilide herbicides (Wilson and Foy, 1989). Large volumes of water (up to 120 litres) can be extracted, where necessary, using a Goulden large sample extractor. This enables lower detection limits ($1\,ng\,l^{-1}$) to be achieved if required (Foster and Rogerson, 1990). Pesticides in water have been extracted by on-line continuous flow liquid- liquid extraction using n-hexane. Chlorinated phenoxy-acids (2,4-D, 2,4,5-T and silvex (fenoprop)) were quantitatively extracted, whereas tetrachlorvinphos, its breakdown product 2,4,5-trichloro-phenol and parathion methyl were only extracted with 60% recoveries (Farran *et al.*, 1990).

Solid phase extraction

This technique has been available for almost a decade, initially with C18 and Florisil adsorption cartridges, though now the range is more extensive and includes bonded phases. The attractions of SPE (e.g. ease of use, reproduci-bility, reduced solvent use) have ensured that this sample preparation procedure now rivals liquid–liquid extraction (Majors, 1993). However, as noted previously, a disadvantage is that cartridges can become blocked where samples contain solid materials. This necessitates use of a reduced volume of sample or its filtration/centrifugation before extraction. A literature review supplied by equipment manufacturers on SPE extraction of pesticides has been made (Roulston, 1993). SPE has become increasingly popular for trapping organic compounds from water matrices (Junk and Richard, 1988). Stelluto *et al.* (1990) have shown that the extraction efficiencies of liquid–liquid extraction and solid phase techniques can be comparable, though there have been few detailed studies (Loconto, 1991) which have statistically evaluated the recovery of selected pesticides using different adsorbent cartridges (e.g. recovery of diazinon on C8 and C18 was 85.3% with a standard error of the mean of 2.3% and 95.5% with a standard error of the mean of 3.7%) rather than the general applicability of the techniques to a wide range of pesticides. Practical problems of how to interface large sample volumes (typically a litre) with the cartridges, to ensure reproducibility in an acceptable time, have been addressed. Auto-mated SPE is growing in popularity and a number of instruments capable of solid phase extraction are now available which can automate column conditioning, sample addition, washing steps and subsequent elution of the sample. Automation of solid phase extraction has been achieved in UK water companies (Naish-Chamberlain and Cooke, 1991) where large sample throughput is required for drinking water monitoring.

The majority of applications of enrichment of pesticides in water and sea water have involved the use of C8 or C18 bonded phase cartridges. Hinckley and Bidleman (1989) obtained satisfactory recoveries (85–120%) for a variety of organochlorine, organophosphate and pyrethroid insecticides at $7-110 \, \text{ng} \, \text{l}^{-1}$ concentrations from sea water or river water with high dissolved organic carbon content. Processing of samples (1–4 litres) through a tandem C8 cartridge system took 0.5–2 hours per sample. However, recovery of pesticides was affected by high $(7.6 \, \text{mg} \, \text{l}^{-1})$ organic content (DOC) (Johnson *et al.*, 1990), salinity and pH (Ahel *et al.*, 1992) in the water samples. C8 columns have also been used for benomyl, carbendazim and aldicarb species (Marvin *et al.*, 1991) though Suprynowicz and Gawdzik (1989) have found that C6 bonded phases achieve higher recoveries for organochlorine pesticides than C8 or C18.

Bagnati *et al.* (1988) applied C18 concentration for the rapid screening of 21 pesticides in water using 0.5 g of C18 material. This approach was subsequently improved by Benfenati *et al.* (1990) using a tandem column approach (0.8 g of C18 and 0.4 g of phenyl bonded silica) for the multiresidue analysis of 50 pesticides. C18 columns were also used in a multiresidue method for extraction of 12 triazole or pyridine pesticides in ground water samples (Bolygo and Atreya, 1991) and for chloroacetanilide and triazine herbicides (Meyer *et al.* (1993). US EPA and SCA in the UK have also included SPE for sample preconcentration in their test methods (US EPA, 1988a; SCA, 1994a). Treatment of water samples with acid prior to solid phase extraction on C8 cartridges was found necessary to retain free acid herbicides (e.g. 2,4-D, dichloroprop and dicamba (Swineford and Belisle, 1989; Bogus *et al.*, 1990). Loconto (1991) has also shown that addition of methanol to the water sample can significantly increase the recovery of organochlorine pesticides. The use of deuterium (isotopic)-labelled internal standards to compensate for any sample losses during extraction (Schuette *et al.*, 1990) or storage (Eadsforth *et al.*, 1991) is also recommended.

Less common adsorbents have been used successfully for extracting pesticides from water samples. These include polyurethane foam for S- and P-containing pesticides (Farag *et al.*, 1986; Farag and El-Shahawi, 1991) and polymeric material PLRP-S for phenoxy-acid herbicides (Geerdink *et al.*, 1991). Chromosorb 102 (Senin *et al.*, 1986), styrene-divinylbenzene absorbents, Wofatit Y77 (Edelmann *et al.*, 1992), Aquapak 440A, Amberlite XAD-2 and a styrene–ethylene dimethacrylate copolymer Separon SE (Svoboda *et al.*, 1991), Amberlite XAD-4 (Keydel and Seiber, 1988), XAD-8 (Fan and Chen, 1991), a mixture of XAD-2 and XAD-7 resins (Mattern *et al.*, 1991b) and Florisil (Wang and Huang, 1989) have also been used.

Mixed mode isolation which uses a combination of functional groups on a resin can be used to simplify the co-isolation and purification of polar and non-polar, non-ionic analytes from water by taking advantage of dual mechanisms of isolation. Mills and Thurman (1992) have used a combination of octadecyl chains and sulphonic acid groups for isolation and purification of triazine herbicides and their metabolites. Another approach is to use two precolumns in series. Two columns, C18 (to remove interferences) and the copolymer PRP-1 (which has a high affinity for polar phenylureas), enabled isolation and detection of phenylurea herbicides down to $0.01 \, \mu g \, l^{-1}$ in a 500 ml sample (Hennion *et al.*, 1990).

A non-specific adsorbent (Carbopack B) made from graphitized carbon black has been demonstrated to be more efficient than C18. It can adsorb basic and acidic pesticides from water at any pH (Di Corcia *et al.*, 1989). Fractionation of pesticides can then be achieved by eluting the cartridge with a range of solvents of different polarity or pH (methanol, methylene chloride/methanol (4:1 v/v) and acidified methylene chloride/methanol (4:1)) (Di Corcia and Marchetti, 1991, 1992; Stelluto *et al.*, 1990). This multiresidue approach has enabled extraction and quantitation of 89 pesticides in municipal and ground water samples below $0.1 \mu g l^{-1}$ (with a few exceptions).

Solid phase extractions can also be performed using sorbent discs. These are comprised of chemically bonded silica particles individually suspended in a densely woven 'web' of micro-PTFE fibrils. The discs (25, 47 or 90 mm diameter) are used with standard filtration equipment. A range of matrices, incorporating C8 or C18 together with styrene divinylbenzene polymeric matrix (for phenolic compounds) is available. Similar recoveries using the discs have been observed in comparison with the more traditional liquid–liquid extraction approach (Hagen *et al.*, 1990; Davi *et al.*, 1992). For samples with suspended particulate matter (leachate samples with clay particulates) filtration and centrifugation is needed to minimize clogging of the discs (Crepeau *et al.*, 1991).

Commercially available bonded silica cartridges have been shown to give several undesirable peaks (interferences) when eluted with solvent, due mainly to the plastic container. Bagnati *et al.* (1988) have overcome this problem by using glass columns packed with C18 material. The presence of resin derived co-extractives, which interfere with low-level analysis by electron capture detectors, has been associated with XAD resin (Hinckley and Bidleman, 1989) and necessitates extensive clean-up of the resin before use. Supercritical fluid extraction has been used in place of solvent for elution of pesticides from C18 material (Tang *et al.*, 1993). Advantages of this approach include the reduced amount of solvent waste and the reduction in analysis time.

Although SPE as a sample treatment process is efficient in improving sensitivity, the selectivity, particularly with the non-polar adsorbents, is not optimum. An alternative and potentially interesting technique for future development is to use pre-columns packed with phases containing immobilized antibodies having high selectivity for the analyte (e.g. immunoaffinity precolumns) (Brinkman and Farjam, 1992).

Organochlorine pesticides have been concentrated using a short trace-enrichment pre-column across the sample loop connection of a Rheodyne 7125 valve (Braithwaite and Smith, 1990) or a C18 pre-column (Marvin *et al.*, 1990). The adsorbent cartridges can also be used in the field by using newly available submersible instrumentation (Hadfield *et al.*, 1992). In this way sampling, extraction and preconcentration are done at the sampling site thus eliminating most contamination and handling problems. In addition, immediate isolation of organics from the aqueous matrix by an adsorbing material can preserve analytes from bacterial attack occurring between the time of sample collection and analysis. The small volume cartridges can be sealed and conveniently forwarded to the laboratory for subsequent elution and analysis. *In situ* solid phase extraction techniques are sometimes employed instead of conventional bottle sampling methods for pesticides (e.g. pyrethroids) which can strongly adsorb to glassware (Hadfield *et al.*, 1989).

A recent and novel application of SPE, which offers scope for future development, is solid phase microextraction (SPME). A solid phase in the form of a fibre is held within a syringe which is exposed to the sample when the plunger is depressed. Pesticides and other organics can be adsorbed from solution onto the fibre and then analysed by thermal desorption. SPME is a fast, sensitive, inexpensive, portable and solvent-free method for extracting organic compounds from aqueous solution. It can attain detection limits of $15\,\text{ng}\,\text{l}^{-1}$ and below for both volatile and non-volatile compounds (Arthur *et al.*, 1992).

Steam distillation

In suitable cases good recoveries of pesticides from water samples have been achieved using simultaneous steam distillation/solvent extraction. This procedure uses less solvent ($< 10\,\text{ml}$) than conventional liquid–liquid extraction, but is time consuming (i.e. 1.5 hours for water sample processing) (Hemmerling *et al.*, 1991).

Volatile pesticides

Techniques such as closed-loop stripping, headspace analysis, and purge and trap are applied to the isolation and determination of volatile organics. Few pesticides are sufficiently volatile to make these techniques feasible. Exceptions, such as methyl bromide and the nematocide D-D (1,3-dichloro-propene/1,2-dichloropropane) can be analysed by purge and trap according to US EPA method 524.2 using a three-stage trap consisting of Tenax, silica gel and charcoal (Eichelberger *et al.*, 1990). Methyl bromide, following conversion to the more electron-capture sensitive methyl iodide, was also determined by headspace GC. The limit of determination was $5\,\text{ng}\,\text{l}^{-1}$ for Italian surface waters (Cirilli and Borgioli, 1986). D-D was also analysed by purging water samples and trapping the components on a Tenax adsorbent resin, before concentrating and analysing by GC-MS (limit of determination $0.16\,\mu\text{g}\,\text{l}^{-1}$) (Merriman *et al.*, 1991).

Bioaccumulation/in situ *sampling*

A number of *in situ* sampling devices and bioaccumulation approaches have been used to abstract and concentrate low levels of pesticides from water columns.

Supported liquid membranes, mounted in a flow system, can be used for selective extraction and enrichment of pesticides (Melcher and Morabito, 1990; Jönsson and Mathiasson, 1992). Herbicides are extracted in their protonated form from acidified water samples, which are pumped through a membrane separator. After passage through the membrane, acids are ionized in a stagnant acceptor phase which is sufficiently alkaline to prevent them re-entering the membrane. In this way an enrichment factor of several hundredfold can easily be attained (Jönsson and Mathiasson, 1992). Membrane systems have been incorporated into field sampling devices to permit integrated sampling and quantitation of herbicides in the lower $ng l^{-1}$ range (Mathiasson *et al.*, 1991).

Other approaches to time-integration concentration of pesticides in aquatic systems has included the use of semi-permeable membrane devices (SPMDs) and use of caged mussels or fish. The feasibility of the SPMD approach for *in situ* monitoring of environmental contaminants has been demonstrated by Huckins *et al.*, 1990) with an isomer of fenvalerate in a littoral enclosure of a small pond. Solvent-filled dialysis membranes accumulate persistent lipophilic pollutants in a similar way to that of aquatic organisms (Södergren, 1990). The use of caged mussels or fish has been widely used in monitoring of organochlorine compounds in receiving waters (Heinonen *et al.*, 1986; Herve *et al.*, 1988; Gutierrez-Galindo *et al.*, 1992).

Prest *et al.* (1992) compared the uptake of pesticides and PCBs by SPMDs and freshwater clams. In general, levels of organochlorine compounds were ~1.6 times higher in clams on a wet weight basis than in the SPMDs and trends in accumulation were similar except where biofouling of the SPMD membranes decreased uptake rates. Gill *et al.* (1992) have compared the uptake of the insecticide flufenoxuron from water using several devices (SPMDs filled with triolein or corn oil, C18 discs or plates) with that by fish. The uptake using the passive *in situ* devices was generally equivalent to that of the fish, although, for ease of handling and analysis, the C18 discs (Empore) were preferred.

Green *et al.* (1986) compared two approaches, a SEASTAR™ *in situ* sampler, which periodically draws large volumes of sea water for chlorinated pesticides analysis through extraction columns filled with absorbent XAD resin, and accumulation of the same compounds in transplanted mussels. The mussels provided a measure of the bioavailability of the chlorinated pesticides, whereas the sampler provided a measure of the water concentration. The efficiency of the SEASTAR device has also been confirmed for other pesticides in water (Sarkar and Sen Gupta, 1989).

Clean-up of extracts

The need to determine very small amounts of pesticides in the presence of interfering substances, sometimes in high amounts (particularly in surface waters), makes the clean-up of water sample extracts necessary in some cases. Adsorption column chromatography employing such adsorbents as Florisil, alumina, carbon and silica gel has been a popular method. These procedures typically require relatively large volumes of solvent to recover the pesticides from the clean-up column and consequently the use of micro-columns has been employed to minimize the solvent volume and the time spent. Disposable cartridges (Superclean^{TM}, Sep-pak^{TM}, Quick-Sep^{TM} and Bond-Elut^{TM} etc.) containing different absorbents have gained increasing application to sample extract clean-up (as with sample extraction).

Florisil and alumina clean-up columns give excellent recoveries for a number (25) of priority pollutants, e.g. organochlorine pesticides and PCBs, extracted from water (Millar *et al.*, 1981). A deactivated Florisil column, using 3% (v/v) methanol in benzene (not now a recommended solvent) as eluant, has been used to clean up dichloromethane extracts in the analysis of triazine herbicides. Limits of detection of $<0.025\,\mu g\,l^{-1}$ could be achieved (Lee and Stokker, 1986).

More recent approaches have included the standardization of a procedure which uses silica gel micro-columns and increasing polarity solvents for separating pesticides, PCBs and other organic pollutants from water extracts into four distinct groups (Leoni *et al.*, 1991). Recoveries of triazines and other herbicides were satisfactory at the 0.2 and 1.0 $\mu g\,l^{-1}$ levels. Aakerblom *et al.* (1990), using a multiresidue method for a range of non-polar and semi-polar pesticides, have applied hydrophobic gel permeation for clean-up of dichloromethane extracts of water samples. Using this approach approximately 80% of the pesticides used in Sweden could be determined.

A clean-up involving pre-column switching for the extracts in order to increase the speed and reproducibility for analysis of cyanazine and bentazone has been developed (Hogendoorn and Goewie, 1989). The method was applied to analyses of these herbicides in surface and drinking water at concentrations down to 1 $\mu g\,l^{-1}$ and for bentazone in drinking water and rain water at 0.01 $\mu g\,l^{-1}$. Van der Hoff *et al.* (1991) replaced column chromatography under atmospheric pressure with alumina absorbent and *n*-hexane as the elution solvent with an automated fractionation using ASPEC (*A*utomated *S*ample *P*reparation with *E*xtraction *C*olumns). This permitted analysis at the ng$\,l^{-1}$ concentration for organochlorine pesticides and pyrethroids in ground, drinking and surface water in a single run.

Pesticide determination

Gas chromatography

GC detectors: Selective gas chromatographic detectors are an essential part of the methods used to analyse pesticides in environmental samples including water. However, several different conventional selective detectors are required for selective detection based on all the heteroatoms (except oxygen) and functional groups common to most pesticides (Table 2). In addition, other spectroscopic and mass spectrometric detectors, e.g. atomic emission, fourier transform infrared and mass selective detectors have been introduced which complement, and in the case of the mass selective detector in part replace, the more conventional detectors.

The value of these detectors for determination of traces of pesticides in environmental samples, including water, lies in the combination of their sensitivity and selectivity. The most sensitive, though not the most selective, detector is the ECD with a sub-picogram detection limit for strongly electron capturing compounds (e.g. chlorinated pesticides). Several of the other detectors, with selectivities as indicated in the table, have detection limits in the low picogram range including NPD in the phosphorus mode, AED depending on the element (e.g. sulphur and tin) and ECLD for chlorinated compounds. At the other end of the sensitivity scale is FTIR where even with strong absorbers, only low nanogram amounts can be detected. This is a similar order of sensitivity to MSD for a full spectrum. However, with MSD operating in the selected ion monitoring (SIM) mode typical limits of detection (LOD) for pesticides are in the range 10–50 pg with good selectivity, especially for ions of higher mass. Further details of the sensitivity, linear dynamic range and selectivity of individual detectors,

Table 2   GC detectors for pesticide determination

Detector	Selectivity
Conventional selective detectors	
Electron capture (ECD)	Halogens, CN, NO_2 and electron-capturing structures
Nitrogen-phosphorus (NPD) (thermionic)	Nitrogen, phosphorus
Electrolytic conductivity (ELCD)	Halogens, sulphur, nitrogen
Flame photometric (FPD)	Phosphorus, sulphur, tin
Newer spectroscopic and mass spectrometric detectors	
Atomic emission (AED)	Tunable, any element except helium
Mass selective (MSD)	Tunable for any compound
Fourier transform infrared (FTIR)	IR absorbing structures

all of which have sensitivities in the picogram to nanogram range are given by Buffington and Wilson (1987).

GC-atomic emission detection: Atomic emission detection (AED) is a sensitive multi-element detection method for gas chromatography which offers several advantages over the more conventional detectors (Lee and Wylie, 1991). For example:

(1) any element except helium can be detected;
(2) atomic identity can be confirmed with recorded spectra;
(3) it is fast since four or more elements can be measured simultaneously;
(4) the response is essentially compound independent for most elements so that concentrations of unknowns can be quantitated.

This permits the use of the detector as a screening tool as well as for trace analysis (Koehle *et al.*, 1990; Sullivan, 1991). Although generally less sensitive than the other detectors previously mentioned, by using standard solvent extraction and concentration techniques a level of $0.1 \mu g \, l^{-1}$ can be achieved for a number of pesticides using several different elements for selective detection (Miller *et al.*, 1991). Organotin compounds are particularly suited to trace analysis ($ng \, l^{-1}$) by this means (Gremm and Frimmel, 1992). So far, GC-AED has not been widely used for the determination of pesticides in water.

GC-fourier transform infrared detection: Gas chromatography combined on-line with spectroscopic detection systems is a powerful technique for the rapid detection and characterization of organic compounds. Mass spectrometric detection systems are capable of distinguishing between homologues and of giving molecular weight and chemical structure information from fragmentation patterns. By contrast, the infrared spectrometer is capable of providing information on the presence of functional groups, positional substitution and can differentiate between isomers, but is less likely to distinguish between homologues. The data available from mass spectrometric and infrared detection systems are therefore complementary and the acquisition of both types of data adds both scope and confidence to the characterization of unknown components.

Recent advances in FTIR technology allow the identification of trace components on-the-fly at the low nanogram level. Despite this lack of sensitivity, many applications of GC-FTIR in environmental analysis have appeared. The US EPA has developed protocols (method 8410) for the GC-FTIR analysis of semi-volatile organics, including some pesticides (e.g. hexachlorobenzene, pentachlorophenol) (US EPA, 1989). Malissa *et al.* (1990) have demonstrated that the minimum identifiable concentration of atrazine in drinking water using the light-pipe technique is approximately

$1 \mu g \, l^{-1}$; levels at the EC limit of $0.1 \mu g \, l^{-1}$ can only be achieved using cryodeposition techniques. Detection at the low picogram level is possible with matrix isolation. GC-FTIR and the sensitivity and selectivity of this combination have enabled the technique to be successfully applied to the analysis of dioxin residues where the need to distinguish between individual isomers is paramount (Gurka et al., 1991). Sensitivity limitations currently prevent more widespread use of this technique for determination of pesticides in water.

GC-mass spectrometric detection: The use of gas chromatography-mass spectrometry (GC-MS) with electron impact (EI) and positive and negative ion chemical ionization (PCI and NCI) as a confirmatory technique for pesticide residues has been reviewed (Barcelo, 1991). GC-MS with EI is particularly useful for confirmation studies and in the analyses of pesticides at sub-$\mu g \, l^{-1}$ concentration levels in water ($0.1 \mu g \, l^{-1}$). The use of chemical ionization (CI) as a softer GC-MS ionization technique is known to be more sensitive than EI for those compounds that undergo vigorous fragmentation.

Chemical ionization GC-MS has been used to quantify triazines in water (base peak M + 1) at the low picogram and high femtogram level in injected extracts (Feigel and Holmes, 1991) and to scan selectively for four classes of pesticides in water to $0.005 \mu g \, l^{-1}$ using selected ion monitoring (Hargesheimer, 1984). Herbicides and their degradation products in surface waters have also been detected using gas chromatography/positive ion chemical ionization (isobutane)/tandem mass spectrometry (Rostad et al., 1989). GC-PCI-MS-MS used in the neutral loss mode to detect specific daughter ions eliminated any interferences observed using GC-EI mass spectrometry in the surface water samples. The detection limit of most of the herbicides was $100 \, pg$ injected (equivalent to $10 \, ng \, l^{-1}$) and the instrument response was linear from $200 \, pg$ to $20 \, ng$. Two alachlor degradation products were detected using the GC-PCI-MS-MS which could not be observed by the GC-IT (ion-trap) MS or GC-PCI-MS systems.

Chlorotriazines (simazine and atrazine), molinate, trifluralin, alachlor and metalochlor have been analysed in water samples from the Elaro Delta using GC-NPD at the $0.01 \mu g \, l^{-1}$ concentration level with further confirmation by GC-MS with EI ionization (Durand et al., 1992). Certain compounds which have low halogen, N or P contents (e.g. mecoprop and MCPA) can be analysed with more sensitivity using GC-mass selective detection (Welter and Ottmann, 1988). Capillary gas chromatography-high resolution mass spectrometry (GC-HRMS) has been used to determine atrazine in water at sub-parts per trillion ($ng \, l^{-1}$) concentration. This technique was needed since the existing detection limits of routine GC, GC-MS and HPLC methods were inadequate for the study. Using EI and operating in the SIM mode atrazine could be detected at $0.2–0.5 \, ng \, l^{-1}$ (Cai et al., 1993).

Ion-trap MS has been used for the determination of several commonly used herbicides, including triazines, metolachlor and alachlor and selected degradation products in surface and ground water. Full scan spectra were obtained on $\leqslant 1$ ng of analyte, though using SIM the detection limit was lower, i.e. 60 pg (Pereira $et\ al.$, 1990). Chemical ionization ion trap MS was also used to detect 20 target pesticides (Mattern $et\ al.$, 1991a). Limits of detection (LODs) were often better than $0.005\ \mu g\,l^{-1}$, though carbaryl, cyanazine, fenamiphos, linuron, pendimethalin and terbufos had LODs of $0.005-0.05\ \mu g\,l^{-1}$.

Use of internal standards in GC-MS methods can improve the accuracy of quantitation of analytes, particularly at low concentrations and for complex matrices. Addition of $[^{13}C_3]$ atrazine to water samples prior to extraction resulted in precision and accuracy at the $1\ ng\,l^{-1}$ level of 15% (relative standard deviation, $n = 8$) and >85% respectively (Cai $et\ al.$, 1993). Tsuchiya and Ohashi (1992) have shown that use of internal standards in GC-MS analysis of pesticides in water improves the reproducibility. Coefficients of variation for 32 pesticides ranged from 2.1 to 16.6% with internal standards and from 4.2 to 29.2% without at the $0.2-0.5\ ng\,l^{-1}$ level. These CVs improved further to 0.5–9.1% and 5.3–14.0% respectively at the $2\ ng\,l^{-1}$ level. Deuterium-labelled internal standards were used by Schuette $et\ al.$ (1990) in an SPE extraction method for alachlor, metolachlor, atrazine and simazine, enabling an accuracy of ±5% at the $0.2\ \mu g\,l^{-1}$ level. Because of its sensitivity and ability to be used selectively, GC-MS has gained wide acceptance both for determination and confirmation of pesticides in water at low concentration.

Thin-layer chromatography

Although not widely used, thin-layer chromatography (TLC) has found favour for the screening for pesticide residues in water because of its simplicity and minimal cost. Separation and visualization systems have been identified for several pesticide groups, including triazines (Lawrenz, 1983), organophosphates (Marutoiu $et\ al.$, 1987) and pyrethroids (Patil $et\ al.$, 1992) with detection limits in the range of $0.1-1\ \mu g$ per spot. Other workers have developed TLC systems for a wide range of pesticides. Ambrus $et\ al.$ (1981), using single solvents and visualization modes including o-toluidine plus KI, p-nitrobenzene-diazonium-fluoroborate, silver nitrate and UV radiation, p-methylaminobenzaldehyde and bioassays with fungi and enzymes, were able to detect 188 pesticides. Similar visualization reagents were used by Gardyan and Thier (1991) for direct quantitation of 150 pesticides, though with more modern techniques for spotting, separation (high performance TLC, HPTLC) and evaluation, including computer-assisted densitometry.

An HPTLC method using automated multiple development (AMD) for multi-component analysis of pesticides in water has been developed by Burger *et al.* (1990). The conventional isocratic elution (of TLC) has been replaced by multiple and stepwise development combined with gradient elution. Screening and confirmation gradients, coupled with reflectance spectroscopy (multiwavelength scanning) and post-chromatographic derivatization, make it possible to detect pesticides in ground water and drinking water supplies to a limit of detection of $20\,\mathrm{ng}\,l^{-1}$. At least 100 pesticides can be checked for their presence on one TLC plate. Improvements in the sensitivity, linearity and speed of the method have been gained by reducing the thickness of the silica gel layers of the TLC plates and by decreasing the gradient step increments (de la Vigne *et al.*, 1991).

High performance liquid chromatography

The application of high performance liquid chromatography (HPLC) to pesticide residue analysis continues to grow, particularly for those compounds which cannot be analysed directly to GC due to their polarity, poor volatility and/or thermal instability (e.g. phenylureas, carbamates). The technique is now being more widely used for the determination of pesticides in water and increased attention on the presence of polar pesticide metabolites in surface and ground waters has also been responsible for the growth in HPLC methods.

HPLC-UV detection: Detectors operating on the principle of monitoring ultraviolet (UV) absorption at a specific wavelength still dominate high performance liquid chromatography as the most widely used detector type. Sensitive methods for analysing pesticides in water using HPLC-UV are available with detection often confined to the region 200–260 nm. For improved sensitivity (though also increased possibility of interference from co-extractives) detection at the lower end of this wavelength region is often used, e.g. for triazines (Xu *et al.*, 1986), pyridate (Macomber *et al.*, 1992) and metolachlor metabolites (Heegemann, 1988), as is the use of large sample volumes for injection (Geerdink, 1991) or trace enrichment (Froehlick and Meier, 1989). The latter has enabled analysis of triazines in water in the range $1-10\,\mathrm{ng}\,l^{-1}$.

Other methods of detection may be more sensitive. Electrochemical detection has been compared with UV for DNOC, dinoseb and dinoterb. Limits of detection for the detector in oxidative (0.1 ng) or reductive modes (0.3–1 ng) were superior to those (2–24 ng) using UV for these compounds (Yao *et al.*, 1991).

Post-column derivatization can be used to enhance the sensitivity, usually by addition of a chromophore to the compound of interest. Determination of carbaryl and its hydrolysis product, α-naphthol, involves hydrolysis followed by dye formation using sodium nitrite and sulphanilic acid and UV monitoring at 506 nm (Tena *et al.*, 1992).

HPLC-fluorescence detection: Conventional fluorescence has been applied for highly selective and sensitive detection of both naturally fluorescent pesticides (e.g. coumarin-based rodenticides) and pesticides and their degradation products that have been derivatized.

Chlorophenoxy-acid herbicides can be detected following post-column derivatization with 9-anthryldiazomethane to sub-μg l^{-1} levels in water samples (Suzuki and Watanabe, 1992). Post-column reaction detection is also employed (US EPA method 531.1) for the detection of N-methyl carbamates, following alkaline hydrolysis and fluorogenic labelling with *o*-phthalaldehyde and mercaptoethanol (Dong *et al.*, 1990). Detection limits of 0.2–0.6 μg l^{-1} in drinking water are attainable. This same procedure forms the basis of the recently published SCA method (SCA, 1994b) where limits of determination are typically <0.04 μg l^{-1} for aldicarb and a range of N-methyl carbamates.

HPLC-electrochemical detection: Electrochemical methods of detection in HPLC have failed to achieve widespread popularity. Although electrochemical detection (EC) offers advantages of selectivity and often higher sensitivity, compared with UV detection, it is only applicable to electrochemically active compounds. Other factors include the complexity of operation of some electrochemical detectors and the need for regular maintenance of electrodes and flow cells. However, advances in the coupling of electrochemical detectors with HPLC are being made (Buchberger, 1990). HPLC-EC has been applied to analysis of carbamate pesticides in river water (Anderson *et al.*, 1985) and to phenylurea herbicides in drinking and surface waters (Boussenadji *et al.*, 1991). The latter confirmed that HPLC-EC (at 1.35 V) is more sensitive for detection of phenylurea herbicides than HPLC-UV (at 254 μm).

HPLC-diode array detection: One of the fundamental weaknesses of liquid chromatographic techniques is that the standard detection devices (UV, fluorescence, electrochemical) have provided limited qualitative or confirmatory information. Retention time is insufficient nowadays to confirm the presence of an unknown and the problem is particularly severe for trace analysis of complex mixtures. Diode array detection (DAD) as an extension of conventional UV absorption detection offers good selectivity and sensitivity and has consequently found wide application in HPLC method development and analysis for peak identification and tracking.

The structured and distinct UV spectral information from DAD has permitted workers to differentiate between chlorotriazines and their metabolites (Durand and Barcelo, 1989), paraquat/diquat (Simon and Taylor, 1989), and phenylurea herbicides (Reupert and Plöger, 1989) and it has been used as a standard method for monitoring surface, ground and raw waters in Germany (Friege and van Berk, 1989).

HPLC-mass spectrometric detection: Although HPLC has been used for the determination of pesticides using UV, electrochemical and fluorescence detection, such methods typically do not provide the necessary structural information to confirm or identify the specific pesticide. The use of MS as an HPLC detector is considered a key element for the future development of most HPLC methods; in particular, the confirmation of analyte identification provided by full scan MS detection is invaluable to any laboratory.

The commonly used LC-MS interfaces include thermospray (TSP), particle beam (PB) and atmospheric pressure ionization (API). Thermospray HPLC-MS has been used to screen for relatively non-volatile pollutants in water (Voyksner, 1987), including a number of polar herbicides and insecticides that are difficult to analyse by GC (Bellar and Budde, 1988). Detection limits were in the range 1–$10\,\mu g\,l^{-1}$ for 29 analytes. Hammond *et al.* (1989) also used HPLC-MS for analysis of phenylurea herbicides in river water. It is a relatively soft ionization technique and adducts for selected pesticides include $[M + H]^{+}$, $[M + H - CH_3]^{+}$ and $[M + H - Cl]^{+}$. By concentrating extracts (1 litre to $100\,\mu l$) and injecting $20\,\mu l$ onto the column, Hammond *et al.* (1989) were able to detect pesticides at concentrations equivalent to $0.1\,\mu g\,l^{-1}$ in water samples in the selected ion monitoring mode (SIM). Barcelo *et al.* (1991) have also identified characteristic base peaks and fragment ions for representative pesticides: carbamates, chlorotriazines, phenylureas, phenoxy-acids, organophosphorus and quarternary ammonium compounds under positive and negative ion mode for thermospray HPLC-MS. Generally, normal-phase eluents gave more structural information and enhanced the response of several compounds when compared to reverse-phase HPLC. Tandem mass spectrometry (TSP-LC-MS-MS) has also been used as a rapid screening method to detect hydroxysimazine and hydroxyatrazine at a limit of detection of $0.03\,\mu g\,l^{-1}$ as part of a triazine ground water monitoring study (Cornacchia *et al.*, 1993).

TSP spectra, however, can suffer from a lack of sufficient structural information as well as fluctuating ion signals and poor reproducibility (Allen, 1992). This has encouraged the use of EI spectra generated from particle beam HPLC-MS for confirmation of pesticide residues. Generally PB-LS-MS is less sensitive than the other approaches mentioned, though Miles *et al.* (1992) have shown that 43 out of 100 relatively polar US National Pesticide Survey analytes have sufficient sensitivity using PB-MS for detection of and spectrum generation from 100 ng or less, which is just sufficient for confirming the presence of these analytes in groundwater at the

$0.1\,\mu g\,l^{-1}$ level. A PB-LC-MS method for chlorinated phenoxy-acid herbicides in water has been developed as an alternative to the original GC method with diazomethane derivatization (Brown *et al.*, 1991). After extraction, the esters are hydrolysed to the free carboxylic acids, separated by reverse-phase HPLC and quantitated by UV with confirmation by PB-MS detection. No thermal degradation of these compounds in a PB mass spectrometer has been observed and full scan EI spectra have been obtained from 20 ng total material.

Atmospheric pressure ionization is considered the most impressive of the various ionization techniques. There are a number of atmospheric pressure ionization (API) interfaces, such as electrospray and heated nebulizer, in which the actual formation of ions occurs outside the vacuum system of the instrument. Electrospray (a soft ionization technique) has been interfaced with an ion-trap mass spectrometer (HPLC-MS-MS) (Lin and Voyskner, 1993). The ability to acquire full scan spectra on picogram quantities of material (e.g. propoxur, carbofuran and aldoxycarb) enables the detection of $1\text{--}10\,\mu g\,l^{-1}$ concentrations of these pesticides by direct injection of $10\,\mu l$ of the water sample into a column equilibrated in 100% water.

Electrospray LC-MS at high flow rates ($2\,ml\,min^{-1}$) has been achieved by adding a simple liquid shield between the sprayer and the ion-sampling capillary of a heated capillary-type API interface (Hopfgartner *et al.*, 1993). This has resulted in improved sensitivity and gives the benefits (e.g. ruggedness, efficiency, large injection volumes) of using standard HPLC columns. Applications have included the analysis of carbamate pesticides at the low nanogram levels (with conventional HPLC gradients) and mexacarbate in pond water at $0.1\,\mu g\,l^{-1}$ concentration using selected ion monitoring (SIM).

API interfaces gave lower detection limits for analysis of carbamates in water in comparison with TSP and PB techniques (Allen, 1992; Pleasance *et al.*, 1992). Doerge and Bajic (1992) evaluated atmospheric pressure chemical ionization (APCI) for analysis of four classes of pesticides (triazines, phenylureas, carbamates and organophosphorus compounds) plus some miscellaneous compounds in ground water. Detection limits in full scan or selected ion monitoring mode varied from 0.8 to 10 ng or 0.01 to 1 ng, respectively, with linear calibration curves. APCI-LC-MS is well suited to multiresidue pesticide analysis; it provides the optimal combination of high sensitivity, exclusive formation of protonated molecular ions and broad specificity across chemical classes.

Coupled chromatographic techniques

Coupled chromatographic techniques (e.g. LC-LC; LC-GC and less usually

GC-LC) may be used in order to optimize sample treatment and/or enrichment and clean-up and consequently are amongst the most sensitive and selective techniques available for the determination of pesticide residues in environmental samples.

The first papers on residue analysis of pesticides with reverse phase LC column-switching dealt with on-line preconcentration of analytes from aqueous samples. More recently, researchers have concentrated on the use of column-switching to clean up selectively sample extracts prior to LC analysis (i.e. LC-LC). The application of LC-LC using pre-columns (for enrichment) and longer analytical columns (for chromatographic separation prior to column switching) for pesticides in water has been reviewed (Hogendoorn et al., 1991a; van Zoonen et al., 1992). Successful applications include determination of iprodione, bromacil, diuron and cyanazine in various waters. Coupled reverse-phase LC-LC has been used for the rapid screening of bentazone in water with direct sample injection (Hogendoorn and van Zoonen, 1992). Recoveries ranged from 86 to 103% with detection linearity over the $0.1-100 \, \mu g \, l^{-1}$ range. For more polar compounds such as chloroallyl alcohol (a 1,3-dichloropropene metabolite) and ethylenethiourea, longer C18 pre-columns are required for adequate retention (Hogendoorn et al., 1990, 1991b). A fully automated pre-column LC-diode array UV system has been developed for monitoring polar pesticides in surface waters from European rivers (Brinkman, 1992). A large number of pesticides were included in the programme with detection limits at the sub-$\mu g \, l^{-1}$ level.

Liquid chromatography adds selectivity to the high efficiency of capillary gas chromatography, thus improving the overall separation potential. The technique involves transferring fractions from LC (up to 1 ml in volume) to GC in an automated routine manner. On-line liquid–solid extraction-GC (or trace enrichment LC-GC) has been used by several workers for analysis of pesticides in water at the $ng \, l^{-1}$ level (Noroozian et al., 1987; Noy et al., 1988). Noy et al. (1988) passed an aqueous solution containing organochlorine pesticides over a pre-column, which was dried by helium-flushing and applying vacuum. The pre-column was then extracted using a small volume of hexane (70 μl) which was introduced into the capillary column. Atrazine was analysed at the $ng \, l^{-1}$ level in water using a reverse-phase LC column and a solvent exchange procedure (Davies et al., 1989). LC-LC methods result in good recovery, lower detection levels than conventional methods and partial elimination of manual sample preparation work. Kwakman et al. (1992) have used Empore membrane extraction discs for trace enrichment in on-line LC-GC for analysis of organophosphorus pesticides below $0.1 \, \mu g \, l^{-1}$. A heart cutting technique (GC-LC) has also been used for analysis of terbuthylazine in water ($0.7 \, \mu g \, l^{-1}$) (Sesia et al., 1990).

Capillary electrophoresis

The application of capillary electrophoresis (CE), particularly isotacho-phoresis and zone electrophoresis, to the separation and determination of organic molecules in a wide range of matrices is rapidly expanding, although there is as yet limited success in the area of pesticide chemistry (Kuhr and Monnig, 1992). Triazine herbicides (e.g. prometryne, terbutryne, desmet-ryne, simazine and atrazine) and their solvolytic products have been separated using capillary zone electrophoresis and detected by UV, though a limiting factor was the sensitivity (LOD of $2\,\mu$mol) (Foret *et al.*, 1990). Other pesticides (e.g. prometon, prometryne, propazine and butachlor) have been determined at lower detection limits (18–52 fmol) using micellar electrokinetic capillary chromatography (Cai and El Rassi, 1992). More recently (Dinelli *et al.*, 1993), two sulphonylureas, metsulfuron and chlorsul-furon have been analysed by CE in water. Using SPE and concentration (10 000-fold from 1 litre to 100 μl), 0.1 μg l^{-1} of each compound could be detected using UV at 214 nm, though the electropherograms indicated the presence of some interfering compounds that arose from the SPE procedure. Isotachophoresis has been used for the detection of pyrethroid insecticides, alphacypermethrin and cypermethrin in water. The insecticides are extracted from water, the extract evaporated and the residue hydrolysed at alkaline pH to yield the degradation products, *cis*- and *trans*-dichlorochrysanthemic acids, which are separated and detected by capillary isotachophoresis, with detection to the 0.1 mg l^{-1} level (Dombek and Stránský, 1992).

The drive to obtain additional chemical information about molecules separated by capillary electrophoresis continues to encourage the coupling of CE with mass spectrometers. On-line CE-MS under tandem MS conditions has given full scan mass spectra from 30 pmol amounts for a range of sulphonylureas (Garcia and Henion, 1992).

Other non-chromatographic techniques

Nuclear magnetic resonance (NMR) and electron spin resonance (ESR) spectrometry: Nuclear magnetic resonance spectrometry is now sufficiently sensitive for the analysis of phosphorus- and fluorine-containing pesticides in soil and food products. Organophosphorus pesticides are widely used in agriculture and approximately 10% contain fluorine, though this atom is most commonly found in more recently synthesized structures. It is feasible to use NMR for screening organophosphorus insecticides in crops and soils at residue levels $\sim$0.1 mg kg^{-1} without clean-up of the extracts with only 30 minutes of acquisition time on a 400 MHz instrument (Mortimer and Dawson, 1991a; Krolski *et al.*, 1992). As the receptivity of the fluorine nucleus is 12.5 times greater than that of phosphorus, it is possible to measure residue levels of fluorinated pesticide at $0.01\,mg\,kg^{-1}$ under similar conditions. Any fluorinated pesticides which contain multiple equivalent fluorine atoms (i.e. CF_3 groups) should be detected with more sensitivity. Trifluoralin has been detected at $1\,mg\,l^{-1}$ directly in liquid samples (i.e. wine) (Mortimer and Dawson, 1991b); a detection limit of $\leq 0.1\,\mu g\,l^{-1}$ in water following typical solvent extraction/concentration steps from water is therefore feasible. The advantages of NMR for residue analysis include the acceptable scan acquisition times ($\sim 30\,min$ cf. typical GC or HPLC chromatography), its selectivity and non-destructive analysis, as well as the time saved as no clean-up is expected to be required. The main disadvantages are that it is compound specific and the detection limit can still be a problem.

Another approach, nuclear double resonance (NDR) has also been used to detect pesticides in water at concentrations in the range $10-100\,\mu g\,l^{-1}$ (Osredkar and Kadaba, 1982). This is not a routine analytical test, though it is suitable for thermally unstable compounds (cf. GC) since the measurements are carried out at low temperatures.

ESR has been used for the detection of the herbicide diquat in water (Sanchez-Palacios et al., 1992). The intensity of the diquat radical generated by its reduction with alkaline sodium dithionite is linear with concentration over the range $1-100\,mg\,l^{-1}$. The method is simple and selective, except if paraquat is present, though not particularly sensitive.

Different pulse polarography: This technique has been applied directly but to a limited extent for the determination of several herbicides in water. Water samples are extracted and cleaned up prior to preparation of a reducible derivative, e.g. nitrosoglyphosphate and its determination by differential pulse polarography. Glyphosphate (Friestad and Bronstad, 1985) and 2,4-D, 2,4-DP, MCPA and MCPP (Lechien et al., 1981) have been determined, though the limits of determination ($\leq 30\,\mu g\,l^{-1}$) were not adequate.

Direct spectrophotometric methods: The majority of analytical methods used for the determination of pesticides in water samples involve separatory techniques prior to quantitation (e.g. GLC, HPLC, TLC). Reports of direct spectrophotometric and spectrofluorometric determinations of pesticides are limited. Although such methods are simple, rapid and only require routine equipment and are therefore attractive for less sophisticated laboratories, they tend to suffer from inferior limits of determination ($\mu g\,l^{-1}$ to $mg\,l^{-1}$ range) and therefore lack of selectivity. A derivatization step (e.g. diazotization) of the parent or a degradation product is often employed.

Colorimetric methods are available for paraquat (Shivhare and Gupta, 1991), methyl parathion (Cruces Blanco and Garcia Sanchez, 1990), diphenyl ether herbicides (Ishikawa et al., 1985) and endosulphan (Raju and

Gupta, 1991). The utility of zero-order and first- and second-derivative ultraviolet (UV) spectrophotometry for identification of some dinitroaniline herbicides has been shown (Traore and Aaron, 1989). Limits of determination are only at the $1 \, mg \, l^{-1}$ level. The second-derivative technique is used to determine diquat in the presence of paraquat in the SCA method for these two herbicides in river and drinking waters (SCA, 1987) to a limit of determination of $0.02 \, \mu g \, l^{-1}$. A simple and sensitive spectrofluorometric method was developed by Capitan et al. (1993) for the determination of thiabendazole residues. By using solid phase preconcentration fluorimetry, the compound can be measured directly on a solid phase support to a limit of detection equivalent to $0.1 \, \mu g \, l^{-1}$.

Chemical sensors: Chemical sensors for environmental monitoring of pesticides are under active development. At present, however, they are of limited value.

Sensors which work by current or potential formation include ion-selective electrodes. In environmental monitoring their combination with an enzyme substrate reaction has some application for the detection of pesticides. In particular, organophosphorus compounds and carbamates, which are inhibitors for cholinesterase, can be detected (Durand et al., 1984; Wollenberger et al., 1991) by correlating the cholinesterase activity with intramembranal pH shifts induced by substrate hydrolysis. Stein and Schwedt (1992) have applied an acetylcholinesterase (AChE) biosensor, consisting of a pH-electrode and an enzyme membrane, to the screening of a number of pesticides in drinking water to the limit of $0.1 \, \mu g \, l^{-1}$.

Applications of chemical mass balances such as the piezoelectric (PZ) transducers have been used for solution control, e.g. with immobilized antibodies for atrazine and parathion detection (Guilbault and Schmid, 1991). Similar detection limits to those achievable with enzyme linked immunosorbent assay (ELISA) (i.e. lower $\mu g \, l^{-1}$ range) have been claimed.

Chemical fibre optical sensor systems (FOCS) would appear to have a bright future as they are profiting from the rapid developments in the communications industry. The detection of fluorescent compounds (e.g. PAHs) is particularly widespread with the advantage of combining time-resolved fluorescence emission spectra with chemometrical methods for on-line and *in situ* analysis of multicomponent mixtures. The permanent installation of a bare fibre configuration or a chemically reactive sensor tip should be an attractive solution for monitoring drinking water wells (Niessner, 1991). Adelhelm et al. (1992) used laser-induced photoacoustic spectroscopy with a UV excitation laser beam (253 and 376 nm) transmitted to a sample cuvette through a $600 \, \mu m$ diameter optical fibre to detect pesticides and polycyclic aromatics. Bioanalytical approaches of FOCS include the combination of cholinesterase inhibition and FOCS detection for organophosphorus compounds and carbamates (Wolfbeis and Koller, 1991; Hoebel and Polster, 1992). These remote sensors have application in the area of continuous monitoring of ground water in wells and drilling holes. The method could give a real-time indication of contamination and would therefore provide considerable time-saving over methods involving conventional sampling and subsequent laboratory analysis. While there has been a rapid rise in the number of published references for pesticides in water, these techniques are still under development.

Bioassay: Bioassays play a crucial role in assessing the actual or potential impacts of pesticides on the natural environment. They can be used to probe the extent to which an ecosystem is being or has been polluted. The use of bioassays as monitoring tools has been reviewed (Kohn, 1980; Maltby and Calow, 1989).

Microbial biosensors containing enbacterial and cyanobacterial species can be used for detecting a wide range of pollutants (Gaisford *et al.*, 1991). Bacterial biosensors incorporating the cyanobacterium *Synechococcus* as the biocatalyst are capable of detecting low concentrations of herbicides which interact with photosynthetic electron transfer chains. Linuron and atrazine could be rapidly detected at $50\,\mu g\,l^{-1}$ in water (Van Hoof *et al.*, 1990). A continuous flow system with the unicellular alga *Chlorella vulgaris* was used to detect and quantify triazine herbicides in water (Weston and Robinson, 1991). Samples and algae were passed through a mixing coil and fluorescence (λ emit. 685 nm; λ excit. 440 nm) measured in a flow cell fluorimeter; simazine concentrations as low as $12\,\mu g\,l^{-1}$ could be detected.

A laser/microbe bioassay has been developed (Felkner *et al.*, 1989) for rapid detection of low quantities of toxic contaminants in potable water. Toxicity to an isogensic set of *Bacillus subtilis* mutants is detected by differential light scattering (DLS) of the laser beam, which measures the response of the bacteria in terms of light intensity at various angles over 180°. Shrinking, swelling or failure of the bacteria to divide over time are the measured parameters of toxicity. The 19 bacteria each differ by only one property, which causes them to be more sensitive than other set members based on the mechanism of toxicity. A unique profile for individual pesticides can be generated to assist with identification.

Immunoassay: One of the driving forces behind the development of analytical methods for the detection of pesticides in the environment is the need for increasingly sensitive, yet simple and rapid, procedures. A dramatic expansion in the application of immunoassay technology to environmental monitoring has been observed in recent years. It was hoped to eliminate the tedious, time-consuming preparative work of conventional analytical methods by exploiting the specific molecular recognition and binding properties of antibodies for their analytes. In addition, the simplicity and versatility of immunoassays lends itself to field test kits, which would be of particular value in the environmental monitoring studies.

A considerable amount of development work has been undertaken by many groups to apply immunoassay techniques, in particular enzyme linked immunosorbent assays (ELISA) to the determination of pesticide residues in environmental samples, especially soil and water (Aston *et al.*, 1992; Van Emon and Lopez-Avila, 1992). The ELISA format is typically based on a 96 well microtitre plate, which has the advantage of allowing some 20–30 duplicate samples to be screened simultaneously in addition to standards and QC checks. Owing to the selectivity of the antibody response, sample preparation can often be minimal. The sensitivities of many assays are such that a limit of determination for residues in water in the range 0.01–$0.1\,\mu g\,l^{-1}$ can be achieved following solid phase extraction from a volume of about 150–250 ml. These features allow ELISA to be used for simultaneously screening larger numbers of samples than can be done with instrumental techniques such as GC or HPLC. The simplicity of the technique allows the construction of field test kits for some compounds.

Current disadvantages of ELISA are the considerable investment of time and facilities needed for assay development and the slow acceptance of immunoassay procedures for generating pesticide data for regulatory use. There is only limited regulatory experience in the consideration/use of immunoassay data (e.g. US EPA) and there are as yet no formulated rules. However, the same criteria as for other analytical methods need to be satisfied with this methodology (Van Emon, 1990):

- fully quantitative
- high specificity
- high reproducibility/repeatability
- requires validation data
- reagents (i.e. antibody) must be made available to other laboratories.

The number of immunoassays that have been applied to the detection of pesticide residues in water samples, although restricted in comparison to traditional chromatographic methods, is steadily increasing. At present over 30 individual pesticides can be analysed. Comparative data show that good comparability of ELISA and instrumental methods is being obtained (Eadsforth *et al.*, 1991 (Figure 5); Rittenberg *et al.*, 1991; Itak *et al.*, 1992), though accuracy and precision may be inferior to the instrumental methods (Table 3, Eadsforth *et al.*, 1991). Common sources of error in drinking water analysis are due to cross-reactivities and matrix effects; ELISA can be prone to false positives (Lee and Richman, 1991) although this is less

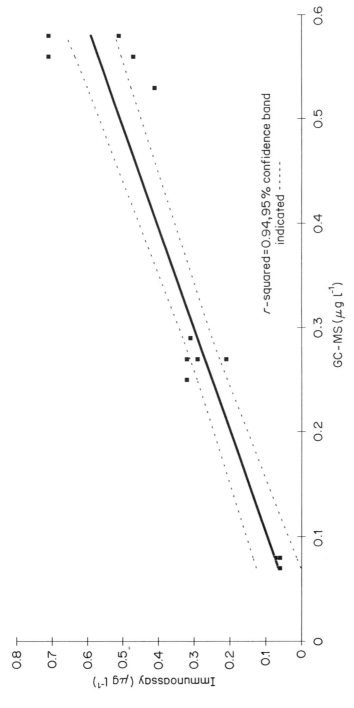

Figure 5 Correlation between GC-MS and immunoassay results for cyanazine in water

Table 3 Cyanazine added to water (μg l^{-1})—comparative analysis, 5 samples per level

Concentration added	–	0.075	0.25	0.60
ELISA	<0.01	0.059	0.29	0.56
		(SD 0.004)	(SD 0.05)	(SD 0.14)
GC-MS	<0.01	0.074	0.27	0.56
		(SD 0.002)	(SD 0.02)	(SD 0.02)

important if the technique is being used to screen samples subject to instrumental confirmation. Validation of ELISA for routine analysis by independent methods and interlaboratory ELISA assessment is considered important (Hansen *et al.*, 1990).

There are some reservations concerning the quality of some kits on the market, but the area of greater concern is the lack of information on antiserum specificity. Moreover, problems of sample matrix effects, which invariably have a significant influence on assay performance, are not always addressed. Antibody quality is a key factor and with advances in antibody production and selection procedures, improvements in the performance of kits are expected (Lee and Richman, 1991; J. P. Aston, 1995, personal communication).

Automation of the competitive ELISA method, which traditionally is carried out on microtitre plates and requires extensive pipetting, has been achieved by transferring to a flow injection system (Kramer and Schmid, 1991).

NATIONAL RESIDUE METHODS

In a number of countries, nationally recognized methods are published which are specifically designed for the determination of pesticides in water. Because these methods are to be used for monitoring purposes they are, where possible, designed as multiresidue procedures. They include as many pesticides as possible, often from a class, that can be analysed by a given method and for which performance data are available. Some of the more widely known series of water methods are listed in Table 4. Some of these, in particular the US EPA 500 series procedures, have gained an internationally accepted status and in this case represent the procedures to be used for compliance monitoring within the country of origin. Such methods are being developed, where possible, with limits of determination to allow national enforcement standards to be met.

Table 4 Sources of national residue methods for water

Country	Source	Comments
Canada	*Methods Manual for the Chemical Analysis of Trace Organics and Pesticides in Environmental Samples*. Part 1 Water 1992	Issued by the Alberta Environmental Centre and published as a consolidated reference for determining organics and pesticides in environmental samples on behalf of the Province of Alberta
UK	*Methods for the Examination of Waters and Associated Materials*	Methods prepared by the Standing Committee of Analysts and published by HMSO
USA	*EPA Methods for the Determination of Organic Compounds in Drinking Water*. EPA/600/4-88/039 December 1988	Methods numbered in the 500 series
	EPA Methods for the Determination of Organic Compounds in Industrial and Municipal Waste Water	Methods numbered in the 600 series
	EPA Test Methods for Evaluating Solid Wastes. SW-846	Methods numbered in the 8000 series. Recent updates 1990

USA, EPA methods

The US EPA 500 series methods provide procedures for several classes of pesticides in drinking water including organochlorine insecticides (OCs), nitrogen- and phosphorus (OP)-containing pesticides, chlorinated acid herbicides, N-methyl carbamoyloximes and N-methyl carbamates. These are detailed procedures largely based on capillary GC with provision for mass spectrometric (MS) determination and confirmation. The carbamates method is different in that it is based on determination by HPLC with fluorescence detection. While the OC and chlorinated acids methods have limits of determination for many of the compounds of $<0.1\ \mu g\,l^{-1}$, limits for the other two procedures are typically closer to $1\ \mu g\,l^{-1}$. These methods are not therefore always directly applicable for monitoring within the EEC.

Some of these methods have also been extensively tested. For example, the GLC electron-capture detection method (EPA method 508) for the determination of organochlorine pesticides in water has been subjected to a joint US EPA/AOAC interlaboratory method validation study (Lopez-Avila *et al.*, 1990) with 11 laboratories. Satisfactory recovery and precision data were obtained in reagent water and finished drinking waters. A similar collaborative study was carried out for nitrogen- and phosphorus-containing pesticides in finished drinking water and the method considered acceptable for all 45 test analytes (Edgell *et al.*, 1991). Edgell *et al.* (1992) later reported on a collaborative study undertaken by 10 laboratories on EPA National Pesticide Survey method 4 to determine the mean recovery and precision for analysis of 18 pesticides and metabolites in low to mid-$\mu g l^{-1}$ concentration range in reagent water and finished drinking water using HPLC.

In addition, EPA 600 series methods include a comparable set of procedures for pesticides in waste water, although being earlier methods they are largely based on GC using packed columns and need up-dating in this respect.

Methods in the 8000 series for wastes include procedures for water and cover OCs, OPs and chlorinated acid herbicides. Recent updates in this series include capillary GC and GC-MS and both of these series provide a good source of multiresidue procedures.

UK methods

The Standing Committee of Analysts (SCA) of the Department of the Environment in the UK issues a series of booklets published by HMSO and entitled *Methods for the Examination of Waters and Associated Materials*, the so-called 'blue book' methods. These booklets are intended to provide guidance on recommended methods for determining the quality of water.

Methods are currently available for similar classes of compounds to those in the USA but, in addition, recent booklets also cover paraquat and diquat, phenylurea herbicides and the important class of synthetic pyrethroids. The determinands normally include those pesticides within a class considered important in the UK and a number of additional booklets including newer compounds are currently in preparation. Depending on the scope of testing, procedures are published as methods, tentative methods or notes. Most of the methods enable a limit of determination close to or lower than $0.1 \, \mu g l^{-1}$ to be achieved and where possible, methods for water in this series are now designed to have limits of determination of the order of $0.01–0.02 \, \mu g l^{-1}$.

Most of the older GC methods including those for OCs, OPs and chlorophenoxy-acidic herbicides are based on the use of packed GC columns. These need updating to use capillary columns as has been done for the equivalent EPA methods. The pesticides panel of SCA draws on the expertise within the UK water industry, pesticide manufacturers, government and private research laboratories to develop and test the published procedures.

Other countries

In Canada, the methods manual from the Alberta Environmental Centre includes several multiresidue methods for water. Again the range of classes covered is similar to those for the EPA and SCA methods but, in addition to phenylurea herbicides, also includes difenzoquat and the more recently introduced herbicide chlorsulfuron. The GC procedures are up to date, being based on capillary techniques. Limits of determination are similar to those quoted for EPA methods and are typically in the range $0.1-1.0\,\mu\mathrm{g}\,\mathrm{l}^{-1}$ except for OCs which are $<0.1\,\mu\mathrm{g}\,\mathrm{l}^{-1}$ and the quaternary ammonium herbicides which are as high as $5-50\,\mu\mathrm{g}\,\mathrm{l}^{-1}$.

There are two sources for multiresidue methods in Germany: the German Research Society (DFG) and the German Standard Methods for Examination of Water, Waste Water and Sludge (DIN). The DFG has issued 13 multiresidue methods for the determination of pesticides in ground and drinking water, 8 of which have a limit of determination of $0.1\,\mu\mathrm{g}\,\mathrm{l}^{-1}$. Several DIN methods have been issued or are under development and all of these are sensitive to the $0.1\,\mu\mathrm{g}\,\mathrm{l}^{-1}$ concentration level.

Advantages and disadvantages

Some of the main advantages and disadvantages of preparing and publishing nationally recognized methods are included in Table 5. Inevitably there is a considerable time lapse between the initial development of a procedure,

Table 5   Advantages and disadvantages

Feature	Positive	Negative
Validity of method	Methods tend to be soundly based and may be extensively tested	Limited test data in some cases
Recognition	Achieve national and sometimes international status	
Speed of publication and updating		Time scale of development and testing tend to be long and methods become dated
Scope	Mostly multiresidue methods enabling numerous analytes to be determined simultaneously	Do not always include all the determinands of current interest
Limit of determination	Designed to meet national requirements	May not be directly transferable between countries, especially for drinking water

often initially for an individual pesticide in a class, and the publication of a fully or even partially (tentative) tested multiresidue method. Consequently, while methods are soundly based, they often tend to become quickly dated. The need to update packed column GC procedures and extend methods to cover additional newer determinands are examples of this. Solid phase extraction (SPE) of water samples has recently been very extensively evaluated for pesticides but is only now beginning to make a significant impact in recognized multiresidue methods where established solvent extraction techniques tend, to some extent because of their general applicability, to continue to be employed.

Although similar techniques are used, the tendency for methods developed outside Europe to be established with higher limits of determination means that not all these methods are directly transferable, especially for drinking water.

Other sources of methods

Attempts are being made to address some of the disadvantages found with existing procedures. A considerable body of technical information is available in the literature and from analytical instrument manufacturers and equipment suppliers. For example, most packed column GC methods have proposed capillary GC alternatives, often with a wider and more up-to-date range of pesticides included within the determination. In addition, SPE procedures and a range of separation (clean-up) techniques have been evaluated for multiresidue applications as discussed elsewhere in this chapter.

Strategies for segregating pesticides into different classes for monitoring purposes have been devised. For example, Moore et al. (1990) analysed for three categories: neutral/basic pesticides (using SPE/GC-MS), phenoxy-acids and phenolics (using SPE/derivatization followed by negative CI GC-MS) and phenylurea and carbamate pesticides (using SPE/LC-MS).

In addition, there is the wider body of literature and range of multiresidue methods for environmental samples, principally crops and soil, from which procedures can be taken and adapted as the basis for water methods. In the case of the corresponding nationally and internationally recognized multi-residue methods for these other matrices, some of the same disadvantages apply, but overall, procedures for this wider range of environmental samples form a valuable source of reference.

INTERNATIONAL RESIDUE METHODS

Work to develop international residue methods for the determination of pesticides in water is being undertaken by technical committees within SEN

(European Committee for Standardisation) and ISO (International Organisation for Standardisation). Procedures are provided for consideration by the countries represented on the committees and may be based on DIN, 'blue book' or other methods developed nationally. UK input to CEN and ISO committees is coordinated by BSI (British Standards Institution). As with the development of national residue methods there is a time lapse before an agreed procedure is available for publication. It is anticipated that CEN and ISO methods for a range of pesticides in water will be available in the near future.

MONITORING

The presence of pesticides in ground water supplies in the late 1970s and early 1980s in the USA (Aharonson, 1987), together with continuing regulatory and public concern over the presence of pesticides in the environment, has resulted in a large number of monitoring studies, initially to assess the condition of ground water, stream/river water and drinking water supplies and more recently to research pesticide volatilization, atmospheric transport and deposition in fog and rain. Multiresidue methods have been extensively applied during the course of these studies.

Since 1979, more than 150 studies of pesticides in ground water have been carried out in the USA by pesticide registrants, state and county governments, US governmental agencies and universities (US EPA, 1988b).

Within Europe, EEC legislation, together with a need to generate information on actual levels of certain pesticides in areas previously unmonitored, has resulted in a number of programmes, e.g.:

- UK   Drinking Water Inspectorate (DWI), 1991
- France   Legrand et al., 1990
- Germany   Iwan, 1988
- Italy   Funari et al., 1990
- Austria   Zouzaneas et al., 1993
- Sweden   Erlandsson et al., 1990
- Norway   Nilsen, 1990
- USSR   Motusinsky and Stroj, 1990
- Croatia   Fingler et al., 1992

CONCLUSIONS

New sampling and analysis approaches, aimed predominantly at more efficient sample analysis, are continually broadening the range of analytical techniques for pesticide trace analysis. Most analytical schemes, whether they are multiresidue or specifically aimed at target compounds, are still based on gas-liquid and high performance liquid chromatography, although new technologies have recently been considered seriously as routine methods. These are aimed either at achieving lower detection limits via improved sensitivity or selectivity of the detection system (e.g. atomic emission detection (GC-AED), immunoassay, capillary electrophoresis, and nuclear magnetic resonance spectroscopy) or confirming the presence of residues with more confidence (e.g. using diode array detection (HPLC-DAD), mass spectrometry (GC- and HPLC-MS) and fourier transform infrared spectroscopy (GC-FTIR)).

Attention is currently being focused on the movement of pesticides and their degradation products in water courses, via leaching or run-off. It is anticipated that, in future, liquid chromatography will play an increasingly important role in the analysis of these more mobile compounds. The use of mass spectrometry as an HPLC detector is considered a key element for the future development of HPLC methods and it is therefore encouraging that there have been promising developments in novel LC-MS interfaces, such as thermospray, particle beam and atmospheric pressure ionization. In particular, the confirmation of analyte identification provided by full scan MS is invaluable to any laboratory undertaking HPLC. Whereas low cost, robust bench top GC-MS instruments are routinely used nowadays for pesticide analysis, there is an urgent need for the equivalent HPLC systems.

There is continued pressure from regulators and the public for the acquisition of pesticides residue data at low level, which requires expensive and time-consuming efforts to avoid contamination during sampling, transport and sample work-up, as well as sophisticated and dedicated laboratory instrumentation for sub-ppb quantitation. There is also an increasing need from those involved in post-market environmental monitoring, site assessment and remediation of contaminated sites for detection methodology aimed at the rapid screening of samples. This is being addressed by the development of such techniques as immunoassays, bioassays, automation of sampling and clean-up (e.g. column switching) as well as the expanding area of *in situ* chemical sensors. Further improvements are envisaged when standard multiresidue methods are extended to include these newer techniques and are applied to a wider range of pesticides.

ACKNOWLEDGEMENTS

We wish to thank colleagues at Sittingbourne Research Centre for help given in preparing this chapter, in particular, Andrew Sherren and Eric Hitchings for development work with cyanazine methods and Richard Stephenson who provided the outline adapted for Figure 1. In addition, we would like to thank David Westwood at the Drinking Water Inspectorate, DoE for advice on work being undertaken to develop international residue methods and also a number of other colleagues for their helpful advice.

REFERENCES

Adelhelm, K., Faubel, W., and Ache, H. J. (1992). 'Fiber optic modified laser induced photoacoustic spectroscopy for the detection of pollutants in solutions', *Springer Ser. Opt. Sci.*, **69** (Photoacoust. Phototherm. Phenom. III), 41–50.

Aharonson, N. (1987). 'Potential contamination of ground water by pesticides', *Pure Appl. Chem.*, **59**(10), 1419–1446.

Ahel, M. A., Evans, K. M., Fileman, T. W., and Mantoura, R. F. C. (1992). 'Solid phase extraction of triazine herbicides from estuarine samples', *Abstr. Pap. Am. Chem. Soc.* (203 Meet. Pt. 1, ENVR 142), American Chemical Society, Washington, DC.

Akerblom, M., Thoren, L., and Staffas, A. (1990). 'Determination of pesticides in drinking water', *Vaar Foeda*, **42**(4–5), 236–243.

Allen, M. H. (1992). 'The evaluation of liquid chromatography/atmospheric pressure ionization/mass spectrometry (LC/API/MS) for the analysis of carbamate pesticides', *Perkin Elmer LC Views*, **Fall 1992**, 7–8.

Ambrus, A., Hargitai, E., Karoly, G., Fulop, A., and Lantos, J. (1981). 'General method for determination of pesticide residues in samples of plant origin, soil and water. II. Thin layer chromatographic determination', *J. Assoc. Off. Anal, Chem.*, **64**(3), 743–748.

Anderson, J. L., Whiten, K. K., Brewster, J. D., Ou, T.-Y., and Nonidez, W. K. (1985). 'Microarray electrochemical flow detectors at high applied potentials and liquid chromatography with electrochemical detection of carbamate pesticides in river water', *Anal. Chem.*, **57**, 1366–1373.

Arthur, C. L., Potter, D. W., Buchholz, K. D., Motlagh, S., and Pawliszyn, J. (1992). 'Solid-phase microextraction for the direct analysis of water: theory and practice', *LC-GC Intl*, **5**(10), 8–14.

Aston, J. P., Britton, D. W., Wraith, M. J., and Wright, A. S. (1992). 'Immunochemical methods for pesticide analysis', in *Emerging Strategies for Pesticide Analysis*, (Eds T. Cairns and J. Sherma), Chapter 15, pp. 309–329, CRC Press, Boca Raton, FL, USA.

Bagnati, R., Benfenati, E., Davoli, E., and Fanelli, R. (1988). 'Screening of 21 pesticides in water by single extraction with C18 silica bonded phase columns and HRGC-MS', *Chemosphere*, **17**(1), 59–65.

Banton, O., Lafrance, P., Martel, R., and Villeneuve, J. P. (1992). 'Planning of soil pore water sampling campaigns using pesticide transport modelling', *Ground Water Monit. Rev.*, **12**(3), 195–202.

Barcelo, D. (1991). 'Applications of gas chromatography – mass spectrometry in monitoring environmentally important compounds', *Trends Anal. Chem.*, **10**(10), 323–329.

Barcelo, D., Durand, G., Vreeken, R. J., de Jong, G. J., Lingemann, H., and Brinkmann, U. A. T. (1991). 'Evaluation of eluents in thermospray liquid chromatography – mass spectrometry for identification and determination of pesticides in environmental samples', *J. Chromatogr.*, **553**(1–2), 311–328.

Bellar, T. A., and Budde, W. L. (1988). 'Determination of nonvolatile organic compounds in aqueous environmental samples using liquid chromatography/mass spectrometry', *Anal. Chem.*, **60**, 2076–2083.

Benfenati, E., Tremolada, P., Chiappetta, L., Frassanito, R., Bassi, G., and Di Toro, N. (1990). 'Simultaneous analysis of 50 pesticides in water samples by solid phase extraction and GC-MS', *Chemosphere*, **21**(12), 1411–1421.

Beynon, K. I., Stoydin, G., and Wright, A. N. (1972). 'The breakdown of the triazine herbicide cyanazine in soils and maize', *Pestic. Sci.*, **3**, 293–305.

Bogus, E. R., Watschke, T. L., and Mumma, R. O. (1990). 'Utilization of solid-phase extraction and reversed-phase and ion-pair chromatography in the analysis of seven agrochemicals in water', *J. Agric. Food Chem.*, **38**(1), 142–144.

Bolygo, E., and Atreya, N. C. (1991). 'Solid-phase extraction for multi-residue analysis of some triazole and pyrimidine pesticides in water', *Fresenius' Z. Anal. Chem.*, **339**(6), 423–430.

Boussenadji, R., Dufek, P., and Porthault, M. (1991). 'Determination of phenylurea herbicides in water using microcolumn high performance liquid chromatography with UV and electrochemical detection', *LC-GC Intl*, **5**(10), 40–43.

Braithwaite, A., and Smith, F. J. (1990). 'Concentration and determination of trace amounts of chlorinated pesticides in aqueous samples', *Chromatographia*, **30**(3–4), 129–134.

Brinkman, U. A. T. (1992). 'The role of column liquid chromatography in hyphenated separation techniques', *Chim. Oggi.*, **10**(11/12), 9–14.

Brinkman, U. A. T., and Farjam, A. (1992). 'Immunoaffinity SPE coupled on-line with HPLC', *Abstr. Pap. Am. Chem. Soc.* (203 Meet., Pt. 1, ENVR 72), American Chemical Society, Washington, DC.

Brown, M. A., Stephens, R. D., and Kim, I. S. (1991). 'Liquid chromatography-mass spectrometry—a new window for environmental analysis', *Trends Anal. Chem.*, **10**(10), 330–336.

Buchberger, W. (1990). 'Trends in the combination of high-performance liquid chromatography and electroanalytical methods', *Chromatographia*, **30**, 577–581.

Buchmiller, R. C. (1989). 'Screening of ground water samples for volatile organic compounds using a portable gas chromatograph', *Ground Water Monit. Rev.*, **9**(3), 126–130.

Buffington, R., and Wilson, M. K. (1987). *Detectors for Gas Chromatography—a Practical Primer*. Hewlett Packard, Avondale, PA, USA.

Burger, K., Koehler, J., and Jork, H. (1990). 'Application of AMD to the determination of crop protection agents in drinking water, Part 1: fundamentals and method', *J. Planar Chromatogr.-Mod TLC*, **3**, 504–510.

Cai, J., and El Rassi, Z. (1992). 'Micellar electrokinetic capillary chromatography of neutral solutes with micelles of adjustable surface charge density', *J. Chromatogr.*, **608**(1–2), 31–45.

Cai, Z., Ramanujam, V. M. S., Giblin, D. E., and Gross, M. L. (1993). 'Determination of atrazine in water at low- and sub-parts-per-trillion levels by using solid phase extraction and gas chromatography/high resolution mass spectrometry', *Anal. Chem.*, **65**, 21–26.

Canova, J. L., and Muthig, M. G. (1991). 'The effect of latex gloves and nylon cord on ground water sample quality', *Ground Water Monit. Rev.*, **11**(3), 88–103.

Capitan, F., Alonso, E., Avidad, R., Capitan-Vallvey, L. F., and Vilchez, J. L. (1993). 'Determination of thiabendazole residues in water by solid phase spectrofluorometry', *Anal. Chem.*, **65**, 1336–1339.

Cavalier, T. C., Lavy, T. L., and Mattice, J. D. (1991). 'Persistence of selected pesticides in ground-water samples', *Ground Water*, **29**(2), 225–231.

Chen, P. H., Van Ausdale, W. A., Keeran, W. S., and Roberts, D. F. (1993). 'GC/MS identification of artifacts formed during sample preparation using US EPA methods 625 and 8270', *Chemosphere*, **26**(9), 1743–1749.

Cirilli, L., and Borgioli, A. (1986). 'Methyl bromide in surface drinking waters. Gas chromatographic determination by the headspace technique', *Water Res.*, **20**(3), 273–275.

Clark, L. (1988). *The Field Guide to Water Wells and Boreholes*. Open University Press, John Wiley & Sons, Chichester.

Clark, L., Gomme, J., Carter, A. D., and Harris, R. (1990). 'WRc/soil survey suction sampler', *Brighton Crop Protection Conference, Pests and Diseases*, **3**, 1011–1016.

Clark, L., Gomme, J., and Hennings, S. (1991). 'Study of pesticides in waters from a chalk catchment, Cambridgeshire', *Pestic. Sci.*, **32**, 15–33.

Cornacchia, J. W., Bethem, R. A., Frier, J. M., and Baln, K. (1993). 'Determination of hydroxysimazine and hydroxyatrazine in well water by high performance liquid chromatography thermospray tandem mass spectrometry (TSP-LC/MS/MS)', Poster presentation, *1st SETAC World Conference, March 28–31, Lisbon, Portugal*.

Crepeau, K. L., Walker, G., and Winterlin, W. (1991). 'Extraction of pesticides from soil leachate using sorbent disks', *Bull. Environ. Contam. Toxicol.*, **46**(4), 512–518.

Cruces Blanco, C., and Garcia Sanchez, F. (1990). 'A kinetic spectrophotometric method to determine the insecticide methyl parathion in commercial formulations and the aqueous environment', *Int. J. Environ. Anal. Chem.*, **38**, 513–523.

Davi, L. M., Baldi, M., Penazzi, L., and Linboni, M. (1992). 'Evaluation of the membrane approach to solid-phase extractions of pesticide residues in drinking water', *Pestic. Sci.*, **35**, 63–67.

Davies, I. L., Markides, K. E., Lee, M. L., Raynor, M. W., and Bartle, K. D. (1989). 'Applications of coupled LC-GC: a review', *J. High Resolut. Chromatogr. Chromatogr. Commun.*, **12**, 193–207.

de la Vigne, U., Jaenchen, D. E., and Weber, W. H. (1991). 'Application of high-performance thin-layer chromatography and automated development for the identification and determination of pesticides in water', *J. Chromatogr.*, **553**(1–2), 489–496.

DFG (Deutsche Forschungsgemeinschaft) (1987). *Manual of Pesticide Residue Analysis* (eds H-P. Their and H. Zeumer), Volume 1, pp 23–27 Weinheim, New York.

Di Corcia, A., and Marchetti, M. (1991). 'Multiresidue method for pesticides in drinking water using a graphitized carbon black cartridge extraction and liquid chromatographic analysis', *Anal. Chem.*, **63**(6), 580–585.

Di Corcia, A., and Marchetti, M. (1992). 'Method development for monitoring pesticides in environmental waters: liquid-solid extraction followed by liquid chromatography', *Environ. Sci. Technol.*, **26**(1), 66–74.

Di Corcia, A., Marchetti, M., and Samperi, R. (1989). 'Extraction and isolation of phenoxy acid herbicides in environmental waters using two adsorbents in one mini cartridge', *Anal. Chem.*, **61**, 1363–1367.

Dinelli, G., Vicari, A., and Catizone, P. (1993). 'Use of capillary electrophoresis for detection of metsulfuron and chlorsulfuron in tap water', *J. Agric. Food Chem.*, **41**, 742–746.

DoE (Department of the Environment) (1990). *Guidance on Safeguarding the Quality of Public Water Supplies*, Department of the Environment and Welsh Office, HMSO, London.

Doerge, D. R., and Bajic, S. (1992). 'Analysis of pesticides using liquid chromato-graphy—atmospheric-pressure chemical ionization mass spectrometry', *Rapid Commun. Mass Spectrom.*, **6**(11), 663–666.

Dombek, V., and Stránský, Z. (1992). 'Determination of alphametrine and cypermethrine in water and soil by capillary isotachophoresis', *Anal. Chim. Acta*, **256**(1), 69–73.

Dong, M. W., Vandermark, F. L., Renter, W. M., and Pickering, M. W. (1990). 'Carbamate pesticides analysis by LC', *Am. Environ. Lab.*, **2**(3), 14–27.

Durand, G., and Barcelo, D. (1989). 'Liquid chromatographic analysis of chlorotria-zine herbicides and degradation products in water samples with photodiode array detection. 1—Evaluation of two liquid–liquid extraction methods', *Toxicol. Environ. Chem.*, **25**, 1–11.

Durand, G., Bouvot, V., and Barcelo, D. (1992). 'Determination of trace levels of herbicides in esturine waters by gas and liquid chromatographic techniques', *J. Chromatogr.*, **607**, 319–327.

Durand, P., Nicund, J. M., and Mallevialle, J. (1984). 'Detection of organophos-phorus pesticides with an immobilized cholinesterase electrode', *Anal. Toxicol.*, **8**, 112–117.

DWI (Drinking Water Inspectorate) (1991). *Drinking Water 1991.* Department of the Environment, HMSO, London, July 1992.

DWI (Drinking Water Inspectorate) (1993). *Further Guidance on Analytical Systems.* DWI information letter 8/93, London, November 1993.

Eadsforth, C. V., Gill, J. P., and Woodbridge, A. P. (1991). 'Sampling and analysis techniques to study environmental fate of pesticides', *Brighton Crop Protection Conference, Weeds*, **3D-3**, 293–300.

Edelmann, B., Dedek, W., Weil, L., and Niessner, R. (1992). 'Preconcentration of insecticides in water using the new polymeric sorbent Wofatit Y77', *Fresenius' Z. Anal. Chem.*, **343**(1), 148–149.

Edgell, K. W., Erb, E. J., Longbottom, J. E., and Lopez-Avila, V. (1992). 'Liquid chromatographic determination of pesticides in finished drinking waters: colla-borative study', *J. Assoc. Off. Anal. Chem. Int.*, **75**(5), 858–871.

Edgell, K. W., Jenkins, E. L., Lopez-Avila, V., and Longbottom, J. E. (1991). 'Capillary column gas-chromatography with nitrogen-phosphorus detection for determination of nitrogen- and phosphorus-containing pesticides in finished drinking waters: collaborative study. *J. Assoc. Off. Anal. Chem.*, **74**(2), 295–309.

Eichelberger, J. W., Bellar, T. A., Donnelly, J. P., and Budde, W. L. (1990). 'Determination of volatile organics in drinking water with USEPA method 524.2 and the ion trap detector', *J. Chromatogr. Sci.*, **28**, 460–467.

Erlandsson, B., Sandberg, E., and Akerblom, M. (1990). 'Residues of pesticides in Swedish drinking water', Poster O8D-37, IUPAC, *Seventh International Congress of Pesticide Chemistry, Hamburg, Aug 5–10th, 1990.*

Fan, D. F., and Chen, X. L. (1991). 'Behaviour of the fungicide MBAMT in water', *Fresenius' Z. Anal. Chem.*, **339**(6), 434–435.

Farag, A. B., and El-Shahawi, M. S. (1991). 'Removal of organic pollutants from aqueous solution. V. Comparative study of the extraction, recovery and chromato-graphic separation of some organic insecticides using unloaded polyurethane foam columns', *J. Chromatogr.*, **552**(1–2), 371–379.

Farag, A. B., El Wakil, A. M., and El Shahawi, M. S. (1986). 'Collection and separation of some organic insecticides on polyurethane foam columns', *Fresenius' Z. Anal. Chem.*, **324**(1), 59–60.

Farran, A., Cortina, J. L., de Pablo, J., and Barcelo, D. (1990). 'On-line continuous flow extraction system in liquid chromatography with untraviolet and mass spectrometric detection for the determination of selected organic pollutants', *Anal. Chim. Acta.*, **234**(1), 119–126.

Feigel, C., and Holmes, W. (1991). 'The determination of triazine herbicides at ultra-trace levels by chemical ionization GC/MS', *Varian Application Note, Number 12*. Varian Associates Inc, Walnut Creek, CA, USA.

Felkner, I. C., Worthy, B., Christiason, T., Chaisson, C., Kurtz, J., and Wyatt, P. J. (1989). 'Laser/microbe bioassay system', in *Aquatic Toxicology Hazard Assessment*, Vol. 12, pp. 95–103, ASTM Special Technical Publication 1027. Tech. Assess. Syst, Inc, Washington DC.

Fingler, S., Drevenkar, V., Tkalcevic, B., and Smit, Z. (1992). 'Levels of polychlorinated biphenyls, organochlorine pesticides, and chlorophenols in the Kupa river waters and in drinking water from different areas in Croatia', *Bull. Environ. Contam. Toxicol.*, **49**(6), 805–812.

Foret, F., Sustacek, V., and Bocek, P. (1990). 'Separation of some triazine herbicides and their solvolytic products by capillary zone electrophoresis', *Electrophoresis*, **11**, 95–97.

Foster, G. D., and Rogerson, P. F. (1990). 'Enhanced preconcentration of pesticides from water using the Goulden large-sample extractor', *Int. J. Environ. Anal. Chem.*, **42**(3–4), 105–117.

Friege, H., and van Berk, W. (1989). 'Pesticides in surface, ground and raw water in North Rhine-Westphalia: measurement strategy, analysis and results', *Schriftenr. Ver. Wasser-Boden-Lufthyg. (Pflantzenschutzm grundwasser)* **79**, 397–414.

Friestad, H. O., and Bronstad, J. O. (1985). 'Improved polarographic method for determination of glyphosate herbicide in crops, soil and water', *J. Assoc. Off. Anal. Chem.*, **68**(1), 76–79.

Froehlich, D., and Meier, W. (1989). 'HPLC determination of triazines in water samples in the ppt-range by on-column trace enrichment', *J. High Resolut. Chromatogr.*, **12**(5), 340–342.

Funari, E., Bastone, A., Bottoni, P., Camoni, I., Carbone, S., and Donati, L. (1990). 'Herbicide contamination of drinking water supplies in Italy', Poster 08D-10, IUPAC, *Seventh International Congress of Pesticide Chemistry, Hamburg, Aug 5–10th, 1990*.

Gaisford, W. C., Richardson, N. J., Haggatt, P. G. D., and Rawson, D. M. (1991). 'Microbial biosensors for environmental monitoring', *Biochem. Soc. Trans.*, **19**(1), 15–18.

Garcia, F., and Henion, J. (1992). 'Fast capillary electrophoresis-ion spray mass spectrometric determination of sulfonylureas', *J. Chromatogr.*, **606**(2), 237–247.

Gardyan, C., and Thier, H. P. (1991). 'Identification and quantitation of currently important pesticides by HPTLC', *Z. Lebensm-Unters. Forsch.*, **192**(1), 40–45.

Geerdink, R. B. (1991). 'Direct determination of metamitron in surface water by large sample volume injection', *J. Chromatogr.*, **543**(1), 244–249.

Geerdink, R. B., Graumans, A. M. B. C., and Viveen, J. (1991). 'Determination of phenoxyacid herbicides in water', *J. Chromatogr.*, **547**(1–2), 478–483.

GIFAP (1990). *Water Quality Monitoring. Site Selection and Sampling Procedures for Pesticide Analysis*, Technical Monograph 13, March 1990, GIFAP, Brussels.

Gill, J. P., Bumpus, R. N., Coveney, P. C., Eadsforth, C. V., Mouillac, A., Rougeaux, J., and Teyras, G. (1992). 'Monitoring the environmental fate of flufenoxuron after use of CASCADE', *Colloque Phyt'eau: Eau-Produits Phytosanitaires Usages Agricoles et Connexes, Versailles, 21–22 Octobre*, pp. 97–101.

Glotfelty, D. E., Seiber, J. N., and Liljedahl, L. A. (1987). 'Pesticides in fog', *Nature*, **325**, 602–605.

Goerlitz, D. F., and Franks, B. J. (1989). 'Use of on-site high performance liquid chromatography to evaluate the magnitude and extent of organic contaminants in aquifers', *Ground Water Monit. Rev.*, **9**(2), 122–129.

Green, D. A., Stull, J. K., and Heesen, T. C. (1986). 'Determination of chlorinated hydrocarbons in coastal waters using a moored *in situ* sampler and transplanted live mussels', *Marine Pollut. Bull.*, **17**(7), 324–329.

Gremm, T. J., and Frimmel, F. H. (1992). 'Determination of organotin compounds in aqueous samples by means of HPGC-AED', *Water Res.*, **26**(9), 1163–1169.

Guilbault, G. C., and Schmid, R. D. (1991). 'Electrochemical, piezoelectrical and fibre-optic biosensors', *Adv. Biosensors*, **1**, 257–289.

Gurka, D. F., Pyle, S. M., Farnham, I., and Titus, R. (1991). 'Application of hyphenated fourier transform-infrared techniques to environmental analysis', *J. Chromatogr. Sci.*, **29**(8), 339–344.

Gutierrez-Galindo, E. A., Flores Munoz, G., Ortega Garcia, M. L., and Villaescusa Celaya, J. A. (1992). 'Pesticides in coastal waters of the Gulf of California: mussel water programme 1987–1988', *Cienc. Mar.*, **18**(2), 77–99.

Hadfield, S. T., Sadler, J. K., Bolygo, E., and Hill, I. R. (1992). 'Development and validation of residue methods for the determination of the pyrethroids lambda-cyhalothrin and cypermethrin in natural waters', *Pestic. Sci.*, **34**(3), 207–213.

Hadfield, S. T., Sadler, J. K., and Hill, I. R. (1989). 'The advantage of solid-phase extraction techniques for water sampling under field conditions. Part I', *Abstr. Pap. Am. Chem. Soc.* (198 Meet. AGRO 1), American Chemical Society, Washington, DC.

Hagen, D. F., Markell, C. G., Schmitt, G. A., and Blevins, D. D. (1990). 'Membrane approach to solid-phase extractions', *Anal. Chim. Acta.*, **236**(1), 157–164.

Hammond, I., Moore, K., James, H., and Watts, C. (1989). 'Thermospray liquid chromatography-mass spectrometry of pesticides in river water using reversed-phase chromatography', *J. Chromatogr.*, **474**, 175–180.

Hanson, P. D., Hock, B., Kanne, R., Krotsky, A., Otst, U., Oehmichen, U., Schlett, C., Schmid, R., and Weil, L. (1990). *DECHEMA Biotechnol. Conf. (1990).* 4 (Pt. A, Lect. DECHEMA Annu. Meet. Biotechnol. 8th, 1990), 31–45.

Hargesheimer, E. E. (1984). 'Rapid determination of organochlorine pesticides and polychlorinated biphenyls (in river water) using selected-ion-monitoring mass spectrometry', *J. Assoc. Off. Anal. Chem.*, **67**(6), 1067–1075.

Harris, G. L., Bailey, S. W., and Mason, D. J. (1991). 'The determination of pesticide losses to water courses in an agricultural clay catchment with variable drainage and land management', *Brighton Crop Protection Conference, Weeds*, **3**, 1271–1278.

Heegemann, W. (1988). 'Detection of traces of pesticides in ground water with reference to use of contact herbicides in maize', *Pflanzenarzt*, **41**(3), 30–32.

Heinonen, P., Paasivirta, J., and Herve, S. (1986). 'Periphyton and mussels in monitoring chlorohydrocarbons and chlorophenols in water courses', *Toxicol. Environ. Chem.*, **11**, 191–201.

Hemmerling, C., Risto, C., Augustyniak, B., and Jenner, K. (1991). 'Some investigations on the preparation of food and environmental samples for residue analysis of pesticides and PCBs by continuous steam distillation', *Nahrung*, **35**(7), 711–719.

Hennion, M. C., Subra, P., Rosset, R., Lamacq, J. Scribe, P., and Saliot, A. (1990). 'Off-line and on-line preconcentration techniques for the determination of phenylureas in freshwaters', *Int. J. Environ. Anal. Chem.*, **42**(1–4), 15–33.

Herve, S., Heinonen, P., Pankku, R., Knuutila, M., Koistinen, J., and Paasivirta, J.

(1988). 'Mussel incubation method for monitoring organochlorine pollutants in water courses. Four year application in Finland', *Chemosphere*, **17**, 1945–1961.

Hinckley, D. A., and Bidleman, T. F. (1989). 'Analysis of pesticides in seawater after enrichment onto C-8 bondend-phase cartridges', *Environ. Sci. Technol.*, **23**(8), 995–1000.

Hoebel, W., and Polster, J. (1992). 'Fiber optic biosensor for pesticides based on acetylcholine esterase', *Fresenius' Z. Anal. Chem.*, **343**(1), 101–102.

Hogendoorn, E. A., de Jong, A. P. J. M., van Zoonen, P., and Brinkman, U. A. Th. (1990). 'Reversed-phase chromatographic column switching for the trace-level determination of polar compounds. Application to chloroallyl alcohol in ground water', *J. Chromatogr.*, **511**, 243–256.

Hogendoorn, E. A., and Goewie, C. E. (1989). 'Residue analysis of the herbicides cyanazine and bentazone in sugar maize and surface water using high-performance liquid chromatography and on-line clean-up column-switching procedures', *J. Chromatogr.*, **475**, 432–441.

Hogendoorn, E. A., Goewie, C. E., and van Zoonen, P. (1991a). 'Application of HPLC column switching in pesticide residue analysis', *Fresenius' Z. Anal. Chem.*, **339**(6), 348–356.

Hogendoorn, E. A., and van Zoonen, P. (1992). 'Coupled-column RPLC (or LC) for the rapid screening of bentazone in water with direct sample injection', *Fresenius' Z. Anal. Chem.*, **343**(1), 73–74.

Hogendoorn, E. A., van Zoonen, P., and Brinkman, U. A. Th. (1991b). 'Column-switching RPLC for the trace-level determination of ethylenethiourea in aqueous samples', *Chromatographia*, **31**, 285–292.

Hopfgartner, G., Wachs, T., Bean, K., and Henion, J. (1993). 'High flow ion spray liquid chromatography/mass spectrometry', *Anal. Chem.*, **65**(4), 439–446.

House, W. A., and Ou, Z. (1992). 'Determination of pesticides on suspended solids and sediments; investigations on the handling and separation', *Chemosphere*, **24**(7), 819–832.

Huber, S. A., Scheunert, I., Doerfler, U., and Frimmel, F. H. (1992). 'Interaction of dissolved organic carbon (DOC) with some pesticides in aqueous systems', *Z. Wasser-Abwasser-Forsch.*, **25**(2), 74–81.

Huckins, J. N., Tubergen, M. W., and Manuweera, G. K. (1990). 'Semi-permeable membrane devices containing model lipid: a new approach to monitoring the bioavailability of lipophilic contaminants and estimating their bioconcentration potential', *Chemosphere*, **20**(5), 533–552.

Hunt, D. T. E., and Wilson, A. L. (eds) (1986). *The Chemical Analysis of Water, General Principles and Techniques*. Second edition. The Royal Society of Chemistry, Alden Press, Oxford.

Ishikawa, K., Kamo, E., Sasaki, A., Kuzuoka, S., Yonekura, Y., Suzuki, K., and Ito, T. (1985). 'Colorimetric method for the determination of diphenyl ether herbicides in water', *Eisei Kaguku*, **31**(2), 72–78.

Itak, J. A., Selisker, M. Y., and Herzog, D. P. (1992). 'Development and evaluation of magnetic pesticide based enzyme immunoassay for aldicarb, aldicarb sulfone and aldicarb sulfoxide', *Chemosphere.*, **24**(1), 11–21.

Iwan, J. (1988). 'Pesticides in ground and drinking water—results of a monitoring program in Germany', *Gesunde Pflanz.*, **40**(5), 208–213.

Johnson, W. E., Fendinger, N. J., and Plimmer, J. R. (1990). 'Solid-phase extraction of pesticides from water: possible interference from dissolved organic material', *Abstr. Pap. Am. Chem. Soc.* (200 Meet. Pt. 1. AGRO 56), American Chemical Society, Washington, DC.

Jönsson, J. A., and Mathiasson, L. (1992). 'Supported liquid membrane techniques for sample preparation and enrichment in environmental and biological analysis', *Trends Anal. Chem.*, **11**(3), 106–114.

Junk, G. A., and Richard, J. J. (1988). 'Organics in water: solid-phase extraction on a small scale', *Anal. Chem.*, **60**, 451–454.

Keydel, S. P., and Seiber, J. N. (1988). 'In situ extraction of water for multiresidue pesticide analysis', *Abstr. Pap. Am. Chem. Soc.* (196 Meet. AGRO 76), American Chemical Society, Washington, DC.

Koehle, H., Karrenbrock, F., and Haberer, K. (1990). 'Atomic emission detector: trace analysis of herbicides in water', *Wasser, Luft Boden*, **34**(6), 16–20.

Kohn, G. K. (1980). 'Bioassay as a monitoring tool', *Residue Rev.*, **76**, 99–129.

Kongovi, R. R. (1981). 'Problem isolation in the analysis of pesticides and herbicides in water. Part 2', *Am. Lab.*, **13**(8), 61–62.

Koskinen, W. C., Otto, J. M., Jarvis, L. J., and Dowdy, R. H. (1992). 'Potential interferences in the analysis of atrazine and deethylatrazine in soil and water', *J. Environ. Sci. Health, Part B. Pestic. Food Contam. Agric. Wastes*, **27**(3), 255–268.

Kramer, P. M., and Schmid, R. D. (1991). 'Automated quasi-continuous immuno-analysis of pesticides with a flow injection system', *Pestic. Sci.*, **32**, 451–462.

Kratochvil, B., and Peak, J. (1989). 'Sampling techniques for pesticide analysis', in *Analytical Methods for Pesticides and Plant Growth Regulators—Advanced Analytical Techniques* (ed. J. Sherma, Vol. 17, Chapter 1, pp. 1–33, Academic Press, San Diego, California, USA.

Krolski, M. E., Bosnak, L. L., and Murphy, J. J. (1992). 'Application of nuclear magnetic resonance spectroscopy to the identification and quantitation of pesticide residues in soil', *J. Agric. Food Chem.*, **40**, 458–461.

Kuhr, W. G. and Monnig, C. A. (1992). 'Capillary electrophoresis', *Anal. Chem.*, **64**, 389R–407R.

Kwakman, P. J. M., Vreuls, J. J., Brinkman, U. A. Th., and Ghijsen, R. T. (1992). 'Determination of organophosphorus pesticides in aqueous samples by on-line membrane disk extraction and capillary gas chromatography', *Chromatographia*, **34**(1/2), 41–47.

Lawrenz, A. (1983). 'Analysis of triazine herbicides in water by thin-layer chromatography', *Acta Hydrochim. Hydrobiol.*, **11**(3), 347–350.

Lechien, A., Valenta, P., Neurnberg, H. W., and Patriarche, G. J. (1981). 'Differential pulse polarography of some herbicides derived from 2,4-dichlorophenoxyacetic acid. II. Determination of herbicide residues in irrigation waters', *Fresenius' Z. Anal. Chem.*, **306**(2–3), 156–160.

Lee, H. B., and Stokker, Y. D. (1986). 'Analysis of eleven triazines in natural waters', *J. Assoc. Off. Anal. Chem.*, **69**(4), 568–572.

Lee, S. M., and Richman, S. (1991). 'Thoughts on the use of immunoassay techniques for pesticide residue analysis', *J. Assoc. Off. Anal. Chem.*, **74**(6), 893.

Lee, S. M., and Wylie, P. L. (1991). 'Comparison of the atomic emission detector to other element-selective detectors for the gas chromatographic analysis of pesticide residues', *J. Agric. Food Chem.*, **39**, 2192–2199.

Legrand, M. F., Costentin, E., and Bruchet, A. (1990). 'Occurence of 38 pesticides in various French surface and ground waters', Poster 08D-09, IUPAC, *Seventh International Congress of Pesticide Chemistry, Hamburg, Aug 5–10th, 1990*.

Leoni, V., Cremisimi, C., Casuccio, A., and Gulliotti, A. (1991). 'Separation of pesticides, related compounds, polychlorobiphenyls, and other pollutants into four groups by silica-gel microcolumn chromatography (application to surface water analysis)', *Pestic. Sci.*, **31**, 209–220.

Lin, H. Y., and Voyksner, R. D. (1993). 'Determination of environmental contaminants using an electrospray interface combined with an ion trap mass spectrometer', *Anal. Chem.*, **65**(4), 451–456.

Loch, J. P. G., and Verdam, B. (1989). 'Pesticide residues in groundwater in the Netherlands. State of observations and future directions of research', *Schriftenr. Ver. Wasser-, Boden-Lufthyg.*, **79**, 349–363.

Loconto, P. R. (1991). 'Solid-phase extraction in trace environmental analysis, Part I — current research', *LC-GC Intl*, **4**(9), 10–15.

Lopez-Avila, V., Wesselman, R., and Edgell, K. (1990). 'Gas chromatographic-electron capture detection method for determination of 29 organochlorine pesticides in finished drinking water: collaborative study', *J. Assoc. Off. Anal. Chem.*, **73**(2), 276–286.

Macomber, C., Bushway, R. J., Perkins, L. B., Baker, D., Fan, T. S., and Ferguson, B. S. (1992). 'Determination of the ethanesulfonate metabolite of alachlor in water by high performance liquid chromatography', *J. Agric. Food Chem.*, **40**(8), 1450–1452.

Majors, R. E. (1993). 'A comparative study of European and American trends in sample preparation', *LC-GC Intl*, **6**(3), 130–140.

Malissa, H., Kreindl, T., and Winsauer, K. (1990). 'GC-FTIRS — applications in organic trace analysis', *Fresenius' Z. Anal. Chem.*, **337**, 843–847.

Maltby, L., and Calow, P. (1989). 'The application of bioassays in the resolution of environmental problems; past, present and future', *Hydrobiologia*, **188–189**, 65–76.

Marutoiu, C., Vlassa, M., Sarbu, C., and Nagy, S. (1987). 'Separation and identification of organophosphorus pesticides in water by HPLTC', *J. High Resolut. Chromatogr. Chromatogr. Commun.*, **10**(8), 465–466.

Marvin, C. H., Brindle, I. D., Hall, C. D., and Chiba, M. (1991). 'Rapid on-line precolumn high performance liquid chromatographic method for the determination of benomyl, carbendazim and aldicarb species in drinking water', *J. Chromatogr.*, **555**(1–2), 147–154.

Marvin, C. H., Brindle, I. D., Singh, R. P., Hall, C. D., and Chiba, M. (1990). 'Simultaneous determination of trace concentrations of benomyl, carbendazim (MBC) and nine other pesticides in water using an automated on-line pre-concentration high performance liquid chromatographic method', *J. Chromatogr.*, **518**(1), 242–249.

Mathiasson, L., Nilvé, G., and Ulén, B. (1991). 'A liquid membrane enrichment technique for integrating field sampling in water applied to MCPA', *Int. J. Environ. Anal. Chem.*, **45**, 117–125.

Mattern, G. C., Louis, J. B., and Rosen, J. D. (1991a). Multipesticide determination in surface water by gas chromatography/chemical ionization/mass spectrometry/ion trap detection', *J. Assoc. Off. Anal. Chem.*, **74**(6), 982–986.

Mattern, G. C., Roinestad, K. S., and Rosen, J. D. (1991b). 'Pesticide analysis in water using ion-trap mass spectrometry', *Abstr. Pap. Am. Chem. Soc.* (202 Meet, Part I, AGRO 60), American Chemical Society, Washington, DC.

Melcher, R. G., and Morabito, P. L. (1990). 'Membrane/gas chromatographic system for automated extraction and determination of trace organics in aqueous samples', *Anal. Chem.*, **62**(20), 2183–2188.

Merriman, J. C., Struger, J., and Szawiola, R. S. (1991). 'Distribution of 1,3-dichloropropene and 1,2-dichloropropane in Big Creek watershed', *Bull. Environ. Contam. Toxicol.*, **47**(4), 572–579.

Meyer, M. T., Mills, M. S., and Thurman, E. M. (1993). 'Automated solid-phase extraction of herbicides from water for gas chromatographic-mass spectrometric analysis', *J. Chromatogr.*, **629**(1), 55–59.

Miles, C. J., Doerge, D. R., and Bajic, S. (1992). 'Particle beam/liquid chromatography/mass spectrometry of National Pesticide Survey analytes', *Arch. Environ. Contam. Toxicol.*, **22**(2), 247–251.

Millar, J. D., Thomas, R. E., and Schattenberg, H. J. III (1981). 'Determination of organochlorine pesticides and polychlorinated biphenyls in water by gas chromatography', *Anal. Chem.*, **53**(2), 214–219.

Miller, J. W., Markell, C., and Wylie, P. L. (1991). 'Analysis of pesticides at the parts-per-billion level in surface water using gas, chromatography with atomic emission detector', Hewlett–Packard Application Note 228–127, Hewlett–Packard Co, Avondale, PA, USA.

Mills, M. S., and Thurman, E. M. (1992). 'Mixed mode isolation and purification of atrazine herbicides from organic and aqueous solutions. *Abstr. Pap. Am. Chem. Soc.* (203 Meet. Pt. 1, ENVR 12), American Chemical Society, Washington, DC.

Mohnke, M., Rohde, K. H., Bruegmann, L., and Franz, P. (1986). 'Trace analysis of some chlorinated hydrocarbon in water by gas-liquid chromatography', *J. Chromatogr.*, **364**, 323–337.

Moore, K., Gibby, S., Fielding, M., and Watts, C. (1990). 'Multi-residue analytical method for monitoring pesticides for compliance with the EC Drinking Water Directive', Poster 08B/54, IUPAC, *Seventh International Congress of Pesticide Chemistry, Hamburg, Aug 5–10th, 1990*.

Mortimer, R. D., and Dawson, B. A. (1991a). 'A study to determine the feasibility of using ^{31}P NMR for the analysis of organophosphorus insecticide residues in cole crops', *J. Agric. Food Chem.*, **39**, 911–916.

Mortimer, R. D., and Dawson, B. A. (1991b). 'Using ^{19}F NMR for trace analysis of fluorinated pesticides in food products', *J. Agric. Food Chem.*, **39**, 1781–1785.

Motusinsky, N. F., and Stroj, A. N. (1990). 'Quality of underground waters under conditions of pesticide uses', Poster 08D-08, IUPAC, *Seventh International Congress of Pesticide Chemistry, Hamburg, Aug 5–10th, 1990*.

Munch, D. J., and Frebis, C. P. (1992). 'Analyte stability studies conducted during the National Pesticide Survey', *Environ. Sci. Technol.*, **26**(5), 921–925.

Naish-Chamberlain, P. J., and Cooke, A. R. (1991). 'Automation of sample preparation for pesticide analysis in water', *LC-GC Intl*, **4**(9), 38–45.

Nations, B. K., and Hallberg, G. R. (1992). 'Pesticides in Iowa precipitation', *J. Environ. Qual.*, **21**(3), 486–492.

Niessner, R. (1991). 'Chemical sensors for environmental analysis', *Trends Anal. Chem.*, **10**(10), 310–316.

Nilsen, H.-G. (1990). 'Monitoring of pesticides in water in Norway', Poster 08D-38, IUPAC, *Seventh International Congress of Pesticide Chemistry, Hamburg, Aug 5–10th, 1990*.

Noroozian, E., Maris, F. A., Nielen, M. W. F., Frei, R. W., de Jong, G. J., and Brinkman, U. A. T. (1987). 'Liquid chromatographic trace enrichment with on-line capillary gas chromatography for the determination of organic pollutants in aqueous samples', *J. High Resolut. Chromatogr. Chromatogr. Commun.*, **10**(1), 17–24.

Noy, Th., Weiss, E., Herps, T., Van Crutchen, H., and Rijks, J. (1988). 'On-line combination of liquid chromatography and capillary gas chromatography. Preconcentration and analysis of organic compounds in aqueous samples', *J. High Resolut. Chromatogr. Chromatogr. Commun.*, **11**(2), 181–186.

Osredkar, R., and Kadaba, P. K. (1982). 'Application of nuclear double resonance technique for detection of pesticides in water at low concentrations', *Bull. Environ. Contam. Toxicol.*, **28**, 513–518.

Parker, L. V. (1992). 'Suggested guidelines for the use of PTFE, PVC and stainless steel in samplers and well casings', in *Current Practice in Ground Water Vadose Zone Investment*, (Eds. D. M. Nielsen and M. N. Sata) pp. 217–229, ASTM Special Technical Publication, STP 1118.

Patil, V. B., Sevalkar, M. T. and Padaliker, S. V. (1992). 'Thin-layer chromatographic detection of pyrethroid insecticides containing a nitrile group', *Analyst*, **117**, 75–76.

Pereira, W. E., Rostad, C. E., and Leiker, T. J. (1990). 'Determination of trace levels of herbicides and their degradation products in surface and ground water by gas chromatography/ion-trap mass spectrometry', *Anal. Chim. Acta*, **228**(1), 68–75.

Pestemer, W., Nordmeyer, H., and Scholz, G. (1989). 'Degradation behaviour of pesticides in porous aquifers', *Schriftenr. Ver. Wasser-, Boden-Lufthyg.*, **79**, 313–327.

Pettyjohn, W. A., Dunlap, W. J., Cosby, R., and Keeley, J. W. (1981). 'Sampling ground water for organic contaminants', *Ground Water*, **19**(2), 180–189.

Pleasance, S., Anacleto, J. F., Bailey, M. R., and North, D. W. (1992). 'An evaluation of atmospheric pressure ionization techniques for the analysis of N-methyl carbamate pesticides by liquid chromatography mass spectrometry', *J. Am. Soc. Mass Spectrom.*, **3**, 378–397.

Poole, S. K., Dean, T. A., Ondsema, J. W., and Poole, C. F. (1990). 'Sample preparation for chromatographic separations: an overview', *Anal. Chim. Acta*, **236**, 3–42.

Prest, H. F., Jarman, W. M., Burns, S. A., Weismuller, T., Martin, M., and Huckins, J. N. (1992). 'Passive water sampling via semi-permeable membrane devices (SPMDS) in concert with bivalves in the Sacramento San Joaquin River Delta', *Chemosphere*, **25**(12), 1811–1823.

Raju, J., and Gupta, V. K. (1991). 'A simple spectrophotometric determination of endosulphan in river water and soil. *Fresenius' Z. Anal. Chem.*, **339**(6), 431–433.

Reupert, R., and Plöger, E. (1989). 'The determination of N-herbicides in ground, drinking and surface water: analytical method and results', *Vom. Wasser*, **72**, 211–233.

Rittenberg, J. H., Fitzpatrick, D. A., Stocker, D. R., and Grothaus, G. D. (1991). 'Development of simple and rapid immunoassay systems for analysis of pesticides', *Brighton Crop Protection Conference, Weeds*, **3D-1**, 281–288.

Rostad, C. E., Pereira, W. E., and Leiker, T. J. (1989). 'Determination of herbicides and their degradation products in surface waters by gas chromatography/positive chemical ionization/mass spectrometry', *Biomed. Environ. Mass Spectrom.*, **18**, 820–827.

Roulston, S. A. (1993). 'Solid phase extraction in foods and related matrices', *Leatherhead Food R.A. Technical Notes, April, 1993.*

Sanchez-Palacios, A., Perez-Ruiz, T., and Martinez-Lozano, C. (1992). 'Determination of diquat in real samples by electron spin resonance spectrometry', *Anal. Chim. Acta*, **268**(2), 217–223.

Sarkar, A., and Sen Gupta, R. (1989). 'Determination of organochlorine pesticides in Indian coastal water using a moored *in situ* sampler', *Water Res.*, **23**(8), 975–978.

SCA (Standing Committee of Analysts) (1987). Determination of diquat and paraquat in river and drinking waters, spectrophotometric methods tentative 1987. *Methods for the Examination of Waters and Associated Materials*, HMSO, London.

SCA (Standing Committee of Analysts) (1994a). 'Phenylurea herbicides (urons), dinocap, dinoseb, benomyl, carbendazim and metamitron in waters 1994', *Methods for the Examination of Waters and Associated Materials*, HMSO, London.

SCA (Standing Committee of Analysts) (1994b). 'Determination of aldicarb and other N-methyl carbamates in waters 1994', *Methods for the Examination of Waters and Associated Materials*, HMSO, London.

Schuette, S. A., Smith, R. G., Holden, L. R., and Graham, J. A. (1990). 'Solid-phase extraction of herbicides from well water for determination by gas chromatography-mass spectrometry', *Anal. Chim. Acta*, **236**(1), 141–144.

Senin, N. N., Filippov, Y. S., Tolikina, N. F., Smol'Yaninov, G. A., Volkov, S. A., and Kukushkin, V. S. (1986). 'Chromatographic determination of some trace organic impurities in natural and waste waters with preliminary adsorption trapping', *J. Chromatogr.*, **364**, 315–321.

Senseman, S. A., Lavy, T. L., Mattice, J. D., Myers, B. M., and Skulman, B. W. (1993). 'Stability of various pesticides on membraneous solid phase extraction media', *Environ. Sci. Technol.*, **27**(3), 516–519.

Sesia, E., Gasparini, G. F., Icardi, M. L., and Dagna, L. (1990). 'Bidimensional chromatography of agrochemical residues in honey and water', *Lab. 2000*, **4**(4), 24–29.

Shivhare, P., and Gupta, V. K. (1991). 'Spectrophotometric method for the determination of paraquat in water, grain and plant materials', *Analyst*, **116**(5), 391–393.

Simmons, N. D. (1991). 'Residues in ground water', *Brighton Crop Protection Conference, Weeds*, **3**, 1259–1269.

Simon, V. A., and Taylor, A. (1989). 'High-sensitivity HPLC analysis of diquat and paraquat with confirmation', *J. Chromatogr.*, **479**, 153–158.

Singmaster, J. A. III (1991). 'Contaminants in aged pesticide grade dichloromethane interfering in herbicide and residue analysis and a method for their removal', *J. Agric. Univ. P.R.*, **75**(4), 329–344.

Smith, S., Jr (1993). 'Pesticide retention by a programmable automatic water/suspended-sediment sampler', *Bull. Environ. Contam. Toxicol.*, **50**(1), 1–7.

Södergren, A. (1990). 'Monitoring of persistent, lipophilic pollutants in water and sediment by solvent-filled dialysis membranes', *Ecotoxicol. Environ. Safety*, **19**, 143–149.

Stein, K., and Schwedt, G. (1992). 'Application of an acetylcholinesterase (AChE) biosenser consisting of a pH-electrode and an enzyme membrane for pesticide screening in drinking water', *Vom. Wasser*, **79**, 211–229.

Steinheimer, T. R. (1993). 'HPLC determination of atrazine and principal degradates in agricultural soils and associated surface and ground water', *J. Agric. Food Chem.*, **41**, 588–595.

Stelluto, S., Marcomini, A., di Corcia, A., Marchetti, M., Capri, S., and Liberatori, A. (1990). 'Method comparison for herbicide determination in water: GC-MS versus HPLC and liquid–liquid versus solid-phase extraction', *Ann. Chim. (Rome)*, **80**(7–8), 369–377.

Sullivan, J. J. (1991). 'Screening in gas chromatography with atomic emission detection', *Trends Anal. Chem.*, **10**(1), 23–26.

Suprynowicz, Z., and Gawdzik, J. (1989). 'Enrichment of trace amounts of organochlorine pesticides from water for GC analysis', *Fresenius' Z. Anal. Chem.*, **334**(7), 662–663.

Suzuki, T., and Watanabe, S. (1992). 'Liquid chromatographic screening method for fluorescent derivatives of chlorophenoxy acid herbicides in water', *J. Assoc. Off. Anal. Chem. Int.*, **75**(4), 720–724.

Svoboda, L., Jandera, P., and Cheracek, J. (1991). 'Study of the sorption of selected pesticides on non-polar sorbents', *Collect. Czech. Chem. Comm.*, **56**(2), 317–333.

Swineford, D. M., and Belisle, A. A. (1989). 'Analysis of trifluralin, methyl paraoxon, methyl parathion, fenvalerate and 2,4-D dimethylamine in pond water using solid-phase extraction', *Environ. Toxicol. Chem.*, **8**, 465–468.

Tang, P. H., Ho, J. S., and Eichelberger, J. W. (1993). 'Determination of organic pollutants in reagent water by liquid–solid extraction followed by supercritical fluid elution', *J. Assoc. Off. Anal. Chem. Int.*, **76**(1), 72–82.

Tena, M. T., Luque de Castro, M. D., and Valcarcel, M. (1992). 'HPLC-postcolumn derivatising-integrated retention-detection system for the determination of carbaryl and its hydrolysis product', *J. Chromatogr. Sci.*, **30**(7), 276–279.

Topp, E., and Smith, W. (1992). 'Sorption of the herbicides atrazine and metolachlor to selected plasticis and silicone rubber', *J. Environ. Qual.*, **21**(3), 316–317.

Traore, S., and Aaron, J.-J. (1989). 'Analysis of trifluralin and other dinitroaniline herbicide residues by zero-order and derivative ultraviolet spectrophotometry', *Analyst*, **114**, 609–613.

Tsuchiya, Y., and Ohashi, N. (1992). 'Determination of pesticides in water by GC/MS using internal standards', *Jpn. J. Toxicol. Environ. Health*, **38**(6), 560–565.

US EPA (1988a). *Methods for the Determination of Organic Compounds in Drinking Water*, EPA-600/4-88/039, December 1988. US EPA, Washington, DC.

US EPA (1988b). *Pesticides in Ground Water Database, 1988 Interim Data*. Office of Pesticide Programs, EPA, Washington, DC.

US EPA (1989). *Method 8410, Rev. 1, SWA 846*, US EPA, Washington, DC.

Valkirs, A. O., Stallard, M. O., Stang, P. M., and Frank, S. (1990). 'Assessment of frozen storage of tributyltin in sea-water samples using hydride derivatisation', *Analyst*, **115**(10), 1327–1328.

Van der Hoff, G. R., Gort, S. M., Baumann, R. A., Brinkman, U. A. Th., and van Zoonen, P. (1991). 'Clean-up by automated solid phase extraction cartridges coupled on-line to capillary GC-ECD; application to the analysis of synthetic pyrethroids in surface water. *3rd Workshop on Chemistry and Fate of Modern Pesticides, Bilthoven, Sept. 2–6*, B30.

Van Emon, J. M. (1990). 'Immunoassay methods—EPA evaluations', in *Immunochemical Methods for Environmental Analysis* (eds J. M. Van Emon and R. O. Mumma), pp. 58–64, Am. Chem. Soc. Symp. Ser. 442, American Chemical Society, Washington, DC.

Van Emon, J. M., and Lopez-Avila, V. (1992). 'Immunochemical methods for environmental analysis', *Anal. Chem.*, **64**(2), 79A–88A.

Van Hoof, F., Van Es, T., Briers, M., Hansen, P., and Pluta, H. (1990). 'The evaluation of bacterial biosensors for water quality monitoring', *Meded. Fac. Landbouwwet Rijkuniv. Gent.*, **55**(4), 1429–1437.

Van Zoonen, P., Hogendoorn, E. A., van der Hoff, G. R., and Baumann, R. A. (1992). 'Selectivity and sensitivity in coupled chromatographic techniques as applied to pesticide residue analysis', *Trends Anal. Chem.*, **11**(1), 11–17.

Voyksner, R. D. (1987). 'Thermospray HPLC/MS for monitoring the environment' in *Application of New Mass Spectrometry Techniques in Pesticide Chemistry* (ed. J. D. Rosen), pp. 146–160, John Wiley and Sons Inc., New York.

Waite, D. T., Grover, R., Westcott, N. D., Sommerstad, H., and Kerr, L. (1992).

'Pesticides in ground water, surface water and spring runoff in a small Saskatchewan watershed', *Environ. Toxicol. Chem.*, **11**, 741–748.

Wang, W. K., and Huang, S. D. (1989). 'Rapid determination of seven herbicides in water or isooctane using C18 and florisil Sep-pak cartridges and gas-chromatography with electron capture detection', *J. Chromatogr.*, **483**, 121–129.

Wauchope, R. D. (1978). 'The pesticide content of surface water draining from agricultural fields—a review', *J. Environ. Qual.*, **7**, 459–472.

Welter, A., and Ottmann, J. (1988). 'Testing for pesticide residues in drinking water and Rhine water by gas chromatography with a mass selective detector', *Lebensmittelchem. Gerichtl. Chem.*, **42**(1–3), 14–15.

Weston, L. H., and Robinson, P. K. (1991). 'Detection and quantification of triazine herbicides using algal cell fluorescence', *Biotechnol. Tech.*, **5**(5), 327–330.

Williamson, A. R., and Carter, A. D. (1991). 'Field studies to determine the potential risk of contamination of ground and surface water by an autumn and spring applied herbicide in oilseed rape and fodder maize', *Brighton Crop Protection Conference, Weeds*, **1**, 477–484.

Wilson, J. S., and Foy, C. L. (1989). 'Analysis of water samples for triazine and acetanilide herbicides from selected watersheds in Virginia', *Proc. Northeast. Weed Sci. Soc.*, **43**, 36.

Wolfbeis, O. S., and Koller, E. (1991). 'Fibre optic detection of pesticides in drinking water' in *Biosensors: Applications in Medicine, Environmental Protection and Process Control* (eds R. D. Schmid and F. Schneller), Volume 13, pp. 221–224, GBF Monographs, VCH, New York.

Wollenberger, U., Setz, K., Scheller, F., Loffler, U., Gopel, W., and Gruss, R. (1991). 'Biosensors for choline, choline esters and inhibitors of choline esterase'. *Sensors & Actuators*, **B4**, 257–260.

Xu, Y., Lorenz, W., Pfister, G., Bahadir, M., and Korte, F. (1986). 'Residue analysis of triazine herbicides in soil: comparison of a capillary gas chromatographic and a high performance liquid chromatographic method', *Fresenius' Z. Anal. Chem.*, **325**, 377–380.

Yao, S., Meyer, A., and Henze, G. (1991). 'Comparison of amperometric and UV-spectrophotometric monitoring in the HPLC analysis of pesticides', *Fresenius' Z. Anal. Chem.*, **339**(4), 207–211.

Zouzaneas, P., Soevegjarto, F., Stadlbauer, H., and Schmid, E. R. (1993). 'Determination of 35 selected pesticides in underground water from Southern Styria, Austria', *Ernaehrung (Vienna)*, **17**(2), 77–82.

Index

Contents–Volume 1

Contents—Volume 2

Contents—Volume 3

Contents–Volume 4

Contents–Volume 5

Contents–Volume 6

Contents—Volume 7

Contents–Volume 8

Contents—Volume 9